COMMANDER
STEVEN HAINES
ROYAL NAVY

Unlawful Combatants

The Changing Character of War Programme is an inter-disciplinary research group at the University of Oxford, and was funded by the Leverhulme Trust between 2003 and 2009.

Unlawful Combatants

A Genealogy of the Irregular Fighter

Sibylle Scheipers

OXFORD
UNIVERSITY PRESS

OXFORD
UNIVERSITY PRESS

Great Clarendon Street, Oxford, OX2 6DP,
United Kingdom

Oxford University Press is a department of the University of Oxford.
It furthers the University's objective of excellence in research, scholarship,
and education by publishing worldwide. Oxford is a registered trade mark of
Oxford University Press in the UK and in certain other countries

First Edition published in 2015

Impression: 1

Published in the United States of America by Oxford University Press
198 Madison Avenue, New York, NY 10016, United States of America

British Library Cataloguing in Publication Data
Data available

Library of Congress Control Number: 2014942232

ISBN 978–0–19–964611–1

Printed and bound by
CPI Group (UK) Ltd, Croydon, CR0 4YY

To my son, Leonard: his generation probably will not live in an age of fewer wars, but may they be less blinded by prejudice towards those who fight in them (and those who get caught up in the fighting).

Acknowledgements

As so often happens, this book arose out of the failure of a research project. In 2006, when I had just finished my PhD, I started working on a post-doctoral project on transatlantic differences regarding approaches to the law of armed conflict in the framework of the 'war on terror'. I wanted to know why officials in the USA had assigned the label 'unlawful combatants' to irregular fighters captured in the 'war on terror' and what their European counterparts thought about it. I wish to thank those who shared their thoughts with me on this topic—from the Pentagon and the State Department, as well as from the British Ministry of Defence, the German Verteidigungsministerium, the French Ministère de la Défense, and NATO. With hindsight, it surprises me how many of these people agreed to speak to me about matters that were already far beyond their political control and that involved legal inquiries and litigations that are still ongoing at the time of writing in 2014. To be sure, they did not tell me much more than that which was already public knowledge at the time. Some were obviously troubled by what was happening and genuinely sympathetic towards my research. Others were probably bewildered, embarrassed, or even annoyed by my questions, since they must have realized that we lived in parallel universes concerning our understandings of the treatment of irregular fighters. For this reason, I appreciate all the more their willingness to meet and to talk with me. Of course, these encounters proved rather fruitless with respect to my immediate research question. But as I considered the meagre fruits reaped from these interviews (and others with officials from the International Committee of the Red Cross (ICRC) and non-governmental organizations (NGOs), who for various reasons could not provide more information either), I was struck by one assumption that all my interlocutors held in common: that irregular fighters are a distinct category of combatants and that they should be treated differently from regular armed forces. I was intrigued by the way in which this assumption was never questioned.

By that time, I had fortunately joined Oxford University's Changing Character of War (CCW) programme. Remarkably, my colleagues at CCW never gave up hope that I would eventually produce a study on unlawful combatants (or if they quietly did, they had the grace to not let me notice it). More importantly, CCW's interdisciplinary set-up was an eye-opener for the problem I was wrestling with: I realized that, in order to understand the emergence of the concept of the 'unlawful combatant', I had to study the modern history of irregular fighters. I am grateful to all the programme's staff, associate faculty, and visiting fellows for listening to my ideas on irregular fighters and for sharing their comments on the tentative beginnings of this book. Hew Strachan and Adam Roberts in particular have been tireless in their support and advice throughout my time at CCW and ever since.

The main part of the book was written after I had been appointed as a lecturer by St Andrews' School of International Relations. I am grateful to both colleagues and students at St Andrews for patiently enduring my obsession with irregular fighters. Special thanks are due to my students in IR 3029 'Irregular Warfare', who were brilliant in challenging my evolving thoughts on irregular fighters while I was teaching them and writing the book in my head at the same time. I also owe thanks to my colleagues for their flexibility in working around my Leverhulme fellowship in 2012 and 2013 and for enabling me to take an additional semester of research leave subsequent to the fellowship.

I am also grateful to the Leverhulme Trust for awarding me a research fellowship (RF-2012-111), without which the book would have taken much longer to finish. It was a precious gift of free time to read, think, and write, which I appreciated more than I can express.

Many people have contributed to the book in a number of ways. Michael Broers, Cian O'Driscoll, Thomas Hippler, Bettina Renz, and Alex Danchev read parts of the typescript and shared their comments. I am particularly grateful to Huw Bennett and Antulio J. Echevarria II, who made time to read the entire typescript before I submitted it to OUP. Richard and Regine Davy were fantastic hosts during my visits to Oxford, never failing to enquire about the progress of the book, to offer their friendship, and to act as adoptive grandparents to my son while I was busy attending conferences. My parents also did their share of babysitting when I had to work. Last but not least, Melisa Salazar did far more for the book than the actual editorial work she undertook with such dedication: she patiently listened to my occasional frustrations with the project over a cup of coffee, put all her heart into improving the typescript in all its aspects, from the big questions to the tiniest footnote, and probably made me sound more intelligent than I am. Needless to say that all remaining flaws are my responsibility alone.

Contents

Abbreviations

4GW	fourth-generation warfare
AK	Armja Krajowa; Home Army
ALN	Armée Libération Nationale
ANA	Afghan National Army
AQM	al-Qaeda in Mesopotamia
AS	Armée Secrète
CCW	Changing Character of War
CDDH	Conférence Diplomatique sur le Droit Humanitaire
CDLL	Ceux de la Libération
CDLR	Ceux de la Résistance
CIA	Central Intelligence Agency
COIN	counter-insurgency
CT	communist terrorist
CTT	*centre de tri et de transit*
DOD	Department of Defense
DOJ	Department of Justice
EAM	Ethniko Apeleftherotiko Metopo; National Liberation Front
EDES	Ethnikos Dimokratikos Ellinikos Syndesmos; National Republican Greek League
EKKA	Ethniki kai Koinoniki Apeleftherosis; National and Social Liberation
ELAS	Ellinikós Laïkós Apeleftherotikós Stratós; Greek People's Liberation Army
EPW	enemy prisoners of war
EU	European Union
FATA	Federally Administered Tribal Areas
FFI	Forces Françaises de l'Intérieure
FLN	Front Libération Nationale
FTP	Francs-Tireurs et Partisans
GPW	Geneva Convention on Prisoners of War
ICG	International Crisis Group

ICRC	International Committee of the Red Cross
IDF	Israeli Defence Forces
ISAF	International Security Assistance Force
ISIS	Islamic State of Iraq and the Levant
KAR	King's African Rifles
MCP	Malayan Communist Party
MNLA	Malayan National Liberation Army
MPAJA	Malayan People's Anti-Japanese Army
NATO	North Atlantic Treaty Organization
NDH	Nezavisna Država Hrvatska; Independent State of Croatia
NGO	non-governmental organization
NKGB	Narodnij Komissariat Gossudarstwennoi Besopasnosti; People's Commissariat for State Security
NKO	Narodny Kommissariat Oborony; People's Commissariat of Defence
NKVD	Narodnyy Komissariat Vnutrennikh Del; People's Commissariat for Internal Affairs
OAS	Organisation Armée Secrète
OEF	Operation Enduring Freedom
OEF-A	Operation Enduring Freedom-Afghanistan
OLC	Office of Legal Counsel
ORA	Organisation de Résistance de l'Armée
OS	Organisation Spécial de Combat
PAM	*pris les armes à la main*
PCF	Parti Communiste Français
PCI	Partido Comunista Italiano
POW	prisoner of war
RAF	Royal Air Force
RMA	Revolution in Military Affairs
RSI	Repubblica Sociale Italiana
SD	Sicherheitsdienst
SIPO	Sicherheitspolizei
SIS	Secret Intelligence Service
SOE	Special Operations Executive
SS	Schutzstaffel
STO	*service du travail obligatoire*
TNC	Transitional National Council

UK	United Kingdom
UPA	Ukrajinska Powstanska Armija; Ukranian Insurgent Army
USA	United States of America
WCA	War Crimes Act
WMD	weapon of mass destruction
ZWZ	Związek Walki Zbrojnej; Union of Armed Struggle

1

Introduction

When the USA launched Operation Enduring Freedom Afghanistan (OEF-A) on 7 October 2001, it was in dire need of local allies in order to achieve the ambitious aims of ousting the Taliban from Afghanistan, capturing Osama bin Laden and the al-Qaeda leadership, and closing all terror training camps in the country. OEF-A was to rely primarily on Western airpower and Special Forces, while local anti-Taliban alliances were supposed to provide the ground force element. Since Afghanistan had been more or less in a constant state of civil war following the withdrawal of the Soviet forces in 1989, there were numerous potential anti-Taliban allies eager to volunteer. Two individuals who were recruited to support the US military operations on the ground were Abdullah Mujahid and Ziauddin, both local strongmen of Tajik ethnic origin from Paktia province and opponents of the Taliban regime. A US Special Forces commander reportedly equipped both with weapons, military gear, and money, and they subsequently fought alongside Western forces in the battle of Shah-i-Kot in the spring of 2002.[1] They were rewarded for their services with local security posts; Abdullah Mujahid became police chief in Paktia province and Ziauddin local army commander. It was alleged that both were corrupt, unprofessional, and constantly embroiled in infighting with other local warlords, and that this infighting in particular at times jeopardized Western operational plans in Paktia. In 2003, both were arrested by US security forces; Abdullah Mujahid was sent to Guantánamo Bay, Cuba; Ziauddin was held in the US prison at Bagram air base. How could this happen? How could these two men, who were initially regarded as Western allies and had proven their commitment in battle, morph into 'unlawful combatants' within a matter of a few months?

This book will show that the above example was not the result of an error of judgement on the part of the USA, committed in the chaos and confusion of

[1] Farah Stockman, 'US behind Afghan Warlords Rise and Fall: At Guantánamo Unruly Chieftains Join Combatants', *Boston Globe*, 12 August 2007.

the first months of the war in Afghanistan. Rather, the example reflects central features that are inherent in the concept of the 'unlawful combatant' itself: the ambiguity of the status of irregular fighters, the political opportunism entangled with categorizing someone as an irregular fighter, and, finally, the stark consequences of such a categorization.

In his *Theory of the Partisan*, Carl Schmitt contended that the distinction between the regular and the irregular is one of the most basic organizing principles of modern war.

> The partisan fights irregularly. But the difference between regular and irregular battle depends on the precision of the regular, and finds its concrete antithesis and thereby its concept only in modern forms of organization, which originated in the wars of the French Revolution.... for a complete history of the partisan, it is important to recognize that the power and significance of his irregularity has [sic] been dependent on the power and significance of the regularity that he challenges.[2]

In contrast to the regular combatant, Schmitt wrote, the irregular fighter 'does not risk simply his life... He knows, and all depends on this, that the enemy regards him as acting outside legality, right and honour.'[3] The marginalization of the irregular fighter is reflected mainly in the modern law of armed conflict.[4] Under the law of armed conflict, irregular fighters such as insurgents, guerrillas, and rebels are largely excluded from the privileges and protections of prisoner-of-war (POW) status. While there was a move towards greater inclusiveness regarding irregular fighters with the adoption of the 1977 Additional Protocols, the debate about the status and rights of 'unlawful combatants' following the terror attacks of 9/11 shows that the law of armed conflict is far from being all-inclusive in this respect. Irregular fighters continue to have a contested legal status. This contested legal status is linked to questions about their legitimacy. Even if most contemporary observers do not think that irregular fighters should be executed on the spot, tortured, or kept incommunicado for long periods of time, they would agree that there is and should be a difference between the way regular and irregular combatants are treated.

Scholarship in the disciplines of strategic studies, security studies, and International Relations more broadly rarely problematizes the dichotomy between the regular and the irregular. On the contrary, since the end of the

[2] Carl Schmitt, *Theory of the Partisan: Intermediate Commentary on the Concept of the Political* (New York: Telos Press, 2007 [1963]), 3.

[3] Schmitt, *Theory of the Partisan*, 35.

[4] Note that Schmitt argued that, with the codification of the Hague and the Geneva laws, many forms of irregulars, or 'partisans' in Schmitt's terminology, were 'equated with regular combatants' (Schmitt, *Theory of the Partisan*, 23). Schmitt is correct in as far as the codified law recognizes militias and spontaneous risings en masse against an invading army as lawful belligerents, but other forms of irregulars remained excluded from the law.

cold war the prevailing tendency was to reproduce it: the 1990s saw a rise in analyses of war that were inspired by the concept of 'globalization'. They equated irregular fighters with 'non-state' actors and argued that these actors were not only at odds with the state system, but also threatened to unravel it. The 'new wars' were depicted as an anomaly, a deviation from the norm of interstate wars that had supposedly prevailed in Europe since the Peace of Westphalia in 1648. The alleged rise of 'non-state' actors after the end of the cold war signified the return of more sinister, quasi-medieval forms of warfare.[5] However, the 'new wars' thesis and related approaches obscured the phenomenon of irregular fighters more than they provided new insights. First, the 'new wars' thesis misrepresented the role that irregular fighters continued to play in the alleged 'old wars'.[6] Since the French Revolutionary and Napoleonic Wars and throughout the nineteenth and twentieth centuries, irregular fighters had played an important role in almost all major conflicts, both in Europe and beyond. Secondly, by equating irregular fighters with 'non-state' actors, it also glossed over the complex and intricate relationship between irregular fighters and the state. Groups of irregular fighters often arose out of state structures, sought and received the support of states, and/or fought with the aim of acquiring statehood themselves.

The recent rise in interest concerning insurgency and counter-insurgency did not change this trend fundamentally. As opposed to the 'new wars' thesis, the counter-insurgency (COIN) literature is aware of the large role that irregular fighters played throughout the history of modern war. Yet, it also approaches irregular warfare as something fundamentally different from the 'norm' of modern war. The title of a monograph written by RAND's COIN expert Austin Long, *On 'Other War': Lessons from Five Decades of RAND Counterinsurgency Research*,[7] is revealing in this respect. It reflects the perception that COIN is organized under fundamentally different conditions and prioritizes

[5] See, e.g., Martin van Creveld, *The Transformation of War* (New York: Free Press, 1991); Mary Kaldor, *New and Old Wars: Organized Violence in a Global Era* (Stanford, CA: Stanford University Press, 2012); Herfried Münkler, *The New Wars* (Cambridge: Polity, 2005). In the US context, the concept of Fourth Generation Warfare (4GW) also played a large role in this debate. See William S. Lind et al., 'The Changing Face of War: Into the Fourth Generation', *Marine Corps Gazette*, 73/10 (1989), 22–6; Thomas X. Hammes, *The Sling and the Stone: On War in the 21st Century* (St Paul, MN: Zenith Press, 2006). See also Chapter 6 for a more detailed discussion of these approaches. Note also that the dichotomy of the irregular ('partisan') and the state can already be found in Schmitt: 'partisan insurgence and partisan groups are for Schmitt the symptoms of a "weak" state', in Gabriella Slomp, 'Theory of the Partisan: Carl Schmitt's Neglected Legacy', *History of Political Thought*, 26/3 (2005), 506.

[6] For a critical debate on the lack of historical awareness in the 'new wars' thesis, see, e.g., Edward Newman, 'The "New Wars" Debate: A Historical Perspective Is Needed', *Security Dialogue*, 35/2 (2004), 173–89.

[7] Austin Long, *On 'Other War': Lessons from Five Decades of RAND Counterinsurgency Research* (Santa Monica, CA: RAND, 2006). The title also alludes, of course, to the bible of modern strategic theory, Clausewitz's *On War*, which only serves to underline the distinction between regular (supposedly 'Clausewitzian') war and COIN.

fundamentally different practices and resources from 'regular' or 'conventional' warfare.[8] This dichotomy even manifested itself in the emergence of 'factions' of military practitioners and experts, in particular in Washington, along the cleavage lines of the COIN question. If they were 'COINdinistas', they advocated far-reaching transformations of the US defence sector; if they were not from COIN disciples, they cautioned against such radical measures.[9]

In contrast to these approaches, this book aims to bring the study of irregular fighters back into the centre of war studies. My argumentation builds upon two discussions: first, I show that irregular fighters were deeply involved in most modern wars, both in Europe and beyond, and that hence the categorical distinction between 'old', 'normal', or 'big' wars and 'new', 'other', or 'small' (irregular) wars is neither accurate nor helpful. Second, this book starts from the assumption that the artificial categorical distinction between 'normal' and 'other' wars is rooted in the marginalization of the irregular fighter throughout the history of modern war. Thus the question that it seeks to answer is how this marginalization of the irregular fighter and his or her exclusion from legal protections came about.

Existing scholarship in this field offers two possible explanations for the exclusion of irregular fighters. The first focuses on its historical origin. It is argued that with the emergence and consolidation of the state system in early modern Europe, states had an interest in acquiring an absolute monopoly over the use of military force. Hence they tried to eliminate all 'non-state' armed competitors. In other words, the political, moral, and legal exclusion of irregular fighters was the logical result of the process of state formation. The second explanation for the marginalization of irregular fighters is based on transhistorical, moral arguments. These claim that irregular fighters are not worthy of the same privileges and protection as regular fighters, since they fight in ways that make them morally reprehensible. First, their irregular tactics put civilians at risk. Second, they fight an unfair fight, as they exploit their opponents' commitment to the rules and customs of war, while they themselves do not abide by them.

The next two sections critically discuss these arguments. They conclude that neither the historical origins thesis nor the transhistorical rational thesis convincingly explains how and why irregular fighters have become marginalized. The third section suggests that it is more fruitful to reconstruct the

[8] See also M. L. R. Smith, 'Guerrillas in the Mist: Reassessing Strategy and Low Intensity Warfare', *Review of International Studies*, 29/1 (2009), 19–37.

[9] The term 'COINdinistas' was introduced by Douglas Porch, 'The Dangerous Myths and Dubious Promise of COIN', *Small Wars & Insurgencies*, 22/2 (2011), 239–57; see also Douglas Porch, *Counterinsurgency: Exposing the Myths of the New Way of War* (Cambridge: Cambridge University Press, 2013).

emergence and development of the irregular as a genealogy of non-deterministic historical events and narratives.

The Rise of the State and the Marginalization of the Irregular

Military historians have explained the rise of the state system in early modern Europe as the direct result of a 'military revolution'. The most prominent of these approaches was an argument put forward by Michael Roberts in the 1950s.[10] According to Roberts, in the period between 1560 and 1660 a 'military revolution' took place in Europe. Its causes were mainly technological—namely, the introduction of firearms into European warfare. At first, the introduction of firearms seemed to decrease the importance of military training. The arquebus and the wheel-lock pistol replaced the crossbow and the longbow. Bowmen had been highly trained specialists. However, the arquebus and the wheel-lock pistol were less accurate and slower to reload than the bows. Specialist training on such inaccurate weapons would have been wasted. The introduction of the Swiss column reinforced this trend to reduce military training, because deep formations such as the Swiss column did not require much drill or discipline to function. On the whole, the advantage that the early firearms conveyed on the battlefield was limited: the first line could fire only once before the opponents had closed up with them.

According to Roberts, the turning point came when Maurice of Orange and King Gustav Adolf of Sweden reformed the use of firearms on the battlefield. They introduced the volley technique, which meant that deep formations were broken up into linear formations of two or three lines. The rate of fire was increased through volley and counter-march, in which the rear line loaded while the front line fired, after which the front line marched to the rear and the next line fired, and so on. While this tactical innovation soon proved superior on the battlefield, it also required higher degrees of drill, discipline, and training. 'The army was no longer to be a brute mass, in the Swiss style, nor a collection of bellicose individuals, in the feudal style; it was to be an articulated organism of which each part responded to impulses from above.'[11]

However, the question was whether the available ways of recruiting armies in the early seventeenth century were able to meet the new requirements for drill and discipline. According to Roberts, they were not. There were two options for recruiting at the time: either to rely on mercenary armies or to

[10] Michael Roberts, 'The Military Revolution, 1560–1660', in Michael Roberts (ed.), *Essays in Swedish History* (London: Weidenfeld & Nicolson, 1967), 195–225.

[11] Roberts, 'The Military Revolution', 198.

raise local militias. Mercenaries were notoriously unreliable and adverse to battle, while local militias were unable to achieve the high levels of training and drill practice required to perform the new tactics. Roberts argued that the use of mercenaries in Europe thus underwent a transformation. While states had previously disbanded mercenaries at the end of every campaigning season and re-enlisted them in the spring, mercenaries eventually remained embodied over the winter. This enabled their principals to train and drill them longer. It was, in effect, the birth of standing armies. According to Roberts, the technological and tactical innovations of the early seventeenth century had mutually reinforcing ramifications at the strategic, the political, and the societal level. Ultimately, it is argued, they led to the emergence of the European state system and, conversely, to the marginalization of irregular fighters:

> The transformation in the scale of war led inevitably to an increase in the authority of the state.... Only a state, now, could supply the administrative, technical and financial resources required for large-scale hostilities. And the state was concerned to make its military monopoly absolute. It declared its hostility to irregular and private armies, to ambiguous and semi-piratical naval ventures.[12]

Roberts's thesis drew profound criticism. While Geoffrey Parker reinforced the notion of a military revolution in early modern Europe, he insisted that it was far less momentous and decisive than Roberts had thought.[13] Parker emphasized that in early modern Europe siege warfare was more important than battles. He argued that, while the 'gunpowder revolution' did have a profound impact on war, the consequent military revolution was neither unilinear nor quick and decisive. Increased firepower threatened to render fortifications vulnerable, but with the introduction of the *trace italienne* a new form of fortification was developed that was less susceptible to attack and easier to defend against bombardment. The result was a new equilibrium between offence and defence in siege warfare. According to Parker, the military revolution regarding tactics and technology of land armies did take place, but it did not have a large impact at first in those European countries where fortifications were prevalent, such as Italy, the Netherlands, France, and Spain, because these fortifications actually reduced the importance of battles.

[12] Roberts, 'The Military Revolution', 204 ff. For a similar argument in the field of International Relations, see Charles Tilly, 'War Making and State Making as Organized Crime', in Dietrich Rueschemeyer, Peter B. Evans, and Theda Skocpol (eds), *Bringing the State Back In* (Cambridge: Cambridge University Press, 1985), 203–21; and Charles Tilly, *Coercion, Capital and European States, AD 990–1990* (Oxford: Blackwell, 1990).

[13] Geoffrey Parker, *The Military Revolution: Military Innovation and the Rise of the West 1500–1800* (Cambridge: Cambridge University Press, 1988); see also Geoffrey Parker, 'The "Military Revolution", 1560–1660—a Myth?', *Journal of Military History*, 48/2 (1976), 195–214.

Importantly, Parker emphasized that the military revolution did not immediately eliminate all 'non-state', 'non-royal', or peripheral actors. Regular troops were indeed being merged into broader field armies, but, according to Parker, small strongholds, fortifications, and garrisons continued to exist. These were mostly manned by local, 'irregular' forces:

> *Der kleine Krieg* was thus as much the staple of early modern warfare in many parts of Europe as pitched battles and protracted sieges. By the side of the spectacular confrontations and campaigns, any complete military history of the age must consider the countless 'small wars' in which numerous concentrations of troops, although often only a few hundred strong, aimed to 'damage the enemy without trying to force a decision by battle'. This complex pattern of warfare, in which guerrilla was as important as guerra, only came to an end with the demolition of the network of strongholds which sustained it.[14]

Others criticized the notion of a 'military revolution' as such. According to David Parrott, the increase in the size of European armies was not limited to the period between 1560 and 1660.[15] Rather, it had already started in the mid-fifteenth century and continued into the mid-sixteenth century. Most of this growth was achieved through an increased reliance on military entrepreneurs—in other words, mercenaries. As the level of this 'outsourcing' grew, successive European governments came to the verge of bankruptcy, as they could no longer finance their wars. According to Parrott, this led to a 'military–fiscal crisis' in mid-sixteenth century Europe.[16] However, in contrast to Roberts, Parrott argued that the result of this was not the abolition of the mercenary system, but a further ratcheting-up of military outsourcing. Large-scale military contractors such as Albrecht von Wallenstein emerged. These offered not only military but also financial services in the framework of something akin to a 'war-on-loan' scheme. Hence they were more accurately military enterprisers rather than mere contractors.

Parrot conceded that this did change after 1648 and that European rulers did try to seize upon the modernizing initiative taken by Gustav Adolf of Sweden to raise state-controlled armies. Yet, this led to a paradox. On the one hand, rulers wished to assert their state sovereignty by gaining a monopoly on the use of armed force. On the other, they were aware that a pure militia model could never achieve the same war-making potential that the mercenary and military entrepreneur model of the preceding period had unleashed. So the question was how to build a powerful sovereign state if eliminating

[14] Parker, *The Military Revolution*, 41.

[15] David Parrott, 'Had a Distinct Template for "Western Way of War" Been Established before 1800?', in Hew Strachan and Sibylle Scheipers (eds), *The Changing Character of War* (Oxford: Oxford University Press, 2011), 48–63.

[16] Parrott, '"Western Way of War"', 57.

mercenaries would severely curtail the state's fighting power. What use was a monopoly on armed forces if that armed force was ineffective? The solution was to rely on the aristocratic elites in the newly formed 'states' to provide the necessary military power. In contrast to Roberts's assertions, Parrott argued that early modern rulers had failed to build up a taxation system that could have provided adequate finances for large regular armies, so they had to create—once again—a private–public partnership in the military realm. This emerged in the form of formal military proprietorship for the unit officers under the façade of centralized state control. This system remained in place throughout the seventeenth and the first half of the eighteenth century. As Parrott concluded:

> Down to the mid-eighteenth century, the response of European rulers to larger-scale, more expensive, and more organizationally demanding warfare was not the growth of the state and the establishment of its monopoly over military force but a series of experiments with various forms of military contracting-out, or, ultimately, private–public partnership.[17]

To complicate matters, the 'armies' of those newly formed 'states' were not only strictly speaking 'parastatal' actors, because they were raised and led by aristocratic elites. They also partly comprised subcontracted, 'irregular' units, which in contemporary parlance meant light infantry and cavalry units whose leaders had licence to operate independently. These were 'the last remainders of the freelance military enterprisers of the sixteenth and seventeenth centuries'.[18] They sometimes formed almost half of the military forces of European states.[19] One of the most outstanding of these early modern 'partisans' was Colonel Franz von der Trenck. He had entered the Russian army in 1737 and served as a private contractor in the war against the Turks, but he was soon accused of bad conduct and insubordination, released of military duty, and subsequently imprisoned. However, an influential patron obtained an amnesty for von der Trenck and offered him a colonel's commission. Von der Trenck raised a unit of irregular fighters, his Pandours Corps, which fought on behalf of the heiress to the Austrian throne Maria Theresa in the War of Austrian Succession (1740–8). Von der Trenck's Pandours were soon accused of having committed atrocities. There were reports that, after they had defeated a Prussian battalion on 31 July 1741 in the Silesian town of Zobten, they had plundered the local population and killed and mutilated Prussian

[17] Parrott, '"Western Way of War"', 58.

[18] Martin Rink, 'The Partisan's Metamorphosis: From Freelance Military Entrepreneur to German Freedom Fighter, 1740 to 1815', *War in History*, 17/1 (2010), 6–36.

[19] In 1740 about 40% of the forces of the Austrian monarch Maria Theresa were light 'irregular' units. See Rink, 'The Partisan's Metamorphosis', 11.

prisoners as well as local inhabitants.[20] Similar scenes were reported from von der Trenck's operations in Bavaria and the Upper Rhine.

Von der Trenck was an extreme figure, and he was tried and imprisoned for his behaviour in 1748. But the arrangement within which he operated was far from unique. The use of 'irregular' light units was widespread in early modern Europe. They were hastily recruited and inadequately paid, which meant that they not only were lacking discipline from the start, but also had every incentive to plunder and loot.[21]

Hence, it is not convincing to argue that the marginalization of irregular fighters was caused by the emergence of the state system in early modern Europe and by the states' attempt to eliminate non-state competitors in the field of military power and war. On the contrary, until the mid-eighteenth century the European state relied on non-state contractors and subcontractors to fill the ranks of its 'armies'. Beyond the question of the accuracy of these 'military revolution' arguments, they also leave important questions unanswered. If states had such a strong interest in the elimination of non-state military competitors, why did they continue to rely on irregular fighters throughout the history of modern war into the twenty-first century? And why did European states often disagree on the exclusion of irregular fighters, in particular during the process of the codification of the law of armed conflict? Most importantly, why did the marginalization of the irregular fighters at the practical level as well as at the level of moral and legal arguments only start, as this book will show, at the end of the eighteenth century, and not at the time of the emergence of the state system in early modern Europe? Arguments about the early modern 'military revolution' cannot answer these questions.

Moral Problems Related to Irregular Fighters

Moral arguments against irregular fighters have a long history in international ethics and just war theory. As James Turner Johnson put it, irregular warfare tends to be 'intractably dirty, from a moral standpoint'.[22] There are two main moral arguments against irregular fighters. First, it is often argued that irregular tactics put civilians at a high risk of suffering from war-related violence. Second, both scholars and practitioners have contended that the exclusion of irregulars from certain privileges and protections is based on reasons of reciprocity: since irregulars do not play by the rules of a fair and lawful fight, they should not be entitled to legal protection either. If these arguments turned out

[20] Rink, 'The Partisan's Metamorphosis', 12.
[21] Rink, 'The Partisan's Metamorphosis', 14.
[22] James Turner Johnson, *Can Modern War Be Just?* (New Haven: Yale University Press, 1986), 55.

to be valid, the marginalization of the irregular could be explained with reference to transhistorical moral reasons.

Irregular Fighters put Civilians at Risk

The allegation that irregular fighters pose an increased risk to civilians points to both direct and indirect aspects of their behaviour and their tactics. Francis Lieber already contended in his 1862 legal treatise on guerrilla bands, produced at the request of United States Union General Henry W. Halleck, that irregular fighters are renowned for pillaging and terror tactics:

> Thus many persons associate the idea of pillage with the guerrilla band, because, not being connected with the regular army, the men cannot provide for themselves, except by pillage, even in their own country.... Others connect with it the idea of intentional destruction for the sake of destruction, because the guerrilla chief cannot aim at any strategic advantages or any regular fruits of victory.[23]

Moreover, as most irregular fighters wear neither a uniform nor other distinctive insignia,[24] they are difficult to distinguish from 'peaceful citizens'. In fact, this was one of the problems that had caused Halleck to seek Lieber's advice on the status and treatment of guerrilla fighters in the first place. As Halleck wrote in correspondence with Lieber: 'The rebel authorities claim the right to send men, in the garb of peaceful citizens, to waylay and attack our troops, to burn bridges and houses and to destroy property and persons within our lines.'[25] Lieber acknowledged this problem: 'Others, again, associate with it [the guerrilla band] the idea of the danger with which the spy surrounds us, because he that to-day passes you in garb and mien of a peaceful citizen, may to-morrow, as a guerrilla man, fire your house or murder you from behind the hedge.'[26]

Such arguments against irregular fighters have not changed much since the American Civil War. In a 2003 paper on the status of terrorists, for instance, John Yoo and James Ho argued that al-Qaeda fighters are 'unlawful combatants', because

[23] Francis Lieber, 'Guerrilla Parties Considered with Reference to the Laws and Usages of War', printed in Richard Shelly Hartigan (ed.), *Lieber's Code and the Laws of War* (Chicago: Precedent, 1983), 39.

[24] Indeed, wearing neither a uniform nor other distinctive insignia is one of the defining features of irregular fighters, according to the legal sources. See Hague Convention IV respecting the Laws and Customs of War on Land and its Annex. The Hague, 18 October 1907, article 1 (2), and Geneva Convention III relative to the Treatment of Prisoners of War. Geneva, 12 August 1949, article 4 (2) (b); both Conventions are printed in Adam Roberts and Richard Guelff (eds), *Documents on the Laws of War* (Oxford: Oxford University Press, 2000), 73, 247.

[25] Henry W. Halleck, quoted in Courtney S. Campbell, 'Moral Responsibility and Irregular War', in James Turner Johnson and John Kelsay (eds), *Cross, Crescent, and Sword: The Justification and Limitation of War in Western and Islamic Tradition* (New York: Greenwood Press, 1990), 104.

[26] Lieber, 'Guerrilla Parties', printed in Hartigan (ed.), *Lieber's Code*, 39.

> They operate covertly by intentionally concealing themselves among the civilian population; they deliberately attempt to blur the lines between civilians and combatants. Most importantly, they have attacked purely civilian targets with the aim of inflicting massive civilian casualties.[27]

However, many of these arguments do not withstand closer scrutiny. First, the assumption that regular armed forces fulfil per se the criteria of distinction is problematic. The relevance of Special Forces within the force structure of Western states is currently on the increase; since their mode of operation often depends on their stealth, they pose a growing challenge to the image of the regular fighter and the requirement of distinction.[28] In addition, the increasing 'civilianization' of the regular armed forces[29] raises the question of where to draw the line between the civilian and the regular combatant. Since this distinction is often blurred, it is unjustified to charge that it is exclusively irregulars who threaten to undermine the category of the civilian and the protections that come with it. Rather than blaming either regular or irregular forces for the difficulties related to the status of civilians, one should acknowledge 'that civilian status and identity are indeed ambiguous and problematic sometimes'.[30]

It is also not clear that the other characteristics that Lieber ascribed to irregular fighters are really distinctive features. For instance, it seems problematic to assert that irregular fighters were more prone to looting since they could not rely on an extensive logistics network, while regular armies could. Before the mechanization of warfare in the first half of the twentieth century, large armies had to requisition and to live off the land. More generally, while the increasing mechanization of warfare did enhance the provision of supplies for regular armies and thus decreased their need for foraging, the history of logistics continued to be 'an endless series of difficulties succeeding each other'.[31] Thus, it seems as if Lieber's insinuation that regular armies do not pillage because they receive a continuous supply from their base was an ideal rather than a reflection of the reality of the American Civil War. Equally, regular troops do not necessarily refrain from the 'intentional destruction

[27] John C. Yoo and James C. Ho, 'The Status of Terrorists', *Boalt Working Papers in Public Law*, 25 (Berkeley: University of California, Berkeley, 2003), 10.

[28] This challenge is not new, however, as Watkin points out the widespread use of irregular forces by the Allies in the Second World War. Cf. Kenneth Watkin, 'Warriors without Rights? Combatants, Unprivileged Belligerents, and the Struggle over Legitimacy', *Humanitarian Policy and Conflict Research Occasional Paper* (Cambridge, MA: Harvard University, 2005), 38. See also Chapter 4.

[29] A prominent example of this trend are UAV 'pilots' who direct attacks against targets, but are located several thousand miles away from the battlefield.

[30] Hugo Slim, *Killing Civilians: Methods, Madness and Morality in War* (London: Hurst, 2007), 268.

[31] Martin van Creveld, *Supplying War: Logistics from Wallenstein to Patton* (Cambridge: Cambridge University Press, 1977), 231.

for the sake of destruction' that Lieber saw as a characteristic feature of irregular fighters.

Neither is the use of human shields, a contemporary charge laid exclusively on irregular fighters. Regular armies such as the Israeli Defence Forces (IDF) have been accused of using Palestinian civilians as human shields in operations conducted in Palestinian population centres.[32] In fact, it is difficult to find any reliable evidence in the history of war that regular troops are less inclined than irregulars to use terror tactics against civilians.[33]

Finally, it is doubtful that the protection of 'civilians' in our contemporary sense was really Lieber's aim. Lieber speaks not of civilians, but of 'peaceful citizens', as the potential illegitimate target of irregular fighters.[34] In fact, it was not until the 1949 Geneva Conventions that the protection of 'civilians' became one of the core concerns of the law of armed conflict. Arguably, already during medieval Christendom the idea had become established that immunity should be granted to certain groups of individuals who were assumed not to contribute to the war effort owing to their social role or occupation, such as women, clerics, and merchants. However, this convention was still far from offering categorical protections for civilians in war.[35] The 1907 Hague Convention did not codify civilian immunity in an unequivocal way either. It contained weak protections for civilians, such as the prohibition of the bombardment of undefended towns. However, civilians in fortified towns could still be targeted, though the attacker ought to warn them if feasible.[36]

In addition, the Hague rules' weak protections for civilians were dependent on these citizens' non-involvement in the conflict. This favoured the invading army, in that it set limits to spontaneous civilian uprisings during occupations and emphasized the 'ancient privileges of soldiers'.[37] Hence, the Hague law was as much about disciplining the civilian into passivity and obedience towards the occupying force as it was about civilians' protections. While the protection of civilians became undoubtedly one of the main concerns of the law of armed conflict after the Second World War, it was not central during

[32] Cf. B'tselem, 'Human Shields' <http://www.btselem.org/english/Human_Shields/Index.asp> (accessed 15 June 2013).

[33] Cf. Slim, *Killing Civilians*; Alexander B. Downes, *Targeting Civilians in War* (Ithaca, NY: Cornell University Press, 2008).

[34] Cf. Lieber, 'Guerrilla Parties', printed in Hartigan (ed.), *Lieber's Code*, 39.

[35] Adam Roberts, 'The Civilian in Modern War', in Strachan and Scheipers (eds), *The Changing Character of* War, 357–80.

[36] Amanda Alexander, 'The Genesis of the Civilian', *Leiden Journal of International Law*, 20/2 (2007), 364; Hague Convention IV, articles 25 and 27, printed in Roberts and Guelff (eds), *Documents*, 78.

[37] Karma Nabulsi, 'Evolving Concepts of Civilians and Belligerents: One Hundred Years after the Hague Peace Conferences', in Simon Chesterman (ed.), *Civilians in War* (Boulder, CO: Lynne Rienner, 2001), 17.

the emergence of the legally codified POW regime in the second half of the nineteenth century.

Thus, irregular fighters were excluded from certain legal protections and privileges *before* the contemporary moral argument arose that they endanger civilians. In its initial stages, the legal exclusion of irregular fighters was mainly meant to serve regular combatants, not civilians. This observation is further supported by the fact that in many cases the measures taken by regular forces to protect themselves from attacks by irregulars lead to more civilian deaths than the actions and tactics of the irregulars themselves.

In fact, the historical record confirms that regular troops' harsh, repressive responses to guerrilla warfare tended to be directed towards the civilian population. From the late eighteenth to the mid-twentieth century a punitive approach towards civilian populations was both commonplace and regarded as legally sanctioned. Hostage-taking, reprisals, and other repressive measures against civilians continued to be a common practice in wars both within Europe and in the colonies until the end of the Second World War.[38] In some cases regular armies became so preoccupied with the dangers posed by guerrilla fighters that they developed an outright obsession with the repression of potential civilian uprisings in occupied territory. Such was the case, for instance, with the 'francs-tireurs' myth, a 'collective delusion' that beset the German armed forces during the early weeks of their occupation of Belgium and northern France in 1914. Between the start of the invasion in early August and the end of October 1914, German troops killed 5,146 civilians with the intention of suppressing an imminent 'people's war', for which, however, there had been very little if any evidence.[39]

Thus, it seems that strong narratives about the threats posed by irregular forces actually *hindered* the development of an unequivocal and protected status of the 'civilian' in the early phases of the codification of the law of armed conflict. This was precisely because occupying forces could never be sure whether civilians were in fact innocent. Plus, even if they were innocent, civilians could still be lawful targets. Reprisals against civilians with the aim of repressing guerrilla warfare remained lawful until after the Second World War, although article 50 of the Hague Convention prohibited collective punishment and hostage-taking in occupied territory. It was only the 1949 Geneva

[38] On the German military's approach to dealing with *franc-tireurs*, guerrilla warfare, and repressive measures against civilians, see Isabel V. Hull, *Absolute Destruction: Military Culture and the Practices of War in Imperial Germany* (Ithaca, NY: Cornell University Press, 2005); on the British internment of civilians in the Second Boer War (1899–1902), see Thomas Pakenham, *The Boer War* (New York: Random House, 1979).

[39] John Horne and Alan Kramer, *German Atrocities, 1914: A History of Denial* (New Haven: Yale University Press, 2001).

Convention IV, the so-called Civilian Convention, that ruled out reprisals against civilians and their property.[40]

Irregulars Undermine Fairness and Reciprocity

The second moral argument against irregulars is that they undermine fairness and reciprocity in war by flouting the legal rules, while taking advantage of the fact that their regular counterparts will most likely abide by them. In this vein, the simplest reason for excluding guerrilla fighters from the privileges of POW status is that they are unlikely, in fact often unable, to reciprocate:

> Others connect with the guerrillero the idea of necessitated murder, because guerrilla bands cannot encumber themselves with prisoners of war; they have, therefore, frequently, perhaps generally, killed their prisoners, and of course have been killed in turn when made prisoners, thus introducing a system of barbarity which becomes intenser in its demoralization as it spreads and is prolonged.[41]

However, this unwillingness or inability to treat prisoners well is, according to Lieber, not the only way in which irregular fighters fall short of reciprocity:

> Others, again, connect the ideas of general and heinous criminality, of robbery and lust with the term, because the organization of the party being but slight and the leader utterly dependent upon the band, little discipline can be enforced, and where no discipline is enforced in war a state of things results which resembles far more the wars recorded in Froissart or Comines, the Thirty Years' War, and the Religious War in France, than the regular wars of modern times.[42]

Two themes are particularly interesting in this quotation: first, that irregular tactics are equated with a 'system of barbarity' and, second, that the flouting of legal rules is seen as the result of a lack of discipline. The alleged barbarity of one's opponents in war has long been a common justification for not applying the laws of war. This fact stands in stark contrast to the description of the 'self' as 'civilized' and restrained by legal rules.[43] Colonial warfare became the most notorious example of the suspension of the law of armed conflict towards irregular fighters, as it was the hallmark of colonialist narratives to declare the local inhabitants of the colonies as 'savages'.[44] Yet the idea that regular troops should not be bound by the laws of war when faced with irregulars did not

[40] Geneva Convention IV, article 33, printed in Roberts and Guelff (eds), *Documents*, 312.

[41] Lieber, 'Guerrilla Parties', printed in Hartigan (ed.), *Lieber's Code*, 39.

[42] Lieber, 'Guerrilla Parties', printed in Hartigan (ed.), *Lieber's Code*, 39–40.

[43] Patrick Porter has coined the phrase 'military orientalism' to describe this phenomenon: *Military Orientalism: Eastern War through Western Eyes* (London: Hurst, 2009).

[44] Cf. Frédéric Mégret, 'From "Savages" to "Unlawful Combatants": A Post-Colonial Look at International Humanitarian Law's "Other"', in Anne Orford (ed.), *International Law and its 'Others'* (Cambridge: Cambridge University Press, 2006), 265–317. However, see also Chapter 5 for a discussion of the relevance of the so-called standard of civilization argument.

specifically emerge from colonial warfare. Rather, this idea developed and transformed as it travelled back and forth between the centre and the periphery.

The alternative (or sometimes complementary) image of the irregular is that of the bandit or ordinary criminal. The term 'brigand' is paradigmatic in this respect, as it captured the ambiguous nature of the irregular as a bandit using guerrilla tactics. It was designed to deny that irregulars were motivated by political grievances and causes. As with the civilized/savage distinction, this image of the irregular depicts him or her as the conceptual opposite of the regular. Irregulars fight only for their own gain, and are therefore unlikely to submit to any kind of discipline. The regular, by contrast, fights for the good of his nation, but not for personal gain, and he does so in an orderly manner, in compliance with the laws of war and under the command of a responsible authority.

Eventually, the criteria of discipline and compliance with legal rules became the cornerstone of the codified law. The 1907 Hague Convention stipulated that, in order to qualify as lawful belligerents, either individuals had to be members of an army of a belligerent party or, if they were members of a militia or volunteer corps, they had 'to be commanded by a person responsible for his subordinates' and 'to conduct their operations in accordance with the laws and customs of war'. Also popular risings en masse against an invading army were lawful, even if they failed to meet the criterion of organization.[45] As with the criteria relating to the principle of distinction, it was initially assumed that members of the regular armed forces would comply with these criteria by definition. They were reiterated in article 4 of the 1949 Geneva Convention III, the Prisoner of War Convention.[46] It was Additional Protocol I that loosened the criteria a combatant had to fulfil in order to qualify for POW status. Article 43, 1 formally abandoned the notion of lawful combatancy and stated that the 'armed forces of a Party to a conflict consist of all organized armed forces, groups and units which are under a command responsible to that Party for the conduct of its subordinates'. Article 44, 3 makes the protections of POW status dependent on the condition that a combatant 'carries his arms openly: (*a*) during each military engagement, and (*b*) during such time as he is visible to the adversary while he is engaged in a military deployment preceding the launching of an attack in which he is to participate'.[47] However, in

[45] Annex to the Hague Convention IV, printed in Roberts and Guelff (eds), *Documents*, 73. The other two criteria—'to have a fixed and distinctive emblem recognizable at a distance' and 'to carry arms openly'—are geared towards the distinction between combatants and civilians, as discussed in the previous section.

[46] Geneva Convention III, article 4, printed in Roberts and Guelff (eds), *Documents*, 245 ff.

[47] Protocol Additional to the Geneva Conventions of 12 August 1949, and relating to the Protection of Victims of International Armed Conflicts (Additional Protocol I). Geneva, 8 June 1977, articles 43 and 44, printed in Roberts and Guelff (eds), *Documents*, 445.

order to ensure that ordinary criminals motivated by greed were not covered by these protections, article 1 of Additional Protocol I makes it clear that the Protocol applies to situations of international armed conflict, including 'armed conflicts in which peoples are fighting against colonial domination and alien occupation and against racist régimes in the exercise of their right of self-determination'.[48] In other words, the irregular has to be a 'rebel with a (just) cause' in order to be recognized as a lawful combatant. This requirement amounts to an elimination of the distinction between *jus in bello* and *jus ad bellum*, which had been a linchpin of the law of armed conflict since its inception.

However, Additional Protocol I is not as widely ratified as the Geneva Conventions, and those of its rules that affect irregular fighters in particular have not yet become customary law.[49] Crucially, the USA, as the most important power in the current international system, is not party to it. Thus it comes as no surprise that contemporary reciprocity-based arguments against irregular fighters and terrorists closely resemble the previously established themes. Echoing Lieber's suspicion that irregulars notoriously fall short of treating their prisoners in a lawful way, John Yoo, for instance, who served as Deputy Assistant Attorney General in the Office of Legal Counsel for the Bush administration, wrote in an article in 2004: 'Without territory, it [al-Qaeda] does not have the resources to provide detention facilities for prisoners, even if it were interested in holding captured POWs.'[50] And in another article co-authored with James Ho, they write that 'Al Qaeda violates the very core of the laws of war. Al Qaeda members are not under the control of a nation-state that will force them to obey the laws of war'.[51] Taken together with the decidedly Manichaean overtones of the Bush administration's justifications for the 'war on terror' as a fight of civilization against barbarism,[52] this strand of the legal discourse on terrorism drew heavily upon nineteenth-century themes of non-compliance with the laws of armed conflict and lack of discipline; these themes have been used to justify partial suspensions of the law of armed conflict towards terrorists. It is important to note, however, that reciprocity-based legal arguments have not been confined to the more 'extreme' legal opinions produced by some individuals in the Bush administration. Rather, their narrative was complemented by their more 'moderate' counterparts, in

[48] Additional Protocol I, printed in Roberts and Guelff (eds), *Documents*, 423.

[49] Currently only article 75 of Additional Protocol I, which offers minimum protections to all categories of persons and slightly extends the rules included in common article 3 to the 1949 Geneva Conventions, is regarded as customary international law; printed in Roberts and Guelff (eds), *Documents*, 463 ff.

[50] John C. Yoo, 'Terrorists have No Geneva Rights', *Wall Street Journal*, 26 May 2004.

[51] Yoo and Ho, 'The Status of Terrorists', 10.

[52] Alia Brahimi, 'Religion in the War on Terror', in Strachan and Scheipers (eds), *The Changing Character of* War, 184–201.

both Europe and the USA, who regarded the 'war on terror' as a law-enforcement operation, and the terrorists themselves as mere criminals, who should not be endowed with combatant status.

According to this argument, the problem with irregulars' lack of reciprocity in armed conflicts is not simply that they appear to be the antithesis of the regular armed forces. The real issue that makes this argument morally compelling is that the resulting fight is unfair, because it puts the regular party at a disadvantage. This was already a vital concern for some of the delegates to the 1899 Hague Conference, one of whom argued that '"civilized nations" might well be put "in a disadvantageous position in time of war with less civilized nations or savage tribes"'.[53] This theme recurs in contemporary debates about irregular fighters and their tactics under the header of 'lawfare'. Lawfare, in the words of an American commentator, is the

> exploitation of real, perceived, or even orchestrated incidents of law-of-war violations being employed as an unconventional means of confronting American military power. Make it appear that the United States is fighting in an illegal or immoral way, and the damage inflicted upon the public support the forces of a democracy need to wage war is as real as any caused by a traditional defeat.[54]

The use of human shields by insurgents or terrorists clearly falls into this category, but at the same time human shields touch upon the theme of irregular fighters' propensity to expose civilians to greater risks. More generally, the concept of lawfare tends to substitute the older notion of the 'civilized nation' fighting against savages with the more contemporary image of a democracy fighting against terrorists, but the moral coding remains the same.

The moral dilemma created in a situation in which one party ignores the laws of war is how the other party or parties ought to react. Is it possible to re-establish a basic degree of fairness by suspending one's own constraints on the use of force during that particular conflict? Does it help to use belligerent reprisals in order to bring the opponent back on the legal track? Or should a regular army stick to the rules and risk being defeated in an unfair fight? At first sight, it seems reasonable to mete out some form of punishment towards irregular fighters by withholding some privileges from them. In particular, excluding them from the protections of the POW regime seems a proportionate as well as discriminate approach—more discriminate, at least, than adopting harsh measures against the civilian population at large owing to the suspicion that they harbour irregular fighters.[55] Suspending aspects of the legal constraints of warfare when dealing with irregulars is supposed to

[53] Quoted in Mégret, 'From "Savages" to "Unlawful Combatants"', 281.

[54] Charles C. Dunlap, 'Lawfare amid Warfare', *Washington Times*, 3 August 2007.

[55] Yet, as studies on the inmates of the Guantánamo detention site show, it is often rather difficult to establish whether a captured person is, in fact, an irregular fighter or just a random

safeguard the law rather than undermine it. After all, it is argued, it is the irregular who does away with legal constraints in the first place: 'the guerrilla parties may be culpable for precipitating rapid moral degeneracy in compliance with the laws on combatant conduct, quickly undoing a long historical endeavor to set limits on killing in war'.[56] This impetus to moral degeneration is believed to be even stronger when the irregular party turns out to be successful in its struggle against a regular opponent, and if its success is at least partly due to its flouting of legal rules.[57]

Yet the seemingly straightforward moral answer to this dilemma is not without its problems. Reciprocity is an ambiguous concept. First, the category of the irregular tends to take on a life of its own, in that it is applied a priori to certain groups of fighters without even considering whether individual members of this group fulfil the criteria for lawful belligerency.[58] If this is the case, it is in effect the regular party who makes the first move towards suspending constraints on the use of force: when particular groups know that the irregular status will be attributed to them per se, they also understand that the laws of war will not be applied to them.

Second, it is important to scrutinize the notion of reciprocity in the framework of this moral dilemma. Can the regular party in the conflict safeguard the system of constraints by suspending certain constraints on the use of force towards the irregular party, or does this undermine this system in the long run? Mark Osiel cautioned that, while the law of armed conflict leaves a large margin for such measures, a cost-benefit analysis in this situation would suggest that upholding the law is more beneficial for the regular party in the long run. Rather than forcing the actual opponent back into the remit of the law, the suspension of constraints could provoke future opponents (regular or irregular) to do away with the rules on the basis of the behaviour they observed in the previous conflict. Hence it is perhaps not so much reciprocity in its narrow bilateral form that counts, but the long-term multilateral or 'systemic' quality of reciprocity.[59]

Finally, and related to the previous argument, suspending certain constraints on the use of force could threaten not only the system of constraints,

civilian. Cf. Mark Denbeaux and Joshua Denbeaux, *Report on Guantánamo Detainees* (South Orange, NJ: Seton Hall University School of Law, 2008).

[56] Campbell, 'Moral Responsibility', 106.

[57] This seems to be one of the implicit or explicit assumptions of the discourse on 'asymmetric warfare'. See, e.g., Ivan Arreguín-Toft, *How the Weak Win Wars: A Theory of Asymmetric Conflict* (Cambridge: Cambridge University Press, 2005).

[58] Many legal scholars criticized the Bush administration for denying the Taliban collectively and a priori the status of lawful combatants. They argued that this decision could be made only with respect to captured individuals, but not with respect to whole groups of fighters. Cf. Joseph Blocher, 'The Guantanamo Three-Step', *Yale Law Journal*, 117/1 (2007), 1–7.

[59] Mark Osiel, *The End of Reciprocity: Terror, Torture, and the Law of War* (Cambridge: Cambridge University Press, 2009), 368.

but the very sense of self-identity of the regular armed forces involved. If their identity and their achievements with respect to the civilization of warfare hinge upon their compliance with constraints on the use of force, the suspension of these constraints could negatively affect their self-perception, even if this suspension is a direct response to the non-compliance by an irregular opponent. This is precisely what seems to have animated American Judge Advocate Generals in their protest against the relaxation of the legal rules on detention and interrogation that their civilian counterparts in the Bush administration implemented after 9/11. In the words of US Air Force Deputy Judge Advocate General Jack L. Rives:

> The cultural and self-image of the US Armed Forces suffered during the Vietnam conflict and at other times due to perceived law of armed conflict violations.... Approving exceptional techniques... even though lawful, may create uncertainty among interrogators regarding the appropriate limits of interrogations, and may adversely affect the cultural self-image of the US Armed Forces.[60]

Hence, the 'cult of the regular' as the law-abiding, civilized, and disciplined belligerent, whose identity is the conceptual and moral opposite of the savage, undisciplined, and thuggish irregular, can be interpreted in both ways: as a justification for the exclusion of the latter from privileges and protections, but also as a source of protection for the irregular. The latter is the case if the positive attributes of restraint, discipline, and compliance with the laws of war are disentangled from the condition of reciprocity between belligerents and reinterpreted in the sense of 'martial honour' and professional virtue of the armed forces. As Osiel suggested, virtue and honour are motivations for certain moral forms of behaviour that are independent from cost–benefit calculations or conditions of direct reciprocity.[61] If the law of armed conflict and compliance with it has, then, developed into such a defining feature of the armed forces at the beginning of the twenty-first century, it has unfolded its disciplinary force in the Foucauldian sense: it 'produces' the identity of the regular armed forces. Ironically, this achievement then makes it impossible for them to suspend legal rules towards their irregular counterparts.

To conclude, moral arguments about irregular fighters fail to explain the marginalization of irregular fighters. Neither the argument that irregulars put civilians at risk nor the assumption that irregulars fight an unfair fight provides a sufficient moral basis for the exclusion of irregulars from legal protections.

[60] Jack L. Rives, Deputy Judge Advocate General, Department of the Air Force, 'Memorandum for SAF/GC', 6 February 2003, in Karen J. Greenberg (ed.), *The Torture Debate in America* (Cambridge: Cambridge University Press, 2008), 380.

[61] Osiel, *End of Reciprocity*, 344 ff.

Method, Argument, and Definitions

As the exclusion of irregular fighters in modern warfare cannot be satisfactorily explained with reference either to a formative historical origin or to transhistorical moral arguments, the book's aim is to explore an alternative route to understanding the dichotomy between the regular and the irregular. The approach that it pursues is a genealogical one. Following the writings of Friedrich Nietzsche and Michel Foucault, the genealogical method has been described as 'an approach that seeks to uncover the conditions under which moral institutions are devised and to interpret the value that these norms themselves possess'.[62] It is hence 'a method specifically concerned with interpreting the origins of moral interpretations'.[63] Above and beyond the origins, however, the genealogical approach is concerned with the question of how certain moral norms and meanings emerge, persist, or change. In contrast to approaches that seek the causes for moral phenomena either in a specific, formative historical mechanism or in their transhistorical rational or moral roots, genealogy implies a non-deterministic form of causation that allows for contingency. In the case of a formative historical origins explanation, a phenomenon X emerges as a necessary and inevitable result of an historical event Y. Applied to the case at hand, this would mean that irregular fighters were excluded as a direct and logical result of the emergence of the modern state system. The relationship between the two would be one of logical necessity, albeit one that was not borne out by empirical events, as we have seen. In the case of moral reasoning and deduction, a phenomenon X can be deduced from a more general moral principle Y. In the case of irregular fighters, the sequence of deduction would be that 'irregular fighters put civilians at risk'—'putting civilians at risk is morally wrong'—'therefore irregular fighters are morally wrong'. Alternatively, the argument could be 'irregular fighters fight an unfair fight'—'fighting unfairly is morally wrong'—'therefore irregular fighters are morally wrong'. Again, the connection between X and Y is a necessary and deterministic one. In contrast to these deterministic forms of causation, the genealogical method presupposes a form of causation that is non-deterministic. The aim of the method is not to explain, but to understand and to interpret. Characteristically, a genealogical approach asks 'how' a particular phenomenon developed, instead of 'why' in the sense of a straightforward mechanism for 'what caused it'.[64]

[62] Richard Price, 'A Genealogy of the Chemical Weapons Taboo', *International Organization*, 49/1 (1995), 85; Jens Bartelson, *A Genealogy of Sovereignty* (Cambridge: Cambridge University Press, 1995), 53 ff.

[63] Price, 'A Genealogy of the Chemical Weapons Taboo', 86.

[64] Price, 'A Genealogy of the Chemical Weapons Taboo', 84.

Since this method is open for contingency and not focused on connecting a phenomenon to a particular deterministic cause, it is particularly well suited for analysing the conditions for continuity and change in the development of norms. As we shall see in the subsequent chapters of the book, the marginalization and exclusion of irregular fighters evolved over time and went through various phases of legal codification and reinterpretation. This evolutionary process in the dichotomy of the irregular and the regular once again casts doubt on the usefulness of deterministic explanations in this case, which by necessity possess a static, almost lawlike character (Y causes X regardless of the circumstances, only a change in Y can result in a change in X). In contrast, the genealogical approach assumes that, 'as a result of the marriage of chance occurrence, fortuitous connections, and reinterpretations, the purposes and forms of moral structures often change in such a way that they come to embody values different from those that animated their origins'.[65] That does not mean that the genealogical approach assumes or even predicts change in normative structures. As we shall see in the subsequent chapters, the modern history of the irregular fighter as an intellectual, moral, and legal concept features large degrees of continuity. However, the genealogical approach explicitly asks how and under what conditions norms persist or change. In this respect the genealogical approach differs from studies that focus on the impact of norms on social behaviour, which tacitly or explicitly imply the persistence of norms.[66]

However, if the persistence of norms is the default assumption of most studies on the impact of norms, normative change seems to be more difficult to explain. What are the possible sources of normative change? First of all, we should remember the seemingly trivial, but often forgotten, reply that Harold Macmillan gave to a journalist who had asked him what was most likely to blow government off course: 'Events, dear boy, events.' With respect to the concept of irregular fighters, the events most likely to trigger reinterpretations (which rarely reached the level of full-blown revisions) were, as one may guess, wars. The evolution of the law of armed conflict was in many respects a reaction to perceived gaps, deficiencies, or unintended consequences of the law, as they became apparent in the course of a specific war and as lawyers and policymakers undertook to address them after the event. Since difficulties in dealing with irregular fighters on the battlefield (or, more likely, in the rear areas) were often identified as the core problem ultimately triggering the

[65] Price, 'A Genealogy of the Chemical Weapons Taboo', 86.

[66] Jeffrey T. Checkel correctly pointed out that the first generation of 'constructivist' International Relations scholarship, meaning precisely those studies that pioneered the study of norms in International Relations, was lacking a theory of agency. Jeffrey T. Checkel, 'The Constructivist Turn in International Relations Theory', *World Politics*, 50/2 (1998), 340. Hence they focused mainly on the persistent nature of normative structures.

codification of the law, it is not surprising that the experience of specific wars had a deep impact on the evolution of the concept of the irregular.

Another possible factor for reinterpretations in the concept of the irregular were technological innovations. As Chapter 3 of this book demonstrates, the introduction of more effective firearms in the mid-nineteenth century had a large impact not only on infantry tactics, but also on the perception of irregular fighters. However, it is important to note that events such as wars and technological innovations do not affect moral and legal institutions in an unmediated way. Rather, this books shows that such factors become meaningful and effective only as they are mediated or filtered through different actors' perceptions and narratives.[67]

Finally political ideologies themselves can be sources for the emergence of moral and legal institutions as well as for their subsequent transformations. Changes in the broader understandings of political legitimacy can affect reinterpretations of existing legal and moral arguments. For instance, the emergence of the cult of the *levée en masse* during the French Revolution posed an intricate political and moral problem with respect to irregular fighters: if the armed forces of revolutionary France were morally vindicated because they represented the pure and unmanipulated desire of the French nation to rush to the defence of the fatherland, how could one morally distinguish them from the counter-revolutionary uprisings in the Vendée and elsewhere? The answer was to claim that the latter were 'irregular fighters' and as such illegitimate. This signified the birth of the modern concept of the irregular.[68] As this defining strategy was fraught with internal contradictions, this emergence of the irregular was by no means an inevitable occurrence; the French Republican's problem could have been resolved in a different way, but it was not. In the twentieth century, as the moral weight of the idea of national self-determination became more widely accepted and widely claimed after the Second World War, this had an important impact on how the concept of the irregular became reinterpreted.

This brings us to the final assumption upon which this study is based: normative structures such as moral and legal institutions do not exist, much less change, by themselves. They need agents to produce, reproduce, and/or reinterpret them. Structure and agency are mutually constitutive: structures 'define' agents in the sense of providing them with possible identities. Thereby they empower or disempower agents.[69] Agents, in turn, animate structures; they bring them to life by using them. At the same time, agents

[67] Cf. Ole Wæver, 'Discursive Approaches', in Thomas Diez and Antje Wiener (eds), *European Integration Theory* (Oxford: Oxford University Press, 2004), 197–215.

[68] See Chapter 2.

[69] Cf. Stefano Guzzini, 'Structural Power: The Limits of Neo-Realist Power Analysis', *International Organization*, 47/3 (1993), 443–78.

introduce contingency into moral structures. In principle, they have the capacity to change them. However, structures tend to be sticky and inert, which means that revolutionary change is improbable, while minor reinterpretations or incremental transformations are more likely to occur. In other words, in order to bring about revolutionary change in established norms, it either takes an extreme maverick figure who is only marginally socialized into the prevailing moral structures of a society or an institution, or the complete overhaul of the surrounding system of political legitimacy. Moreover, all degrees of change, from small reinterpretations to major transformations, come about as a result of disagreement between major political actors. This means that the present study interprets the legal codification of the status of the irregular as a result of political power struggles, which take the form of definitional struggles. In this context, contingency not only means that normative structures can change in principle, but also that, at each 'critical juncture' in the evolution of a particular norm, alternative positions were debated. The subsequent chapters highlight these 'critical junctures' in the trajectory of the concept of the irregular and demonstrate the extent to which they were characterized by political power struggles.

The material upon which this study is based consists entirely of legal texts and secondary sources. This approach is not without its problems. In many cases military history focuses on big battles and famous strategic thinkers. This has unsurprisingly been influenced by and has reproduced the marginalization of irregular fighters. If military history applies itself to the study of irregular warfare, it often does so in a way that disconnects the phenomenon from the 'regular' parts of wars. Moreover, the assessments of military historians often echo the established stereotypes about irregular fighters, which this study intends to problematize. The way in which this analysis deals with these problems is to rely on as broad a selection of secondary sources as possible and to pursue every clue that could unravel the established narratives about irregular fighters. Needless to say, this approach will have missed numerous empirical facts that could have been revealed only through thorough archival research. Nonetheless, conducting archival research on a phenomenon as ubiquitous as irregular fighters over a span of several centuries would have rendered this study unmanageable. No doubt this study will be only a first step, and future works will complement, reinforce, or challenge the argument put forward here.

The central argument of the book is that the puzzle of the dichotomy between the irregular and the regular can be resolved by investigating its historical evolution. It first emerged during the French Revolutionary Wars, as the distinction between 'the nation in arms' and popular armed revolts became ambiguous but crucial for the emerging authority of the nation state. Since then, it was reinterpreted during and after crucial historical episodes, such as the American Civil War and the Franco-Prussian War, the Second World War and the wars of decolonization. Hence, the dichotomy between the irregular and the regular can be understood neither in terms of its linear, formative historical origin, nor in terms of any rational or moral purpose. Instead, it is best explained as a process of historical contingencies in which various states (and other actors) pushed for various forms of distinctions between regular and irregular fighters at various times, and these distinctions developed into morally and legally accepted norms. These norms, in turn, affected states' attitudes towards irregular fighters. My main concern regards this complex interplay of political power and the power of moral and legal norms.

The focus of the book is predominantly Eurocentric. This choice was deliberate, as most of the efforts towards the codification of the law of armed conflict, in particular up to the Second World War, originated in Europe and North America. However, Chapter 5 on colonial warfare and the wars of decolonization explicitly highlights the interplay between the treatment of irregulars in the centre and in the periphery. It shows that both aspects were deeply intertwined. However, until the second half of the twentieth century, voices from the periphery did not feature largely as agents in the process of the legal codification of the exclusion of the irregular. The intention is by no means to diminish the relevance of non-Western approaches to the problem of irregular fighters. The conclusion will re-evaluate the Western trajectory from an alternative, non-Western approach and ask to what extent the latter could serve as a corrective to this trajectory.

However, before we can start with the analysis, it is necessary to clarify the subject of the study—in other words, to define what an irregular fighter is. One important limitation of this book is that it focuses exclusively on particular irregular fighters: partisans, guerrillas, rebels, insurgents, and terrorists. Thus it leaves aside other categories of combatants that are sometimes subsumed under the general header of irregular fighters, such as pirates and mercenaries. While pirates and mercenaries underwent a process of exclusion that was similar and at times parallel to that of guerrillas, insurgents, and terrorists, it was subject to different bodies of law. As this study aims to trace the exclusion of the irregular as an effect of the codification of the law of armed conflict from the Lieber Code to The Hague and Geneva, covering pirates and mercenaries as well would have made the study too comprehensive. While such an undertaking is not necessarily unfeasible, it would have meant sacrificing depth for breadth.[70]

[70] For a treatment of the development of legal and moral norms on mercenaries, see Sarah Percy, *Mercenaries: The History of a Norm in International Relations* (Oxford: Oxford University Press, 2007). For a comprehensive account on mercenaries and pirates, see Janice E. Thompson, *Mercenaries,*

What is an irregular fighter? The existing literature offers four characteristics that might serve as components of a definition: the use of irregular tactics, the adoption of the weaker position in an 'asymmetric' conflict, the intensely political character, and the lack of proper authority. These four characteristics will be discussed below. As we shall see, none of them provides a sufficiently distinctive feature that would allow us to define the irregular fighter. This section will hence suggest a different approach to the problem of defining irregular fighters.

The use of irregular tactics is the most commonly mentioned characteristic of irregular fighters and often serves as the main definitional criterion. In his latest heavyweight volume entitled *Invisible Armies*, Max Boot, for instance, defines guerrilla warfare as 'the use of hit-and-run tactics by an armed group directed primarily against a government and its security forces for political or religious reasons'.[71] Apart from hit-and-run tactics, the tactical repertoire of irregular fighters is said to include ambushes and surprise attacks and a general refusal to give battle, terrorist tactics such as assassinations, hostage-taking and suicide attacks, and the use of 'human shields'. Some accounts also attribute the use of 'extreme savagery' specifically to irregulars.[72]

However, the definition of the irregular on the basis of the tactics that they use is problematic. Irregular tactics do, of course, exist, and irregular fighters use them very often, but they are not the exclusive preserve of irregular fighters. Regular armies, for instance, have used ambushes and surprise attacks during all periods of history. Neither is increased mobility a distinctive feature of irregular fighters. As we shall see in the next chapter, the mid-eighteenth century saw a proliferation of doctrinal writings on the use of light infantry and light cavalry units that praised the advantage conveyed by their increased mobility and lower vulnerability. Fighting battles is frequently regarded as the pinnacle of 'regular' warfare, but there were and still are important episodes in the history of modern warfare during which the commanders of regular armies did everything to avoid battles. In the eighteenth century Maurice de Saxe famously remarked: 'I do not favor pitched battles, especially at the beginning of a war, and I am convinced that a skillful general could make

Pirates, and Sovereigns: State-Building and Extraterritorial Violence in Early Modern Europe (Princeton: Princeton University Press, 1996).

[71] Max Boot, *Invisible Armies: An Epic History of Guerrilla Warfare* (New York: Norton & Company, 2013), p. xxii. See also John Arquilla, *Insurgents, Raiders, and Bandits: How Masters of Irregular Warfare have Shaped our World* (Lanham, MD: Ivan R. Dee, 2011), 3 ff.; Robert B. Asprey, *War in the Shadows: The Guerrilla in History* (London: Macdonald and Jane's, 1976), p. xi; Lewis H. Gann, *Guerrillas in History* (Stanford, CA: Hoover Institution Press, 1971), 1. Carl Schmitt lists two tactical criteria in his definition of the 'partisan': 'irregularity' and 'increased mobility' (*Theory of the Partisan*, 22). Cf. also Ian F. Beckett, *Modern Insurgencies and Counterinsurgencies: Guerrillas and their Opponents since 1750* (London: Routledge, 2001), 2; Walter Laqueur, *Guerrilla Warfare: A Historical and Critical Study* (New York: Transaction, 1998), p. xvi.

[72] Gann, *Guerrillas in History*, 22.

war all his life without being forced into one.'[73] Recent US air power doctrine placed a premium on 'rapid dominance' and the strategic paralysis of the opponent, which, if successful, would also imply the avoidance of battle.[74] Finally, as the debate over state terrorism demonstrates, the use of terrorist tactics is also not the sole preserve of irregular fighters.[75] As the previous section showed, neither is the use of human shields limited to irregular fighters. Definitions of the irregular fighter based on tactics are hence not sufficiently distinctive of this phenomenon to be useful.

Implicit in the tactical definition of irregular fighters is often the notion that irregular fighters are typically involved in 'asymmetric warfare'. The typical 'asymmetric adversary' in modern times is a 'sub-state actor' fighting against a powerful state actor.[76] Characteristically, asymmetric warfare has been described as the ability of weaker opponents to use 'the adversary's advantages against them'.[77] However, as one critic remarks:

> the idea of using the adversary's advantage against them is nothing new—it is a principle of all strategy, from Sun Tzu through Machiavelli to Napoleon.... In other words, if asymmetric war is defined in terms of disparities in either military technologies or in tactics, then this would suggest that all warfare is to some extent asymmetric.[78]

But, if 'asymmetric warfare' is itself an unclear concept that fails to distinguish between different forms of warfare, it cannot possibly serve as the basis for a definition of the irregular fighter.

A third approach to the definition of the irregular fighter emphasizes his or her profoundly political character. In the words of John Ellis:

> Guerrilla warfare is not even to be regarded simply as a military phenomenon. The logic of its very nature often means that it is just as much a political activity as a military one.... Because the guerrilla army has to be built from the bottom upwards, its members have to be persuaded to fight rather than being coerced by existing administrative and repressive structures. The guerrilla leaders, therefore, need to show clearly that it is in the people's interest to fight, by appealing either

[73] Quoted in John A. Lynn, *Battle: A History of Combat and Culture* (New York: Basic Books, 2003), 130.

[74] Cf. Harlan Ullman and James Wade Jr, *Shock and Awe: Achieving Rapid Dominance* (Washington: National Defense University, 1996).

[75] See, e.g., Robert Jackson, *Contemporary State Terrorism: Theory and Practice* (London: Routledge, 2010).

[76] See, e.g., Rod Thornton, *Asymmetric Warfare: Threat and Response in the 21st Century* (Cambridge: Polity, 2007), 3.

[77] Christopher Bellamy, '"Tools of Ill-Omen": The Shifted Conflict Paradigm and Reduced Role of Conventional Military Power', *Cambridge Review of International Affairs*, 15/1 (2002), 152.

[78] Yves Winter, 'The Asymmetric War Discourse and its Moral Economies: A Critique', *International Theory*, 3/3 (2011), 493 ff.

> to feelings of nationalism and xenophobia or to the wish for social and economic changes after the war.[79]

However, this argument is not unanimously shared among scholars. Paul Collier, for instance, challenges the notion that irregular fighters in civil wars are fighting for political motives. According to Collier, the predominant motive is 'greed' rather than 'grievance'.[80] While Collier's argument is reminiscent of the denunciation of irregular fighters as mere criminals and should be treated with caution, others have pointed out that, in civil wars, actors on the ground typically have mixed motives, as 'political' motivations to fight become tied up with personal motives such as revenge, gain, and so on.[81] Moreover, what is problematic about this argument is that it seems to imply that the motivations of regular soldiers are 'non-political' or 'apolitical' in comparison. However, a plurality of regular soldiers exists, ranging from local militias through conscripts to professional forces. They have a range of different political or non-political reasons to enlist and to fight. Some authors even suggest that strong political and ideological convictions among regular forces enhance their combat performance.[82] Last, but not least, the assumption that irregular fighters act in a particularly 'political' way leaves open the question of what 'political' precisely means in this context.

Finally, irregular fighters are sometimes defined as those actors in a conflict who take up arms without 'right authority'.[83] The problem with this criterion is, of course, that, in almost every armed conflict involving irregular fighters, the question of who has the 'right authority' is at the heart of the conflict. Quite often the question of 'right authority' can be decided only with hindsight. When irregular fighters succeed at deposing the government forces they fight against and become the new government, as happened in Libya in 2012, they acquire in the process the position of 'right authority'. Sometimes irregular fighters include the defeated and scattered forces of a dethroned monarch, as was the case during the Peninsular War, when Napoleon's army fought against the remnants of the army of King Ferdinand of Spain. Most of the *ancien régime* European states certainly regarded King Ferdinand as possessing the 'right authority' and not Napoleon. The wars of decolonization provide further examples for contested authority as the main bone of contention in

[79] John Ellis, *From the Barrel of a Gun: A History of Guerrilla, Revolutionary and Counter-Insurgency Warfare, from the Romans to the Present* (London: Greenhill Books, 1995), 32. See also Schmitt, *Theory of the Partisan*, 22.

[80] Paul Collier and Anke Höffler, 'Greed and Grievance in Civil War', *Oxford Economic Papers*, 56/4 (2004), 563–95.

[81] Stathis N. Kalyvas, *The Logic of Violence in Civil War* (Cambridge: Cambridge University Press, 2006).

[82] Omer Bartov, *Hitler's Army: Soldiers, Nazis, and War in the Third Reich* (Oxford: Oxford University Press, 1992).

[83] Cf. Watkin, 'Warriors without Rights?', 13.

irregular warfare. Some of the accounts of irregular warfare today seem to substitute the just war theory notion of 'right authority' simply with the authority of the state as such. In this view, irregular fighters are distinctly 'non-state' actors. However, as mentioned above, the concept of the 'non-state' actor is not particularly helpful, not only because it fails to capture the complex relationship that irregular fighters tend to have with states and statehood, but also because it lumps together a plurality of different actors, ranging from terrorists to mercenaries to humanitarian aid organizations.

In sum, the irregular fighter seems impossible to define accurately. The irregular fighter's chameleon-like character is both cause and effect of his or her exclusion from legal protections. Because their manifestations are manifold and difficult to pin down, the law defines irregular fighters *ex negativo* as all those who fail to meet the criteria for combatant privilege. And because the law is based on such a definition *ex negativo*, the contours of the irregular remain fuzzy and unclear. But this legal definition also means that the concept of the irregular fighter emerges from a hierarchical dichotomy in which the 'regular' is taken as the regulative ideal and the 'irregular' as its deviant and morally inferior conceptual opposite. Pitting the regular against the irregular first of all orders the battlefield. It 'opposes the realm of reasoned understanding against a realm of anarchical, threatening Otherness'.[84] The threatening quality of the irregular evolves from two overlapping features. First, the irregular is the conceptual opposite of the regular. He or she is not the heroic, honourable, and honest soldier fighting for the good of his or her nation, but the ruthless and cruel criminal acting out of self-interest. Second, irregulars challenge the basic legal categories: they do not distinguish themselves from civilians and thus threaten the neatly ordered battlefield as such. Moreover, they challenge established legal distinctions, most importantly the distinction between *jus in bello* and *jus ad bellum*, by bringing the question of their potentially just cause back in. They are lacking discipline in two ways: they are not part of a chain of command, nor do their tactics fit the conceptual battle formation.

Hence the impact of the concept of the irregular fighter can be felt mainly at the ideational level of armed conflict. As Patrick Porter observed: 'Armed conflict is an expression of identity as well as a means to an end.'[85] While opponents in a war mobilize their material resources in order to prevail on the battlefield, they cannot do so without at the same time harnessing ideas and attitudes about themselves and their enemies. Opponents who are identified as not belonging to the same category of actors as the self, in other words,

[84] Richard K. Ashley, 'Untying the Sovereign State: A Double Reading of the Anarchy Problematique', *Millennium: Journal of International Studies*, 17/2 (1995), 202.

[85] Porter, *Military Orientalism*, 1.

opponents who are perceived as the 'Other', are characteristically assigned attributes of both inferiority and superiority.[86] They are seen as inferior, because they are regarded as 'savage', 'uncivilized', and 'undisciplined'. At the same time, their negative attributes can be perceived as an advantage: 'uncivilized' behaviour can translate into an increased martial spirits and fighting prowess. In this context, the introduction of hierarchical dichotomies, such as that between the regular and irregular, are a means of taming the ambiguity of the 'Other' and of establishing a certain degree of order on the battlefield. Conversely, at times when the identity of the regular armed forces is in crisis—for instance, owing to changes in recruitment patterns or military technology—harnessing the irregular as the 'conceptual opposite' of the regular can contribute to re-establishing a sense of self-identity of the regular armed forces.

Against this background, this study proposes the following definition of the irregular fighter: *the concept of the irregular fighter is an ideational weapon of war. Charging an opponent with being an irregular fighter has important ramifications in the battle of ideas and definitions. It is a means of denouncing someone for fighting an allegedly unfair fight, for fighting for the wrong motives, and/or for having no right to take up arms.*

However, defining the irregular as an intellectual artefact also implies that the subsequent chapters will take the *concept* of the irregular rather than the 'real' manifestations of irregular fighters as its main empirical object. One consequence of this approach is that the category of the irregular often remains underspecified with respect to its constituent parts and subcategories. The reason for this is twofold. First, as mentioned above, it is one of the essential characteristics of the discourse on irregular fighters that it tends to lump together different groups of actors, thereby implying an artificial homogeneity where none exists. Second, the 'real' manifestations of irregular fighters are difficult to observe, since all available empirical material tends to be discursive by nature. The only way in which definitional hegemonies can be 'deconstructed', to use this once radical but nowadays almost old-fashioned sounding term, is to trace instances of dissent and the voicing of alternative opinions. The subsequent chapters may thus frustrate readers who had hoped for more 'order' within the broad category of the irregular fighter. Instead, the chapters will highlight the dynamic with which the stereotypical application of the label 'irregular' (or any of its substitutes such as 'brigand', 'bushwhacker', 'guerrilla', or 'terrorist') affected political, strategic, operational, and tactical decisions in modern wars. Thus it focuses on what the concept of the irregular *does* rather than who the irregular *is*. Whenever possible, the

[86] Porter, *Military Orientalism*, 42.

study will also cite alternative opinions on the rationality of such labelling—for instance, the plea of some contemporaries of the American Civil War, including Francis Lieber, to differentiate between various categories of irregulars which Union general Henry Halleck, however, chose to ignore.

Structure of the Book

The book is structured along the lines of major historical steps in the genealogy of the irregular fighter. Only the fourth chapter on colonial wars and wars of decolonization will slightly break with this historical narrative in order to trace how dynamics between the centre and the periphery interacted in the trajectory of the concept of the irregular fighter.

The next chapter focuses on the 'making' of the irregular fighter in the second half of the eighteenth century. The making of the irregular is at once characterized by continuity and change. On the one hand, the middle of the eighteenth century saw a heightened interest in 'irregular warfare' as part of the tactical repertoire of states. On the other, the French Revolution profoundly changed the notion of the irregular. It was the emergence of the idea of the nation as the collective wielder of military power and its convergence with the state that triggered the marginalization of the irregular. The nationalization of war led to an increased relevance of the problem of irregular fighters, as conscription was the main trigger of counter-revolutionary uprisings. But it also posed profound questions about the legitimacy of the nationalized armed forces and their opponents: if the Army of the French Revolution was effectively the nation spontaneously rushing to the 'defence' of the fatherland, what about the popular uprisings *against* the revolution? Were they not similarly spontaneous expressions of the will of the people? The answer given in the eighteenth century was that the counter-revolutionaries were illegitimate because they were 'irregular'. The early concept of the irregular tended to function as a smokescreen upon which political protagonists projected all sorts of negative traits also held by the early nationalized armies themselves, in particular their indiscipline and propensity to plunder; in doing so, the nationalized armies symbolically 'purged' themselves of the more sinister aspects of the revolution.

The third chapter traces the emergence of the law of armed conflict and the way in which it codified and 'froze' the dichotomy between the regular and the irregular. It explores how two central wars of the nineteenth century, the American Civil War and the Franco-Prussian War, dispelled the hope that the Congress of Vienna had successfully re-established the pre-revolutionary European system along with its rules and customs of war. These two wars provoked profound confusion both in Europe and in the USA over the question of who qualified as a legitimate combatant. Thus they evidenced the need for a new and lasting legal system. The emergence of this new legal system, however, was not a consensual development. Rather, it was the focus of great international contestation, since it touched upon power interests such as the rights and duties of occupying powers and the legitimacy of a *levée* en masse in occupied territory. The creation of a legal regime had a strong impact on the distribution of power within the international system, for instance, on the balance of power between great powers and small states. The latter insisted that 'irregular' levies were a legitimate means of defence, while the former equated irregulars with an unnecessary obstacle to their large armies. Apart from tracing the legal development, Chapter 3 also shows how technological innovations such as the introduction of more accurate firearms had an impact on both the image of the regular soldier and its conceptual opposite, the irregular.

Chapter 4 explores the relevance of irregular fighters in the Second World War.[87] Mainstream accounts of the Second World War tend to depict it as a regular, 'interstate' conflict par excellence. However, resistance movements and partisans played a large role on the western, eastern, and south-eastern fronts. Even if most of these irregular armed groups did not have a large impact on the strategic outcome of the war, their involvement left an important political legacy. This chapter discusses German anti-partisan warfare from a perspective of continuity in the history of anti-irregular warfare. It puts forward the argument that, while the German forces' ideological dispositions such as racism, anti-Semitism, and *Lebensraum* ideology doubtlessly played a role in the escalation of violence, in particular on the eastern front and in the Balkans, existing scholarship has not paid sufficient attention to anti-irregular prejudices as a source of unrestrained behaviour. It also highlights an aspect that is frequently forgotten in the literature on the Second World War—namely, the extent to which the Second World War signified the return of the irregular as the auxiliary of regular forces. The chapter concludes by highlighting that the Second World War brought a moral re-evaluation of the irregular, as the anti-German resistance movements undoubtedly became morally vindicated after the war. However, this re-evaluation had only an indirect impact on the reinterpretation of the law of armed conflict after the war. The 1949 Geneva Conventions focused instead on protection for civilians, while introducing only minimum protection for actors in non-international armed conflicts in common article 3 of the Conventions.

[87] The book skips the First World War, because it was one of the few conflicts in which irregular fighters did not play a significant role, at least in the European theatres of war. While Europe's peripheries saw irregular fighting in the framework of the Arab Revolt (1916–18) and the Russian Civil War (1917–22), these instances did not trigger major developments in the historical trajectory of the irregular fighter.

The fifth chapter enquires into the role irregular fighters played in wars in the colonies and into the effect these wars had, in turn, on the development of the concept of the irregular fighter. It was necessary to focus on a limited selection of wars: the Haitian Revolution, the Anglo-Zulu War, the Boer War, the war in German South West Africa, the Maji Maji uprising, the Malayan Emergency, the Mau Mau uprising, and the Algerian war. The chapter demonstrates that the treatment of irregulars in wars in the colonies did not differ fundamentally from the practices used in wars in the centre. Moreover, this chapter traces how the concept of the irregular, re-evaluated after the Second World War, impacted on the wars of decolonization in the second half of the twentieth century. It argues that, even if this re-evaluation had not immediately led to a reassessment of the concept of the irregular in the framework of the law, in particular the Geneva Conventions, it did empower irregular fighters in the colonies. Ultimately, this empowerment was reflected in the 1977 Additional Protocol to the Geneva Conventions.

The sixth chapter addresses the emergence of the irregular as the 'unlawful combatant' in the framework of the so-called war on terror. In spite of the ubiquitous assertion by the Bush administration that the 'war on terror' was a new kind of conflict in which the established legal rules would no longer apply, the legal arguments on which the notion of the 'unlawful combatant' was based demonstrate a conspicuous continuity with established exclusionary mechanisms towards irregular fighters. The chapter not only aims to highlight the internal inconsistencies of the 'unlawful combatants' concept, but also argues that the concept is at odds with two other manifestations of irregular fighters at the beginning of the twenty-first century: the irregular as the target of drone attacks and other instances of targeted killing and the irregular as the 'rebel' auxiliary in the West's grand strategies.

The conclusion summarizes the main arguments of the book and highlights their implications for two sets of issues: the question whether the law of armed conflict is outdated, as has been frequently claimed since 9/11, and the issue of how the cooperation between the state and irregular forces in the form of auxiliaries ought to be assessed in the framework of Western strategy-making in the twenty-first century. It concludes with a critical evaluation of the dynamic dichotomy between the regular and the irregular as conceptual opposites and argues that the twenty-first-century revival of the irregular as 'unlawful combatant' had to do not so much with the strategic context of the 'war on terror', but rather with an identity crisis of Western armed forces.

2

The Making of the Irregular Fighter, 1740–1815

In the course of the eighteenth century, the concept of the irregular slowly evolved, as did the concept of regular warfare. At first, these two mutually referencing terms referred to merely tactical issues, and irregular warfare was not yet equated with illegitimate warfare. The concept of illegitimate war did exist, but it mainly covered two forms of war that were not necessarily 'irregular' warfare: peasants' uprisings and wars in the peripheries of Europe or in the colonies. In the first case the illegitimacy of the enemy was based on class, in the second on race. The use of irregular tactics or particularly savage ways of fighting did play a role in denouncing these kinds of opponents as illegitimate; however, their 'irregular' ways of fighting were not the primary reason for their perceived illegitimacy. Rather, their particular ways of fighting were interpreted through the lens of their perceived inferiority as 'peasants' or members of 'uncivilized races'.[1]

A number of military treatises on irregular tactics were published in the mid-eighteenth century. They mainly reflected a backlash in tactical thinking against an increasing trend towards heavier forces, more rigid formations, and a reliance on firepower. Against this trend, authors such as Turpin de Crissé emphasized the advantages of light infantry and cavalry: their manœuvrability, their capacity to surprise and ambush the opponent, and their comparatively low vulnerability. Rulers and strategists in Europe soon came to

[1] Martin Luther's denunciation of the German Peasants' Revolt (1524–6) is a paradigmatic example of the pre-eighteenth-century perspective on peasants' uprisings. While Luther did charge the peasants of having committed atrocities, this charge was framed by two other accusations that targeted the peasants' social origin rather than their ways of fighting: Luther criticized that the peasants had broken their oath of loyalty to their rulers and that they had misinterpreted biblical statements about the equality of all humankind and about the collective good as applying to worldly matters and not only spiritual questions. Martin Luther, 'Wider die räuberischen und mürderischen Rotten der Bawren', *Weimarer Ausgabe*, 18 (Weimar: Böhlau, 1908 [1525]), 357–61. On the question of race and the so-called standard of civilization in relation to the perceived illegitimacy of military opponents, see Chapter 5.

value irregular troops as indispensable auxiliaries to the regular armed forces, although they did not accord them an independent role. Hence, what came to be seen in the nineteenth century as an intolerable challenge to the 'norm' of regular warfare was still seen in the mid-eighteenth century as a vital aspect of European states' force structures. In other words, the irregular initially emerged from the state, and not in opposition to the state, as the equation of irregular fighters with 'non-state' actors implies. This chapter will trace the emergence of the irregular in three overlapping steps: the emergence of irregular auxiliary forces (1740–83), the creation of militia systems and the *levée en masse* (1777–97), and the reconceptualization of the irregular as the conceptual opposite of the regular soldier (1793–1815).

Auxiliary Forces: Useful 'Savages'

Irregular Warfare: European or Indian?

According to common wisdom, the late seventeenth and the eighteenth centuries were the heyday of limited and restrained warfare, at least within Europe. The era is largely seen as a period of respite between the religious conflicts of the first half of the seventeenth century, ending with the Peace of Westphalia, on the one hand, and the nationalistic fervour of the French Revolutionary Wars, and later the unlimited aims of Napoleonic warfare, on the other. Contemporary witnesses such as the Swiss lawyer Emerich de Vattel corroborated this impression when he wrote in 1758 that 'the humanity with which most nations in Europe carry on their wars at present cannot be too much commended'.[2] Often this account of limited war is linked to the idea that the 'military revolution' in early modern Europe was characterized by an increase in state control over the use of military force.[3] The emergence of the modern European state system and the transformation of war were seen as mutually reinforcing: 'War makes the state and the state makes war,' in Charles Tilly's words.[4]

The argument underpinning this idea is that, with increasing troop numbers and the move towards standing armies, the state had to become more efficient at providing the resources necessary to wage war. Standing armies

[2] Emmerich de Vattel, *The Law of Nations; or, Principles of the Law of Nature Applied to the Conduct and Affairs of Nations and Sovereigns* (New York: AMS Press, 1982 [1758]), 482.

[3] Michael Roberts, 'The Military Revolution, 1560–1660', in Michael Roberts (ed.), *Essays in Swedish History* (London: Weidenfeld & Nicolson, 1967), 195–225; Geoffrey Parker, *The Military Revolution: Military Innovation and the Rise of the West, 1500–1800* (Cambridge: Cambridge University Press, 1988). For an extensive discussion of the 'military revolution' thesis, see also Chapter 1.

[4] Charles Tilly, *Coercion, Capital, and European States, AD 990–1990* (Oxford: Blackwell, 1992), 65.

received better and lengthier training. Thus both their maintenance and their training were more costly than before. It was argued that, since states needed to be more efficient with their costly military resources, they were less likely to sacrifice large numbers of troops in battle, thus fostering a more restrained approach to war. Increasing state control over the military enabled absolutist rulers to use military force for narrowly defined dynastic interests rather than religious convictions. Better training, increasing drill, and the deconfessionalization of war aims also meant more discipline: warfare was less likely to slip into slaughter. Alternatively, some authors stressed that, despite the rise of the absolutist state, its rulers still had to rely on cooperation with the societal elite of aristocrats in order to be able to afford efficient and professional armies.[5] This approach highlights the importance of the aristocratic code of honour. From this perspective, war was a highly ritualized undertaking, shaped by prevailing aristocratic attitudes about politeness and courtesy. This is epitomized in an anecdote about the 1745 Battle of Fontenoy during the War of Austrian Succession; it is said that the British commander Sir Charles Hay offered his French counterpart Count d'Anterroches the opportunity for the French forces to fire the first round, which the latter politely declined.[6] But, regardless of whether the accounts of war in the late seventeenth and early eighteenth centuries stress technological, institutional, or cultural aspects, the era is often depicted as restrained and highly disciplined—in other words, as the prime example of limited 'regular' warfare.

If irregular warfare plays a role at all in the historiography of the eighteenth century, it seems to be located outside Europe, in the peripheries and in the colonies, and is often depicted as an anomaly. Colonial war in North America has frequently been discussed from this perspective. John Shy, for instance, argued that British colonists' confrontation with a different kind of opponent—Native Americans, also in coalition with French and Spanish Catholics—led to the emergence of a particular 'American way of war', one that was much closer to the early seventeenth century's unrestricted warfare than to the limited way of war that had come to characterize eighteenth-century Europe: 'The effect was to perpetuate, and in some areas of English settlement even to intensify, the quality of the seventeenth-century military experience.'[7] More generally, John Grenier held that the presence of the 'frontier' had a profound impact on the development of a specifically American combat culture. Because British colonists in North America were faced

[5] Cf. Jeremy Black, 'Introduction', in Jeremy Black (ed.), *European Warfare, 1453–1815* (Houndmills: Palgrave Macmillan, 1999), 1–22.

[6] David A. Bell, *The First Total War: Napoleon's Europe and the Birth of Modern Warfare* (London: Bloomsbury, 2007), 35.

[7] John Shy, 'The American Military Experience: History and Learning', *Journal of Interdisciplinary Research*, 1/2 (1971), 213.

either with Native Americans fighting in an irregular way or with the French using Native American auxiliaries, they developed a way of war that focused on the annihilation of the opponent and left no room for moderation or restraint: 'In the frontier wars between 1607 and 1814, Americans forged two elements—unlimited war and irregular war—into their first way of war.'[8] Others argued that American combat culture developed in a dualistic way. On the one hand, it was strongly influenced by European style warfare and the idea of 'honours of war' and restraint. However, this approach was essentially reserved for white opponents. Native Americans, on the other hand, met with far more ruthless treatment.[9]

The unifying theme of these accounts is that irregular warfare occurred only on the margins of the eighteenth-century international system, while the centre—Europe—had developed more restrained and civilized ways of fighting. Irregular warfare, in other words, occurred either as the shadow of the past or as the spontaneous emulation of the opponent. That is, faced with a barbarian and backward opponent, Europeans had no other choice but to fall back upon long-abandoned tactics and practices. Preferably, they charged auxiliary forces with this task. In this vein, George Washington commented on the British use of Native American auxiliary forces at the beginning of the Seven Years War in America (French and Indian War, 1754–63): 'Indians are the only match for Indians, and without these, we shall always fight upon unequal terms.'[10]

Yet the French and Indian War had far more irregular aspects than the use of Native American auxiliaries or the spontaneous recourse to irregular tactics by the French or the British. British commander Lord Loudoun remarked in 1756 that 'whoever is Superior in irregulars has an infinite advantage over the other side; and must greatly weaken, if not totally destroy them [the irregulars of the other side] before they can get to the Point where they can make their Push'.[11] The competition for irregular fighters occurred only partly in the form of the recruitment of Native American auxiliaries by both sides. The British also sought to gain advantage by deploying irregular troops of English colonists, most notorious among them Robert Rogers's 'Rangers'. Last but not least, the British 'irregularized' parts of their regular troops by breaking up rigid lines and by using tactics that combined fire and manœuvre. In order to become

[8] John Grenier, *The First Way of War: American War Making on the Frontier, 1607–1814* (Cambridge: Cambridge University Press, 2005), 10.

[9] Ian K. Steele, *Warpaths: Invasions of North America* (Oxford: Oxford University Press, 1994), 136.

[10] Quoted in Robert B. Asprey, *War in the Shadows: The Guerrilla in History* (London: Macdonald and Jane's, 1976), 104.

[11] Quoted in Grenier, *The First Way of War*, 126–7.

more proficient in irregular warfare, part of the British regular forces was sent to train with Rogers's Rangers.[12]

It was also not the case that the Europeans adopted irregular measures only because of their confrontation with the colonial Other. Rather, officers serving in the French and Indian War were often familiar with irregular tactics from their experience in Europe. The common historical narrative tends to conceal that mid-eighteenth-century Europe deployed irregular warfare as a tactical repertoire alongside regular tactics. Moreover, this practical experience with irregular warfare in Europe was reflected in a number of publications on the subject by military writers such as Turpin de Crissé, Thomas Antoine le Roy de Grandmaison, Capitaine de Jeney, Count de la Roche, Roger Stevenson, Johann von Ewald, Andreas Emmerich, Georg Wilhelm Baron von Valentini, Gerhard von Scharnhorst, Friedrich Leopold Klipstein, and Carl von Decker.[13]

English and French officers stationed in eighteenth-century America had already encountered three different forms of 'irregular war' in Europe: first, peasant uprisings, second, people in arms such as the Jacobite rebellions in the Scottish Highlands, and, finally, irregular troops acting as auxiliaries to regular forces—for instance, the Pandours, Croats in Austrian service during the War of Austrian Succession, and Cossacks serving with the forces of the Russian empire during the Seven Years War.[14] These auxiliary troops tended to originate from the border regions of Europe and were used to face non-European opponents: Turkish opponents in the case of the Pandours and hussars from the Balkan region and Tatar opponents from the steppe border region in the case of the Russian Cossacks. The history of these irregular troops often dated back to the fourteenth or fifteenth century, but it was in the mid-eighteenth century that their value was discovered in warfare in the centre of Europe. In the course of this process, the names of these troops frequently evolved from an ethnic towards a tactical meaning, as was the case with the hussars, for instance: while the name initially referred to a specific tribe of horsemen in medieval Hungary, it later came to designate units of light cavalry that virtually all European states deployed.[15] But irregular troops that supported a state's defence against invasion could also be found in border regions *within* Europe, such as the border between France and Catalonia as well as France and Piedmont. Both Catalonia and Piedmont had tolerated the existence of more or less autonomous local militias—the *miquelets* in the case of Catalonia and

[12] Peter E. Russell, 'Redcoats in the Wilderness: British Officers and Irregular Warfare in Europe and America, 1740 to 1760', *William and Mary Quarterly*, 35/4 (1978), 629–52.

[13] Beatrice Heuser, 'Small Wars in the Age of Clausewitz: The Watershed between Partisan War and People's War', *Journal of Strategic Studies*, 33/1 (2010), 142.

[14] Armstrong Starkey, *European and Native American Warfare, 1675–1815* (London: UCL Press, 1998), 47.

[15] Cf. Heuser, 'Small Wars in the Age of Clausewitz', 143.

the *barbetti* in Piedmont—because they had risen against French attacks throughout the eighteenth century.[16]

Irregular troops played a large role in the War of Austrian Succession (1740–8). The Battle of 1745 Fontenoy, for instance, which is mostly cited as an example of regular warfare and aristocratic combat culture, saw large use of irregulars.[17] Some argued that the irregular auxiliary troops that were used alongside regular units were in fact a new phenomenon, and not just an innovative form of light infantry or cavalry detachments: 'Croats and Pandours became notorious for a new and higher level of brutality in warfare.... These troops drew no distinction between war and robbery or between combatants and non-combatants.'[18] However, historical accounts of the impact of the newly emerging 'irregular' units vary in their assessment of the perception of contemporaries. According to Armstrong Starkey, 'British officers who encountered the Pandours portrayed them as a murderous mob who mocked the laws of war';[19] this claim seems to be corroborated by others' accounts of the behaviour of Colonel Franz von der Trenck's Pandour regiment.[20] These accounts also challenge the assumption that aristocratic leadership automatically equals a highly ritualistic and restrained combat culture. In contrast, Peter Russell remarked with respect to British encounters with the French *Grassins* that the British 'were not bewildered by the new French tactics, nor did they describe them as unfair or barbarous'.[21]

Hence, European officers in mid-eighteenth-century Europe had many opportunities to familiarize themselves with irregular warfare and to come to appreciate its potential. In fact, from the beginning of colonial warfare in North America, irregular tactics were imported from Europe. The practice of 'feed fights', for instance, which was intended to deprive the opponent of their means of subsistence, had been used during the Irish rebellions before it was deployed in sixteenth-century colonial warfare in North America: 'the ruthlessness and cruelty displayed in abundance in these conflicts [wars in colonial America] were not the fault of the Native Americans; the English conquerors of Ireland did not have to be taught how to be ruthless and cruel'.[22] British officers serving in the French and Indian War had often directly encountered irregular warfare in Europe before this: Lord Loudoun,

[16] Michael Broers, *Napoleon's Other War: Bandits, Rebels and their Pursuers in the Age of Revolution* (Oxford: Peter Lang, 2010), 54.

[17] Russell, 'Redcoats in the Wilderness', 634.

[18] Starkey, *European and Native American Warfare*, 50; cf. Walter Laqueur, 'The Origins of Guerrilla Doctrine', *Journal of Contemporary History*, 10/3 (1975), 341–82.

[19] Starkey, *European and Native American Warfare*, 50.

[20] See Chapter 1.

[21] Russell, 'Redcoats in the Wilderness', 634.

[22] Harold E. Selesky, 'Colonial America', in Michael Howard, George J. Andropoulos, and Mark R. Shulman (eds), *The Laws of War: Constraints on Warfare in the Western World* (New Haven: Yale University Press, 1994), 84.

commander-in-chief of the armed forces in North America from 1756 to 1758, for instance, had seen the 1745–6 Jacobite uprising.[23] Henry Bouquet, who led the British troops to victory over Native American forces in the Battle of Bushy Run in 1763, had had first-hand experience with irregular warfare, having served with the Savoyard army against the French in 1743.[24] Thus, what many scholars described as 'Indian warfare' on closer inspection turned out to be a variety of European warfare.

Conversely, what Europeans encountered as 'Indian warfare'—that is, the conduct of Native Americans on the battlefield—was an adaptation to the new weapons technologies that Europeans had brought to America. Native American warfare before the arrival of the gun had been mostly limited, ritualized, and rather low in mortality. The introduction of more deadly weapons forced the native population to adapt their tactics and to move towards guerrilla warfare. In the words of one contemporary: 'Since the Introduction of Fire Arms [the Iroquois] no longer fight in close bodies, but every man is his own General.'[25]

The Irregular as the 'Savage'

A vital question is how irregular warfare in the colonies was perceived in moral and legal terms. Clearly, the denunciation of the enemy as 'savage' and 'barbarian' played a certain role, in particular with respect to Native American opponents.[26] General James Abercromby, who followed Lord Loudon as commander-in-chief in the American theatre, reassured the French that the British would treat them with utmost humanity should they be captured on the battlefield. However, Native Americans were apparently excluded from this treatment, as it was assumed that they themselves did not treat their captives humanely.[27] 'Indian warfare' was considered cowardly and 'unsporting'.[28] General Jeffrey Amherst lamented that the 'cowardice of these Barbarians is so great'.[29] It was also Amherst who first introduced the idea of using smallpox

[23] Russell, 'Redcoats in the Wilderness', 630 ff.

[24] Russell, 'Redcoats in the Wilderness', 630 ff.; Matthew C. Ward, '"The European Method of Warring is not Practiced Here": The Failure of British Military Policy in the Ohio Valley, 1755–1759', *War in History*, 4/3 (1997), 252.

[25] Sir William Johnson quoted in James Axtell, *The European and the Indian: Essays in the Ethnohistory of Colonial North America* (Oxford: Oxford University Press, 1981).

[26] Cf. Selesky, 'Colonial America', 66.

[27] Linda Colley, *Captives: Britain, Empire and the World, 1600–1850* (London: Jonathan Cape, 2002), 185.

[28] Stephen Brumwell, *Redcoats: The British Soldier and the War in the Americas, 1755–1763* (Cambridge: Cambridge University Press, 2006), 208.

[29] Brumwell, *Redcoats*, 204.

against Native Americans.[30] Yet it is not entirely clear whether the exclusion of some fighters from military constraint was based on racial prejudice alone or on the fact that they used irregular tactics. Brigadier James Wolfe's 1759 General Orders, for instance, 'forbade swearing and scalping "except when the enemy are Indians, *or Canads* [Canadians] *dressed like Indians*"'.[31] This kind of disguise was apparently practised by some regular French troops; and, in this case, it seems as if Wolfe intended to target the tactic rather than the opponent's racial identity.

Native Americans were also perceived as undisciplined. The Fort William Henry massacre was the most noteworthy incident in this respect. In August 1757 French forces and their Native American allies forced the British troops at Fort William Henry to surrender. The terms for surrender offered parole to the British; however, in the end, the Native American auxiliaries assaulted the fort and massacred the British. It is contested whether this was indeed an incident in which the French lost control over their Native American auxiliaries or whether the French commander General Marquis de Montcalm was complicit in the massacre.[32] The British commander-in-chief, Lord Loudoun, protested to the French after the massacre and threatened that he would suspend all restraints even towards French soldiers if French officers failed to control the behaviour of their native auxiliaries: 'I chuse to carry on the war with strict Faith between the Nations and the greatest Humanity to the Particulars,—*Whatever Troops you bring into the Field are to me French*, therefore if any Part of them break through the Rules of War, will immediately lay me under the disagreeable necessity to Treat the whole of your People in the same manner.'[33]

However, Native American auxiliaries were not the only irregular forces involved in the French and Indian War. Another category of irregulars were colonial troops such as 'Rogers's Rangers'. They, too, were renowned for being 'chronically unruly and consequentially unreliable' and were 'composed of some of the most rugged individuals from an individualistic frontier'.[34] But, as in the case of the Native American auxiliaries, it is not entirely clear whether the Rangers' behaviour was due to indiscipline or whether it was based on orders by the British. Robert Rogers himself claimed that in 1755 British

[30] However, it is not clear whether he considered the use of smallpox to be vindicated by the nature of the opponent as 'savage', or whether this type of biological warfare was used because of its deniability, or maybe both. Elizabeth A. Fenn, 'Biological Warfare in Eighteenth-Century North America: Beyond Jeffery Amherst', *Journal of American History*, 86/4 (2000), 1564, 1574.

[31] Asprey, *War in the Shadows*, 106, emphasis added.

[32] Ian K. Steele, *Betrayals: Fort William Henry & the 'Massacre'* (Oxford: Oxford University Press, 1990), 78–128; Francis Jennings, *Empire of Fortune: Crown, Colonies, and Tribes in the Seven Years War in America* (New York: Norton & Company, 1988), 119.

[33] Quoted in Starkey, *European and Native American Warfare*, 102, emphasis added.

[34] John K. Mahon, 'Anglo-American Methods of Indian Warfare, 1676–1794', *Mississippi Valley Historical Review*, 45/2 (1958), 265.

General Shirley had instructed him to 'distress the French and their allies, by sacking, burning, and destroying their houses, barns, barracks, canoes, battoes, &c. and by killing their cattle of every kind; and at all times to endeavour to way-lay, attack and destroy their convoys of provisions by land and water, in any part of the country'.[35] It is hence unclear whether European commanders perceived the lack of discipline among the irregular forces as a problem or whether they tolerated or even encouraged it—possibly all of the above at different times and under different officers.

If imposing discipline among the irregulars was perceived as a problem, however, many contemporaries seem to have regarded it as a resolvable one. Unless the racist perception of the opponent played too large a role, rules and regulations were seen as an appropriate remedy against indiscipline. Many of the military treatises on irregular warfare that had been published in the mid-1750s in Europe essentially aimed at professionalizing irregular warfare.[36] At the same time, contemporaries were aware of the potential trade-off between discipline and effectiveness through spontaneity and the shock effect that undisciplined and unpredictable behaviour is likely to induce. Robert Rogers, for instance, produced a set of 'Standing Orders' for his Rangers that aimed to be sufficiently flexible to safeguard the advantages of irregular warfare but at the same time allowed for a certain degree of effective command.[37] The dilemma of imposing discipline without sacrificing spontaneity and 'popular ownership' of military force would prove to become even more intractable with the emergence of militias and the *levée en masse*.

The uncertainty with which contemporaries in the mid-eighteenth century approached the issue of irregular warfare, legitimacy, and discipline shows that this was a period of transition: the older mechanism of denouncing enemies as 'savage' because of their ethnic or racial identity was gradually complemented by the denunciation of irregular fighters based on the tactics that they use. During this period of transition both categories often overlapped, as was the case with the Native American auxiliaries, and were deliberately intermingled. Moreover, the theme of the irregular as the 'savage' persisted as part of the discourse on irregulars up to the twentieth century and proved particularly important in colonial wars.

The transformation in the perception of irregular warfare was also reflected in the legal discourse of the time. Published in 1758, Vattel's *The Law of Nations* echoed in many respects the essence of the legal thinking of his time, for instance, when he equated irregular tactics with 'savage',

[35] Quoted in Jennings, *Empire of Fortune*, 199.

[36] Starkey, *European and Native American Warfare*, 52.

[37] For details on the 'rules for ranging', see John Ellis, *A Short History of Guerrilla Warfare* (New York: St Martin's Press, 1976), 50.

non-European opponents.[38] But this book also included important innovations compared to previous legal thinking, which had been heavily influenced by Hugo Grotius. In contrast to the latter, Vattel allowed for the possibility of a rightful rebellion against tyranny and cautioned that, in such cases, the ruler should show clemency towards the rebels. The right to rebellion as such was not new; it could be traced back to thinkers such as the late medieval Italian lawyer Bartolus de Saxoferrati, but also to Locke and Montesquieu. Moreover, the right to tyrannicide was an established theme in legal scholarship at the time. However, while this tradition had focused on the just cause as the source of legitimacy for a rebellion, Vattel's innovation consisted in linking the rebels to the potential of statehood:

> If rebels in a civil war succeeded in establishing their rule over a certain territory and creating an organization similar to a state, then the government of a third party state could recognize them as a belligerent party. That was perceived as an anticipation of possible statehood, as recognition of a potential state or state government.[39]

Hence Vattel was the first scholar who explicitly acknowledged the complex and ambiguous nature of the relationship between irregular fighters and the state.[40]

Revolutions and the Issue of Inclusion or Exclusion of Irregular Fighters

On the eve of the American and the French Revolutions, a number of different categories of fighters could be excluded from legal and moral protections in war. First, there were the traditional categories of 'savages', rioting peasants, and rebels, although, as argued above, the exclusion of rebels became increasingly contentious to the extent that Enlightenment thinking promoted the idea that resistance against tyranny could be regarded as justified. The

[38] Vattel, *The Law of Nations*, 482.

[39] Carl Schmitt, *The* Nomos *of the Earth in the International Law of the* Jus Publicum Europaeum (New York: Telos Press, 2003 [1950]), 168. For more details on the legal practice of recognizing insurgency or belligerency, see Stephen C. Neff, *War and the Law of Nations: A General History* (Cambridge: Cambridge University Press, 2005), 250 ff.

[40] Note that Karma Nabulsi placed Vattel firmly in the 'Grotian' tradition of war, which saw war as the sole preserve of the state: *Traditions of War: Occupation, Resistance, and the Law* (Oxford: Oxford University Press, 1999), 36, 64. However, Vattel did introduce the possibility that rebellions could grow into civil wars, in which two equally legitimate belligerents fought each other: 'In [the] case of true civil war, the two parties were deemed to constitute, de facto, two distinct nations. Consequently, the conflict between them was equivalent to a war between fully independent states' (Neff, *War and the Law of Nations*, 255). Hence Vattel's approach is a paradoxical one in that he allows for rebellions and reaffirms prevalence of the state at the same time.

exclusion of 'savages', as we have seen, was not clear-cut any more either, since military campaigns in the eighteenth century had shown that natives in the colonies could be very useful as auxiliaries, in spite of the evident problems of imposing discipline on them. With the American Revolution and the French Revolution, the traditional categories and structures changed for good. New lines of exclusion emerged. The French Revolution, in its early radical phase, toyed with the idea of excluding fighters along the lines of their nationality. The American Revolution ended up in an armed conflict that was partly a colonial war and partly a civil war. As the revolutionary 'rebel' emerged, so did the counter-revolutionary 'loyalist'—a 'rebel' from the point of view of the Americans fighting for independence. Finally, the figure of the irregular as the 'brigand' emerged: the fighter who uses irregular tactics, but has no political aims. The phenomenon itself was not new; in fact, it would be fair to say that all fighters had always had mixed motives for going to war, and pillaging and personal gain had always been prominent among them. However, by focusing the accusation of feigned political motivations upon irregular opponents, regular forces could undermine the formers' political legitimacy in public.

At the same time, with the emergence of revolutionary militia armies and the *levée en masse*, it became difficult to uphold general reservations against the arming of peasants. On the contrary, the new national armies drew their strength from bringing all strata of society into arms. However, reservations about the dangers linked to general armament persisted. Once again, enhancing the discipline of the troops was seen as the best means to confront the pitfalls of the *levée*. Hence, the nationalization of war to include peasants paradoxically led to a greater professionalization of the armed forces. However, this process also furthered the exclusion of irregular fighters on the grounds that they violate the standards of fighting that were said to characterize professional armed forces.

Uncertainty about what categories of fighters should be viewed as legitimate and thus entitled to protection was closely linked to the politicization of military issues characterizing both the American Revolution and the French Revolution. These two aspects turned out to be mutually reinforcing. Indecision about the status and privileges of fighters often led to the breakdown of prisoner exchange cartels. Such breakdowns, in turn, resulted in the imprisonment of enemy fighters for long periods of time, something that had not been practised thus far, and that most military parties were not prepared for. Prisoners thus often became central reference points in an already heated political atmosphere that tended towards radical rhetoric and Manichean world views. Moreover, certain traditional practices such as the impressment of prisoners into one's own armed forces or exchange for ransom were difficult to square with the revolutionary rhetoric of the soldier–citizen.[41] This trend towards politicization stood in marked difference to the continuing and eventually even increasing professionalism of the armed forces, in particular the officer corps, that characterized both revolutionary America and France.

Towards a Nationalistic Paradigm of Exclusion? The Case of the French Revolution

The question of who should be viewed as a legitimate fighter and thus enjoy certain protections was a clear point of conflict between politics and the military. The paradigmatic example for this tension was the French National Convention's announcement in 1794 of a 'no-quarter' policy[42] towards all British and Hanoverian soldiers encountered on the battlefield. From a political perspective, the 1794 decree is an indicator of the radicalization of the revolution. The revolutionaries had initially emphasized the lawful conduct of the French armed forces, and in particular the lawful treatment of prisoners. In this vein, in 1792 the National Assembly had adopted a decree to the effect that, 'first, prisoners of war are safeguarded by the Nation and under the special protection of the Law; second, that unjustifiable severities or insults, violence, or homicidal assaults committed against prisoners of war will be punished by the same Laws, in just the same way as if those excesses had been committed against Frenchmen'.[43] While this initial decree emphasized the inclusive idea that Frenchmen and foreign soldiers were equal, the 1794 decree was inspired by a more Manichean world view that demarcated the revolutionary nation against other nations.[44]

But why did that exclusionary mechanism target the British and their Hanoverian allies in particular? As Tim Blanning pointed out, Britain had long been France's 'other old enemy', besides Austria.[45] Arguably, after the 1763 Treaty of Paris that had officially ended the Seven Years War, enmities between France and Britain had cooled down. However, tensions had picked

[41] For more details on the shift in the treatment of prisoners during the French Revolutionary and Napoleonic Wars, see Sibylle Scheipers, 'The Status and Protections of POWs and Detainees', in Hew Strachan and Sibylle Scheipers (eds), *The Changing Character of War* (Oxford: Oxford University Press, 2011), 394–409.

[42] A 'no-quarter' policy refers to the practice of killing one's surrendering battle opponents instead of sparing them as POWs, which would implicate 'providing them with quarters for habitation'.

[43] Quoted in Geoffrey Best, *Humanity in Warfare: The Modern History of the International Law of Armed Conflicts* (London: Weidenfeld & Nicolson, 1980), 78.

[44] See also Thomas Hippler, *Citizens, Soldiers and National Armies: Military Service in France and Germany, 1789–1830* (London: Routledge, 2007), 77.

[45] T. C. W. Blanning, *The Origins of the French Revolutionary Wars* (London: Longman, 1986), 45. According to Gunther Rothenberg, the British were an even more serious enemy of France than Austria: Gunther Rothenberg, 'The Origins, Causes, and Extension of the Wars of the French Revolution and Napoleon', *Journal of Interdisciplinary History*, 18/4 (1988), 777.

up again considerably with the 1787 competition for control and influence in the United Provinces. Blanning's point is to stress the continuities between the *ancien régime* and revolutionary France in the field of foreign policy, in particular in the realm of enemy perceptions. Indeed, picking up on this familiar image of Britain as the enemy, the 1794 decree expressed an unprecedentedly high level of hostility. In previous conflicts prisoner exchange cartels between the British and the French had functioned remarkably well. The Convention of Écluse (1759–62) that had regulated the prisoner exchange between the two countries during the Seven Years War had been regarded as a highly successful arrangement. It stipulated that prisoners had to be exchanged either man for man or for ransom within fifteen days of their capture, and it applied globally. Although individual soldiers may have violated this agreement in the heat of battle, this did not reflect the intentions of their commanders. As Ferdinand of Brunswick, commander of the Hanoverian army, made clear in an order issued in June 1760:

> It is once and for all to be a General rule with the Army, that during the course of the campaign, if the Fortune of War throws into our hands any prisoners, as soon as they are in our power and from the moment they have been disarm'd, they are to be treated with no Severity, but all possible civility, such as circumstances permit of and is decent and becoming amongst civilized people to be shown them.[46]

Despite this long record of the French and the English regarding each other as *iustus hostis*, as enemies on an equal moral and legal footing, the 1794 decree of the French National Convention depicts the English as 'enemies of humanity', 'assaulting the law of peoples and threatening to annihilate freedom'.[47] As Robespierre explained: 'Those who make war on a people to halt the progress of liberty and destroy the rights of man must be attacked, *not as ordinary enemies, but as assassins and rebel brigands*.'[48] By depicting the English as enemies of humanity, the revolutionaries portrayed France, in turn, as being representative of the humanity in all of mankind.[49] Britain was denounced as the 'the modern Carthage', an empire ruled by an oligarchy

[46] Quoted in Reginald Savory, 'The Convention of Écluse, 1759–1762: The Treatment of the Sick and Wounded, Prisoners of War, and Deserters of the British and French Armies during the Seven Years War', *Journal of the Society for Army Historical Research*, 42/170 (1964), 72.

[47] Bell, *The First Total War*, 143; Erich Pelzer, '"Il ne sera fait aucun prisonnier anglais ou hanovrien": Zur Problematik der Kriegsgefangenen während der Revolutions- und Empirekriege (1792–1815)', in Rüdiger Overmans (ed.), *In der Hand des Feindes: Kriegsgefangenschaft von der Antike bis zum Zweiten Weltkrieg* (Cologne: Böhlau Verlag, 1999), 189.

[48] Quoted in Bell, *The First Total War*, 143–4, emphasis added.

[49] Best, *Humanity in Warfare*, 85; Hippler, *Citizens, Soldiers and National Armies*, 98: 'Being French thus meant representing universal justice, and the enemy to be fought was the enemy of universal justice.'

that oppressed the 'ordinary' British people and the colonial residents alike.[50] According to David Bell, this annihilationist rhetoric evolved paradoxically out of the ideational foundations in pacifism, which had been endorsed by the French revolutionaries at the beginning. From this perspective, revolutionary war was a defensive war against the enemies of peace. This enemy position of the British became all the more intolerable in this case considering the similarities between the British constitutional tradition and the French revolutionaries' aspirations. The British 'revolutionary heritage made their enmity look like betrayal'.[51]

British rhetoric was comparably fierce; this was not only a reaction to the 1794 decree, but had started earlier. Edmund Burke, for instance, believed that 'the nature of Jacobinism was pillage' and that, 'as an attack on property, its tendency was to subvert civilisation itself'.[52] In general, the British perceived this war with France as a 'war of principle' just as much as the French did, although they did not go quite as far as to propose punitive measures on captured French soldiers. Interestingly, both sides made use of the vocabulary of irregular warfare when denouncing the opponent: Robespierre referred to 'rebel brigands' and Burke equated Jacobinism with 'pillage'. This demonstrates the extent to which ambiguity concerning different categories of fighters and their rights went hand in hand with the politicization of the customs and laws of war.

Crucially, however, the 1794 decree of the French National Convention was never implemented on the battlefield. In spite of the Army of the French Revolution's involvement in developing measures to break up the structures of the *ancien régime*'s armed forces, it was never sufficiently politicized to put into practice the more radical ideas that emanated from the political centre. On the contrary, officers in the field often negotiated decentralized, bilateral arrangements regarding prisoners in order to improve their fate.

The American War of Independence: From 'Rebel' to 'Belligerent'

This kind of tension between radicalized political rhetoric in the political centre and cooperation between the enemy troops on the ground was also characteristic for the American War of Independence (1775–83). King George III had initially declared all American troops traitors, because they rebelled against the legitimate authorities. However, since the British pursued a negotiated settlement with the Americans, they deliberately refrained from

[50] Geoffrey Best, *War and Society in Revolutionary Europe, 1770–1870* (Leicester: Leicester University Press, 1982), 84.
[51] Bell, *The First Total War*, 143.
[52] Philip Schofield, 'British Politicians and French Arms: The Ideological War of 1793–1795', *History*, 77/250 (1992), 194.

adopting a systematically harsh approach, which would have consisted in executing captured American troops as rebels. Instead, in a more conciliatory spirit, they often exchanged their troops or released them on parole. Towards the end of the conflict, the term 'rebel' was no longer used in the official discourse; the British referred to the Americans simply as 'the enemy'.[53] Yet the British government never promulgated an official policy concerning the treatment of American prisoners, nor did it ever enact a formal prisoner cartel with them: exchange and release on parole remained decentralized and were organized by military commanders on the ground. Thus, the treatment of captured Americans varied among individual commanders, whereas some regarded these prisoners as quasi-regular captives, and others as instigators of domestic disorder.[54] American officers, however, fiercely objected to 'quartering American prisoners in the public jails, where they were not on all occasions properly separated from common criminals'.[55]

Since a formal exchange cartel had not been arranged, Americans were holding a large number of British prisoners for almost the entire duration of the American War of Independence. One major example of this dynamic arose out of the Battle of Saratoga on 7 October 1777, where British General John Burgoyne surrendered his army of more than 5,000 troops. In the aftermath of the battle, the British tried to set up a convention, the Saratoga Convention, in order to regulate the exchange or release of the prisoners. These efforts were stalled by the American Congress, which insisted that the British delegate in charge of negotiating the convention ought to recognize American independence first.[56] However, that did not happen; the convention was never formally set up, and the 5,000 British prisoners, who came to be known as the 'Convention Army', stayed in American captivity until the end of the conflict in 1783.

Despite the uncertainty about American troops' legal status and the resulting high level of discretion for commanders and soldiers on the ground, atrocities towards Americans were rare. There were reports that some British units refused to accept surrender, but, at the same time, American troops were charged of pretending to surrender in order to get within closer range of the opponent.[57] Generally, however, the British did not necessarily view the Americans as illegitimate, simply because they represented themselves as a

[53] Betsy Knight, 'Prisoner Exchange and Parole in the American Revolution', *William and Mary Quarterly*, 48/2 (1991), 214. See also Stephen Conway, 'From Fellow-Nationals to Foreigners: British Perceptions of the Americans, circa 1739–1783', *William and Mary Quarterly*, 59/1 (2002), 90: 'Prisoner exchanges and the granting of paroles to Continental army officers indicated that the Americans, while routinely labelled "Rebels", were in practice being granted belligerent status.'

[54] George L. Coil, 'War Crimes of the American Revolution', *Military Law Review*, 82 (1978), 185.

[55] Coil, 'War Crimes of the American Revolution', 187.

[56] Knight, 'Prisoner Exchange', 191.

[57] Coil, 'War Crimes of the American Revolution', 182.

'nation in arms'. Indeed, eighteenth-century European armies such as the British one were not as different from the revolutionary American armed forces, as some commentators claim. First, European armed forces consisted of a mixture of mercenaries, European volunteers, and European conscripts.[58] Second, the model of militia forces was a well-known English tradition, dating back to the sixteenth century. And, third, the American forces consisted only partially of militias.

Actually, the Continental army, established in 1775, was far more professional and relevant for military operations than the militias were. Despite the fact that 'a significant element of American resistance consisted not in regular operations but in thousands of episodes of civil disobedience and active opposition throughout the vast area of thirteen states',[59] conventional operations were far more relevant. This was not only due to their greater military effectiveness, but also to the American claim to belligerent status: 'By being militarily conventional, American Revolutionaries created at least the illusion of unified purpose, military strength, and political respectability, both at home and abroad.'[60] George Washington himself, commander of the Continental army, was one of the main driving forces behind this increasing professionalization. He had great reservations regarding militias and complained about 'their undependability, their indiscipline, [and] their cowardice under fire'.[61] Hence, while militias provided an important recruitment pool for the Continental army, they did not play a significant role in military operations overall. Moreover, even if part of the American military operations consisted of irregular warfare, the British could not really take offence at that, since they themselves used irregular units and tactics—as they had also done in previous conflicts on the American continent. For instance, the British deployed the 'British Legion' and the 'Tarleton Raiders', which both combined cavalry and light infantry regiments. The 'Tarleton Raiders' were under the command of the infamous officer Banastre Tarleton, who gained his reputation in the 1780 Battle of Waxhaws, during which he allegedly ordered his men to kill a unit of surrendering Virginian Continentals under the command of Colonel Abraham Burford.

Aside from the strong conventional self-identity of the American revolutionary forces, two other factors may have played a role in preventing the

[58] Peter Paret, 'The Relationship between the American Revolutionary War and European Military Thought and Practice', in Peter Paret (ed.), *Understanding War: Essays on Clausewitz and the History of Military Power* (Princeton: Princeton University Press, 1992), 29.

[59] Paret, 'The Relationship between the American Revolutionary War and European Military Thought and Practice', 31.

[60] John Shy, 'The American Revolution: The Military Conflict Considered as a Revolutionary War', in Stephen J. Kurtz and James H. Hutson (eds), *Essays on the American Revolution* (Chapel Hill, NC: University of North Carolina Press, 1973), 155.

[61] Shy, 'The American Revolution', 216–17.

conflict from slipping into atrocities. First, as already mentioned, the wide reception of Enlightenment thought, and, in particular, of Vattel's ambivalent rehabilitation of the rebel, almost certainly played a role. Secondly, the prevailing reconciliatory spirit in this conflict was facilitated by a sense of commonalities between the British and the American revolutionaries, in particular the perception of 'common blood', 'common Protestantism', and a 'common commitment to constitutional securities'.[62]

Apart from the British colonial troops (and their Hessian allies), the American militias, and Continentals, there were two other groups of fighters involved in the American War of Independence. Both were treated worse than the former. Although Native Americans were referred to as 'savages', they were used as auxiliaries by both sides (just as in the French and Indian War). Their status as 'savages' seemed to warrant harsh treatment, such as at the 1782 'Gnadenhutten massacre', when American militia forces seized over ninety Native Americans of the Lenape tribe of the Ohio Country and massacred them on the suspicion that they had taken part in raids in Pennsylvania. As no disciplinary action was taken against the militia units who carried out the massacre, it seems that American commanders perceived the usual constraints on the use of armed force as inapplicable to Native Americans. A journal entry by a American officer read accordingly: 'I should not think it an affront to the Divine Will to lay some effectual plan, either to civilize, or totally extirpate the race... their friendship... is of no longer duration than while we are in prosperity and the impending rod threatens their destruction.'[63] Here, the 'savage', in parallel with the bandit, was depicted as incapable of committing to genuine political causes. Instead, their motivation was reduced to self-preservation and material gain.

Although the American militia is often perceived as the prime exponent of this line of thinking owing to their constant exposure to frontier warfare, some British commanders also expressed their doubt regarding whether they could ever successfully impose discipline on their Native American allies.[64] This kind of depiction of the 'savage' as the undisciplined fighter was common among British soldiers, who also transferred this theme to the irregular. Still, some Americans tried to hold British officers accountable for the lack of discipline of Native American forces under British command.[65] Such ambivalences seem to reflect the genuine difficulties in establishing whether indiscipline was actively sanctioned from above, whether commanders just turned a blind eye, or whether they were plainly unable to impose orders. It seems

[62] Conway, 'From Fellow-Nationals to Foreigners', 76.
[63] Quoted in Coil, 'War Crimes of the American Revolution', 181.
[64] Coil, 'War Crimes of the American Revolution', 193.
[65] Coil, 'War Crimes of the American Revolution', 195.

plausible that in many cases indiscipline was a manifestation of deliberately 'vain command', with the added value that its deniability made reprisals less likely.

The second group that stood outside the category of legitimate combatants were the so-called bandits operating between the front lines. They were often loyalists who had been driven from their property by Patriots.[66] Their motives ranged from political inclinations over vengeance to economic necessity.[67] The American revolutionaries did not regard them as legitimate combatants and insisted that loyalists be extradited to their individual states for criminal prosecution, thus also reinforcing the revolutionaries' claim to being in charge of domestic law enforcement.

Professionalization of the 'People in Arms'

The trajectory of the American War of Independence reflects a process of rapid professionalization of the 'rebel forces'. A decade later, the development of the French Revolutionary Army confirmed this trend. Like the American militia, the French Revolutionary Army initially made efforts to change the traditional structure of the armed forces, most importantly by electing officers from the ranks, but this practice was soon discontinued. What remained, however, was the opening of military careers to members of the middle classes. Until then, access to the officer corps had been restricted to the aristocracy.[68] Eventually, this led to a more meritocratic system, in which promotion depended on skills and performance. Hence, rather than hindering professionalization of the armed forces, the revolution actually bolstered it in the long run.[69]

Arguably in the early years of the Army of the French Revolution, the soldiers followed the revolutionary political rhetoric and 'depicted themselves as citizen–soldiers fighting slave soldiers'.[70] Despite the fact that European armies consisted of a mixture of volunteers, conscripts, and mercenaries, the French rhetoric implied that their opponents' armed forces consisted of 'hirelings' 'who served for money' and posed inherent dangers to political freedom as such.[71] Yet this attitude hardly survived first contact with the enemy on the battlefield:

[66] Harry M. Ward, *Between the Lines: Banditti of the American Revolution* (Westport, CT: Praeger, 2002), p. x.

[67] Ward, *Between the Lines*, p. 242.

[68] Best, *War and Society in Revolutionary Europe*, 21, 70.

[69] See also Alan Forrest, '*La Patrie en danger*: The French Revolution and the First *Levée en masse*', in Daniel Moran and Arthur Waldron (eds), *The People in Arms: Military Myth and National Mobilization since the French Revolution* (Cambridge: Cambridge University Press, 2003), 30.

[70] Hippler, *Citizens, Soldiers and National Armies*, 165.

[71] Best, *War and Society in Revolutionary Europe*, 77. The difference between the citizen–soldier fighting for liberty and the slave soldier was questionable for two other reasons. First, in early debates in the French Constituent Assembly there were a number of voices arguing that

> By the end of 1794 the experience of combat had eradicated the differences between professional and citizen soldiers and instead *created a distinction between the troops who had been under fire and the politicians who had not.* By 1795 professional leaders, regulars, and citizen-soldiers turned neoprofessional, regained control and soon returned to the traditional customs and usages between soldiers.[72]

This trend towards greater professionalization of the French armed forces continued, and even increased under Napoleon.

In line with this trend towards professionalization, the 1794 decree on the 'no-quarter' policy for British and Hessian soldiers was rescinded in the same year. More generally, the forceful political rhetoric did not seem to have much impact on the battlefield as long as the opposing armies recognized each other as belonging to roughly the same category: regular soldiers. Hence, as in the example of the American Revolution, neither revolutionary ideology nor the 'nationalization of war' led to a lasting gap between the French conscript army and other European armed forces. On the contrary, conscript armies soon found themselves under heavy pressure to become more professional. As a consequence, the revolutionary conscript army animated purely by a nationalistic spirit never really seemed to exist.

David Bell argues that the increasing professionalization of the military—and the emergence of a 'culture of war' associated with this—led to a removal of constraints in the conduct of war. He refers to the French Revolutionary and Napoleonic Wars as the 'first total war[s]'. According to Bell, 'total war' is a category of modern wars that is characterized by the unrestrained use of military force (aiming at the extermination of the opponent and making no difference between combatants and non-combatants), fused with a radical political agenda (such as a revolution).[73] However, Bell goes too far by establishing a connection between increasing military professionalism and the loss of restraint in warfare. Crucially, the examples of 'total war' that he quotes are mainly taken either from the repertoire of revolutionary political rhetoric that was never implemented on the battlefield, such as the 1794 decree on British soldiers, or from specific campaigns against insurgents and guerrillas such as the war in the Vendée (1793–6) or the Peninsular War (1808–14).[74] Hence loss

conscription actually *undermined* the freedom of the individual. Second, this rhetoric had already emerged before the 1793 *levée en masse*—that is, at a time when the French Revolutionary Army itself consisted of a mixture of volunteers and conscripts. See Peter Paret, 'Conscription and the End of the *Ancien Régime* in France and Prussia', in Paret (ed.), *Understanding War*, 57 ff.

[72] Gunther Rothenberg, 'The Age of Napoleon', in Howard, Andreopoulos, and Shulman (eds), *The Laws of War*, 89, emphasis added.

[73] Bell, *The First Total War*, 8 ff.

[74] See also Jeremy Black, 'Book Review of "The First Total War: Napoleon's Europe and the Birth of Warfare as We Know It"', *English Historical Review*, 123/502 (2008), 765.

of restraint was typical for the French army's behaviour towards irregulars, but not for encounters with regular opponents.

Arguably the French Revolution also led to a decrease of military restraint off the battlefield, but this effect had nothing to do with militarism per se. Rather, it was due to the sheer size of the French conscript army, which overwhelmed French military logistics capacities to such an extent that foraging and pillaging in occupied territory reached new dimensions.[75] Therefore, the impact of war on the local population in the areas of operations or in occupied territory—for 'civilians' in our contemporary parlance—became far more palpable.[76]

The problem of indiscipline was already inherent in the concept of the 'people in arms'. Both the American militias and the French National Guard (the local militia in France created in 1789) had a reputation for indiscipline.[77] The respective conscript armies in both countries embodied a certain sense of distinction from these rather raw and untrained militia forces, thus adopting an identity that seemed to be closer to their opponent's professional armies than to their domestic militia forces, although the latter remained an important recruitment pool for the conscript armies. This distinction was all the more important, because the main characteristic feature of regular armed forces—discipline—largely remained an aspiration.

But the issue of indiscipline was not the only one that played a role in distinguishing between militia forces and conscript armies. More fundamentally, the idea of the 'people in arms' could refer both to the conscripted armed forces *of a state* and to the spontaneous uprising of the people *against the state* (or against the occupying forces of another state). 'The word "levée" has several meanings: it signifies both "levy" and "uprising". Recruitment of troops, which is one of the main prerogatives of central power, and revolt are put on the same level. The oxymoron both affirms and denies state power.'[78]

Hence, *the nationalized regular armies and the irregular fighters have precisely the same intellectual roots*. Therefore, the only way for the state to safeguard its authority over the people in arms lies in imposing discipline. It was exactly this paramount importance of discipline that led the emerging conscript armies down the road of rapid professionalization and to a renewed emphasis

[75] Best, *Humanity in Warfare*, 89 ff.

[76] Best, *Humanity in Warfare*, 96 ff.

[77] Hippler, *Citizens, Soldiers and National Armies*, 63. Beatrice Heuser argues that Clausewitz was well aware of this issue: 'Clausewitz recognised that... the French Revolution, as well as Napoleon, as well as the Spanish and Tyrolean reactions were wars that had become "the cause of the people", and that were therefore fought all the more vehemently. The enthusiastic mass armies were thus only the extreme expression of the fanatical terrorist or the small band of *guerrilleros* writ large' ('Small Wars in the Age of Clausewitz', 157 ff.).

[78] Heuser, 'Small Wars in the Age of Clausewitz', 81.

on hierarchy and obedience—at least on the battlefield.[79] Discipline was also to become the main symbolic distinctive feature between the regular and irregular, and allegations of the lack thereof were soon to become the linchpin of the exclusionary rhetoric and measures against irregulars.

'Without discipline, no glory':[80] Irregulars in the French Revolutionary and Napoleonic Wars

With the French Revolutionary and Napoleonic Wars the illegitimacy of the irregular moved to the centre stage. French forces encountered irregular opponents both within France and in the occupied territories. New practices on the treatment of irregular fighters evolved and were reflected in an exclusionary discourse towards them.

The Vendée and the Emergence of the Irregular as the 'Brigand'

The uprising in the Vendée, a rural region in western France, was just one of several counter-revolutionary revolts within France. Most notably, the Chouannerie caused much grief to the Republic. The Chouannerie resistance was much smaller and patchier than the Vendéen uprising, which started as a mass revolt and turned into a guerrilla war after its more conventional efforts were crushed by republican forces. Still, it seemed that the Chouannerie proved more resilient to republican efforts at suppression in the long run. The Chouannerie began in 1792 in the districts of Vendée militaire, Upper Brittany, Lower Normandy, Anjou, and Maine, reached its peak in 1795, and was subsequently subjugated by the republican forces, only to flare up again in 1799.[81] The federalist uprising in the Midi after 1793 provides a third example of counter-revolutionary efforts.[82] Some sources of these uprisings were similar, such as opposition to conscription, religious motives, and alienation with the new regime. Other sources were different, depending on the region; for example, federalism was an important motive in the south, where the movement originated from the urban centres, while royalism and religion were

[79] The perceived link between the risks related to arming the people and the need to impose discipline would become even more acute in the case of Prussian military reform after 1814; see Heuser, 'Small Wars in the Age of Clausewitz', 190 ff.

[80] Napoleon Bonaparte, quoted in Martin Boycott-Brown, 'Guerrilla Warfare *avant la lettre*: Northern Italy, 1792–97', in Charles J. Esdaile (ed.), *Popular Resistance in the French Wars: Patriots, Partisans and Land Pirates* (Basingstoke: Palgrave Macmillan, 2005), 50.

[81] Howard G. Brown, *Ending the French Revolution: Violence, Justice, and Repression from the Terror to Napoleon* (Charlottesville, VA: University of Virginia Press, 2006), 238 ff.

[82] See Colin Lucas, 'The Problem of the Midi in the French Revolution', *Transactions of the Royal Historical Society*, 5th ser., 28 (1978), 1–25.

more important in the north, where the movement arose from the rural regions. All these uprisings, however, forced the republican government to find an answer to the counter-revolutionary challenge, and hence helped shape the emergence of the concept of the 'brigand' as the predominant label for the irregular fighters.[83] The Vendée will be discussed in greater detail, as it presented the most dangerous threat to the Republic, but also because its transition from a mass revolt to a guerrilla war illustrates the variety of roles irregular fighters may play.

French republican historiography written in the nineteenth century treated the wars in the Vendée as an 'insurrection [that] could only have stemmed from a conspiracy, from an "infernal plot" hatched by priests and nobles'.[84] In contrast, contemporary conservative historians tended to refer to the Vendée as a spontaneous rebellion motivated by the peasants' royalist and Catholic convictions.[85] In the 1960s, this tradition of ideologically tainted historiography on the Vendée was superseded by accounts that stressed the economic and structural causes of the uprising and emphasized their roots in pre-revolutionary, long-term economic disparities between the centre and the peripheries of France, but also between the different localities in the Vendée itself.[86] While these accounts were more 'objective' in their approach to the historical facts, they also tended to play down the relevance of religious and political convictions, which transformed individual grievances into a large movement and which provided a source of motivation for combat.[87] An additional problem with these accounts is that they were prone to downplay the novelty of the political and military dynamic that developed in the

[83] The term 'brigand' in the sense of an 'unjust belligerent' had first been used in a legal context by Christian Wolff in his 1750 *Jus gentium methodo scientifica pertractarum.* Wolff defines brigands as 'those who seek for war as an end in itself' and argues that they 'are to be compared to robbers', thereby introducing a criminal element into the definition. Quoted in Dan Edelstein, 'War and Terror: The Law of Nations from Grotius to the French Revolution', *French Historical Studies*, 31/2 (2008), 229. Vattel uses the term 'banditti' to refer to combatants who wage 'illegitimate and informal wars, or rather predatory expeditions, undertaken either without lawful authority or without apparent cause, as likewise without the usual formalities, and solely with a view to plunder.' He goes on to explain that 'a nation attacked by such sort of enemies is not under any obligation to observe towards them the rules prescribed in formal warfare. She may treat them as robbers' (Vattel, *Law of Nations*, 319 ff.).

[84] Claude Petitfrère, 'The Origins of the Civil War in the Vendée', *French History*, 2/2 (1988), 190.

[85] Petitfrère, 'The Origins of the Civil War', 191.

[86] Charles Tilly, *The Vendée: A Sociological Analysis of the Counter-Revolution of 1793* (Cambridge, MA: Harvard University Press, 1976); Marcel Faucheux, *L'Insurrection vendéenne de 1793: Aspects économique et socieaux* (Paris: Imprimerie nationale, 1964); Paul Bois, *Paysans de L'Ouest: Des structures économiques et sociales aux options politiques depuis l'époque révolutionnaire dans la Sarthe* (Le Mans: M. Vilaire, 1960).

[87] 'The shortcomings of the religious policy espoused by the Constituent Assembly, exacerbated by the lack of understanding and intolerance of the local authorities, were undoubtedly the most important cause of the Vendean rebellion' (Petitfrère, 'The Origins of the Civil War', 205).

Vendéan uprising, portraying it instead as 'one of Europe's last great rural rebellions', in line with peasant rebellions such as the 1358 Jacquerie.[88]

Fighting in the Vendée began in March 1793. It was preceded by a general sense among various groups in the region that they were politically estranged from the centre. However, the immediate trigger for the insurrection was the attempted *levée* of an additional 300,000 men for the Army of the French Revolution in February 1793. Resistance to conscription brought together an alliance of peasants, nobles, clergy, as well as some artisans and shopkeepers, who all rose up against the central republican authorities. In the first few months, the Vendéan uprising managed to gather momentum. Its opponents consisted almost entirely of National Guard troops, whose military effectiveness soon proved to be limited. The Vendéans achieved several victories and captured weapons and equipment, including several pieces of heavy artillery. They formed the Royal and Catholic Army, 'an unstable assembly of local units, capable of protracted defence and of short but massive efforts'.[89] It lacked training and discipline, and its members tended to return to their homes after major engagements, but it managed to achieve some successes. By June 1793, the Royal and Catholic Army controlled large parts of the Vendée.

Soon after the uprising, the Republic dispatched 45,000 regular troops and guardsmen to the region. However, the republicans proved fickle with respect to their approach to quelling the revolt: the Convention's first response was to pass a law, according to which all rebels would face the death penalty and their property would be confiscated by the state. In May 1793, this approach was replaced by a more conciliatory effort to separate the rebel leaders from the populace, and to convince the latter to support the Republic.[90] On 1 August 1793, the Convention returned to the punitive approach and called for 'the complete physical destruction of the Vendée'.[91] On 23 December 1793, republican troops under General Kléber eventually defeated and in fact annihilated the Royalist and Catholic Army in the Battle of Savenay. Despite the fact that this reduced the threat to the Republic to a few scattered rebels in the Vendée, the republican government started to implement an even more ruthless 'pacification strategy' in the wake of the Battle of Savenay, possibly motivated by the desire to discourage all future rebel attempts in the region and to set an example for the entire country. This pacification strategy was carried out by General Turreau, who replaced Kléber at the end of 1793, and his so-called *colonnes infernales*, small detachments of regulars that roamed the countryside

[88] Cf. Tilly, *The Vendée*, p. v; cf. Bell, *The First Total War*, 163.

[89] Gunther Rothenberg, *The Art of Warfare in the Age of Napoleon* (London: Batsford, 1977), 118.

[90] Peter Paret, *The Vendée, 1789–1796* (Princeton: Center for International Studies, 1961), 37 ff.

[91] Donald M. G. Sutherland, *The French Revolution and Empire: The Quest for a Civic Order* (Oxford: Blackwell, 2003), 222.

in search of rebels, but more often than not adopted a wholly indiscriminate approach by killing all suspect individuals and by destroying the local populace's property.[92] Jean-Baptiste Carrier, Jacobin deputy in Nantes, gained particular notoriety for his indiscriminate punishment and deterrence policy. Between October 1793 and February 1794, when he was finally recalled to Paris and guillotined, he had ordered the killing of thousands of Vendéan prisoners and suspects, most infamously by drowning them in the Loire.

How did the French republic's first military encounter against other French fighters, as opposed to foreign army soldiers, contribute to the making of the irregular? Upon closer inspection, what is striking about the Vendée is not the extent of the differences between the republican forces and the Vendéans, but the similarities between them. Arguably, discipline among the troops of the Royalist and Catholic Army was low, and the army never constituted itself as a permanent fighting force.[93] However, desertions were also an issue on the republican side.[94] In terms of tactical differences between the opponents, the Vendéans did use ambushes and terror tactics, but they were also capable of using—though with limited success—more conventional tactical approaches such as advancement in columns and deployment into line of battle when approaching the republican forces. Conversely, republican troops did use conventional tactics, but they also emulated the peasants' tactics and broadened their tactical repertoire as a result of their encounter with the Vendéans—for instance, by integrating light infantry in the line battalions.[95] Regarding the adoption of terror tactics, in particular, the regular republican forces probably surpassed their 'irregular' opponents by the end of the campaign. This trend towards emulating the opponent is echoed in a remark by republican general François-Nicolas Salomon in June 1793: 'Since this is a war of brigands, we must become brigands ourselves. We must, for a time, forget all military rules.'[96] Terror tactics were sanctioned by the political regime and were only partly a spontaneous escalation on the part of the military. The Terror itself and the anarchy it produced gave rise to the myth of the pillaging and marauding soldier of the *armées révolutionnaires*. The narrative of the undisciplined soldier survived the regime of the Terror, and the Thermidorean reaction gave Salomon's statement an ironic confirmation in the sense that 'the term "brigand" in fact had changed sides. Reserved in Year II for Chouans and royalist rebels in the interior, it had become in the months of the Thermidoreans . . . the semi-official, indeed almost required description for former terrorists.'[97]

[92] Paret, *The Vendée*, 54 ff. [93] Paret, *The Vendée*, 31 ff. [94] Paret, *The Vendée*, 35.
[95] Paret, *The Vendée*, 10, 34. [96] Quoted in Bell, *The First Total War*, 170.
[97] Richard Cobb, *The People's Armies* (New Haven: Yale University Press, 1987), 488.

When it comes to the assessment of the motives for combat, ideology and material interests were equally important for *both* the Vendéans and the volunteers of the *armées révolutionnaires*. As Petitfrère pointed out, 'if the upper bourgeoisie, artisans and shopkeepers offered their services for the defence of the Revolution, it was because they had derived some benefit from the new order'.[98] Conversely, although the peasants had suffered from the effects of the revolutionary tax reforms, material motives were not the sole driving force in the uprising. Religious belief was equally, if not more, important, in particular because it engendered in them the will to sacrifice.[99] As for the regular republican forces, their motives were equally mixed. Nationalism certainly played a role, although it was not the overwhelming and unanimous motive that the official French revolutionary iconography would have us believe.[100] To a certain extent the campaign in the Vendée supplied new material for this iconography by producing the requisite cultural images of military heroes. A particular 'cult of martyrdom' developed around the deaths of two young volunteers, Joseph Bara and Agricol Viala, who had been killed by Vendéan peasants. They became the object of popular artistic displays of heroic death for the national cause.[101] The official rhetoric of the revolution underlined the nationalist motivation of the republican forces by referring to them as 'volunteers', although they consisted of conscripts who were forced to serve.[102]

Despite—or rather *because of*—the actual similarities between the regular and the irregular forces, the political discourse surrounding them emphasized, or sometimes even constructed, differences. Victor Hugo later depicted the Vendéans as a threatening and invisible opponent of the republican forces: 'Invisible battalions watched and waited. These unseen armies snaked under and around the republican armies, leapt out of the earth in an instant only to disappear, bounded up numberlessly and vanished, everywhere and nowhere, an avalanche one moment, dust the next.'[103] According to a deputy of the National Convention, 'all the present inhabitants of the Vendée are dedicated rebels; it is the women, and the boys and girls over twelve who are the cruellest'.[104] Crucially, this emphasized the alleged difference between the regular forces, who recruited only men, and the irregulars, who supposedly knew of no such differentiating measures.

The experiences with irregular movements such as the Chouannerie, the Vendéan uprising, and the counter-revolution in the Midi had two lasting effects that persisted during the Napoleonic era and had a lasting impact on

[98] Petitfrère, 'The Origins of the Civil War', 198.

[99] Petitfrère, 'The Origins of the Civil War', 205.

[100] Forrest, '*La Patrie en danger*', 26.

[101] Bell, *The First Total War*, 141 ff.

[102] Forrest, '*La Patrie en danger*', 26.

[103] Quoted in Forrest, '*La Patrie en danger*', 166.

[104] Bell, The First Total War, 174.

French operational conduct in the Napoleonic empire. First, the deployment of 'flying columns' that had proven successful within France became a staple of French counter-insurgency campaigns under Napoleon. Secondly, the republican regime's legal response to the challenge of counter-revolution led to a conflation of military, criminal, and political elements in the figure of the irregular fighter. Already in 1792, the Convention had resorted to military commissions that were tasked with adjudicating on both counter-revolutionary resistance and crimes under civil law.[105] The trend towards militarizing civil justice increased in the following years. Challenged by persisting anarchy, the Directory adopted the law of 29 nivôse VI (18 January 1798), which stipulated the death penalty for acts such as highway robbery and housebreaking and determined that, if the offence had been committed by a group of more than two, it would be considered brigandage, and the offenders would have to face trial by a military court. In effect, the government equated common criminals with counter-revolutionary irregulars. What was more, the laws on brigandage were also used against political opponents who were not engaged in violent resistance but were participating in the 'normal' constitutional political process. The militarization of justice thus achieved a politicization of common-law criminals, but at the same time it criminalized counter-revolutionary irregulars and political opponents alike.[106] This conflation of irregulars, criminals, and political opponents not only survived into the Napoleonic era and was applied in the wider empire; it also had a formative effect on the concept of the irregular fighter, in that it merged the rebel with the common criminal.

Rebellion and Guerrilla War in the Empire

Resistance to taxation and to more effective and centralized state power more generally was not confined to France itself. French expansion in the 1790s and later under Napoleon triggered opposition in many conquered and occupied territories, such as the Belgian revolts against conscription in 1798–9, the successful 1799 revolt in Calabria, persistent resistance in Piedmont 1800–7, and several insurrections such as those in 1806 in the Kingdom of Naples, 1808 in Tuscany, and 1809 in the Tyrol and in the Papal States, as well as, most infamously, guerrilla warfare in the Peninsular War 1808–14. Resistance often culminated in attacks and campaigns by local armed groups in those territories. In some cases, irregulars—or at least troops that were perceived as irregulars by the French—already played a role in the initial defence against invasion, as was the case during Napoleon's invasion of Russia in 1812.

[105] Brown, *Ending the French Revolution*, 143 ff.
[106] Brown, *Ending the French Revolution*, 218.

The accounts regarding the nature and the role of irregulars in these territories are beset by the same historiographic problems as the Vendée uprising. The role of guerrilla bands in the Peninsular War is a paradigmatic example for the political contestations surrounding irregulars. Quite unsurprisingly, Spanish sources from the nineteenth century emphasized the positive image of the guerrilla. In the words of the Spanish historian Enrique Rodríguez Solis, 'the guerrilla were [*sic*] the nation in arms . . . They were both soldiers and citizens . . . They abandoned their family and home and gave their lives for the fatherland with . . . joy in their hearts.'[107] The English writer Robert Southey emphasized the guerrillas' honest political motivations by stressing that they were taking 'vengeance for their murdered countrymen' and attempting to deliver 'their country from the violent usurpation' by the Napoleonic forces.[108] Contemporary military observers such as the British officer William Napier underscored the military effectiveness of the Spanish irregular *partidas*.[109] His view was reiterated by subsequent generations of British military historians such as Charles Oman and David Chandler.[110] French historiography shared the assessment that the *partidas* were a very effective military instrument, but denounced them as well as 'cowards, savages and murderers', thus denying their political motives and their political legitimacy.[111]

The case of Russian resistance against Napoleon provides a further example of how irregular fighters can become instrumental in the propagation of a particular view of national history and identity. Soviet historiography depicted the Russian partisans' resistance in 1812 as 'an integral part of the class struggle against the feudal system', whereas Russian historians after the cold war referred to it as proof of Russian patriotism and heroism.[112] Again, while conflicting 'partisan' historiographies are a problem that besets all military history, political contestations tend to be even more acrimonious in the historiography of irregular movements, since basic questions of the identity and the legitimacy of the combatants are at stake.

Indeed, other occupied states in the Empire also created national myths surrounding their own irregular fighters. However, rebels and guerrillas actually did not arise because of nationalistic sentiments, but rather, as mentioned before, because of Napoleon's efforts to strengthen centralized power, especially in the form of taxation and conscription. Resistance against French

[107] Quoted in Charles J. Esdaile, *Fighting Napoleon: Guerrillas, Bandits and Adventurers in Spain, 1808–1814* (New Haven: Yale University Press, 2004), 8.

[108] Quoted in Esdaile, *Fighting Napoleon*, 9.

[109] Esdaile, *Fighting Napoleon*, 13.

[110] Charles Oman, *A History of the Peninsular War* (London: Greenhill, 1995–9), vi. 252 ff.; David G. Chandler, *The Campaigns of Napoleon* (London: Folio Society, 2002), ii. 240.

[111] Esdaile, *Fighting Napoleon*, 17.

[112] Janet Hartley, 'The Patriotism of the Russian Army in the "Patriotic" or "Fatherland" War of 1812', in Esdaile (ed.), *Popular Resistance in the French Wars*, 183.

invasion and occupation generally occurred in places where the state had traditionally been rather weak, local communities were close-knit, and religious ties and traditions strong.[113] Their movements were often more regional and local rather than national. Yet the opposing state authorities often created a myth that fit all too well into nineteenth-century intellectual discourses, regarding not only irregular fighters' legitimacy, but also their important role in military efforts. The Spanish case acquired a particularly exemplary character in this respect. Military and political observers in many other European countries pointed to the role of the 'guerrillas'[114] in the Peninsular War as proof that irregulars could successfully fight against a 'superior' army of regular forces.[115] Carl von Clausewitz found the Spanish example highly instructive as he struggled to find ways to turn around the crushing Prussian defeat against the French in 1806.[116]

French military activities in the Iberian Peninsula began with an attack on Portugal in October 1807 under General Junot, aided by the Spanish army that at the time was allied with Napoleon. The aim of these operations was to prevent British commerce with continental Europe by closing off Portuguese ports to British ships. In 1808, under the pretext of reinforcing the Franco-Spanish army occupying Portugal, French forces invaded and occupied northern and central Spain, and Napoleon installed his brother Joseph on the Spanish throne. While Napoleon's expansionist move across the Peninsula did not meet much military opposition initially, it soon triggered resistance both from the regular Spanish forces returning from Portugal and from irregular groups that were mainly local in character. This, in turn, gave Britain an opportunity to become involved in the struggle by bringing in some 15,000 land forces under General Wellesley. After reinforcing his troops in the Peninsula over the summer of 1808, Napoleon achieved a decisive victory over the regular Spanish forces. However, this did not solve the problem of the irregular resistance efforts. If anything, these efforts had increased after the annihilation of the Spanish army.

[113] Michael Broers, *Europe under Napoleon, 1799–1815* (London: Arnold, 1996), 105, 109 ff.; Charles J. Esdaile, 'Patriots, Partisans and Land Pirates in Retrospect', in Esdaile (ed.), *Popular Resistance in the French Wars*, 6.

[114] Similar to the whole ideational complex surrounding irregular warfare, the term 'guerrilla' was subject to reinterpretation in the eighteenth and early nineteenth centuries. While, in the early seventeenth century, it referred to a dispute between private citizens that was punishable by the state, in the mid-eighteenth century it was synonymous with the tactical concept of 'petite guerre' that authors such as Turpin de Crissé tried to develop. During the Peninsular War, it acquired the meaning of an 'armed struggle of civilians, organized in irregular bands, against an invading enemy and/or against a government that was unjust or, as in Spain, illegitimate' (Vittorio Scotti-Douglas, 'Regulating the Irregulars: Spanish Legislation on *la guerrilla* during the Peninsular War', in Esdaile (ed.), *Popular Resistance in the French Wars*, 139.

[115] Scotti-Douglas, 'Regulating the Irregulars', 176.

[116] Hippler, *Citizens, Soldiers and National Armies*, 201.

Napoleon's initial approach to the irregular challenge consisted in flying columns—a device that had already been developed in the Vendée. From the French point of view, repressing resistance from irregular groups was essentially an exercise in hunting down 'brigands'. As noted above, the notion of 'brigands' had been a 'universally accepted term of abuse for political opponents' in the French revolutionary discourse. It was now reduced back to its narrower meaning aimed at emphasizing the criminal aspect and at denouncing any sincerely political motive. While it could be applied to a whole range of categories of individuals ('irregulars, freedom-fighters, militiamen'[117]), the objective was to deny the opponent political legitimacy. And, since 'brigands' did not have any political legitimacy, but were fuelled by criminal motives, the customs of war did not apply to them. Hence, the French forces were allowed to shoot them without trial. Alternatively, the *gendarmerie* could detain them extra-legally. This approach had already been used in the insurgencies in northern Italy, in particular during the summer revolt of 1809. However, in Italy, the French authorities had still used some form of legal process, even if the outcome for captured irregulars was often the same: 'Identifying resistance as "brigandage" allowed the authorities to place its perpetrators outside the due process of the law and try them in military or special courts where punishments were harsher.'[118]

Arguably, in the Peninsular War harsh treatment and executions were partly the result of a downward spiral of indiscipline on both sides, since the *partidas*, themselves, had committed atrocities towards French soldiers. 'Murder and torture were common, the French retaliating in kind.'[119] Yet the brutal acts committed by both sides during the Peninsular War were more than a spontaneous slippage into slaughter: they were endorsed by the authorities on either side as legitimate ways of fighting the opponent in this specific conflict.[120] Whereas the Spanish perspective defended Spanish irregulars' use of atrocities as a measure of last resort against foreign occupation, the French saw these acts as a confirmation that the 'brigand' was a savage. The figure of the brigand hence oscillated between the ordinary bandit who had no genuine motive but greed, and the savage who committed unspeakable atrocities.[121]

117 Alan Forrest, 'The Ubiquitous Brigand: The Politics and Language of Repression', in Esdaile (ed.), *Popular Resistance in the French Wars*, 37.

118 Alexander Grab, 'State Power, Brigandage and Rural Resistance in Napoleonic Italy', *European History Quarterly*, 25/1 (1995), 40.

119 Rothenberg, 'The Age of Napoleon', 96.

120 Not only had the French authorities permitted harsh measures against the Spanish irregulars; in the 1809 *Instrucción*, the Spanish *junta central* even encouraged its troops to inflict 'all the evil and the damage possible on them, with the understanding that this service will be considered as rendered to the Nation, and rewarded in proportion to its extents and results' (quoted in Scotti-Douglas, 'Regulating the Irregulars', 144).

121 This ambivalence in the French discourse on brigandage had already developed in the Vendée Wars; see Forrest, 'The Ubiquitous Brigand', 39.

This also obscured the differences among the various *partidas* operating in Spain; some of them, such as the *partida* led by Espoz y Mina in Navarre, had treated captured French soldiers as POWs in order to emphasize their claim to belligerent status. Mina issued a 'no-quarter' order to his men only in 1811, when he felt that the ferocity of the French counter-insurgency effort had escalated beyond all limits. After a phase of reprisal killings on both sides, Mina's measure proved successful insofar as the French General Abbé stopped treating guerrillas as criminals.[122]

As we have already seen in the case of the Vendée, neither the depiction of irregulars as heroic freedom fighters nor their denouncement as criminals or savages reflects their reality accurately. While self-confessed motives of irregulars tend to be as problematic as historiographic accounts on 'partisans', a look at the social structure and composition of irregular bands may provide some insights into their potential aims and motivations. The Spanish *partidas* were not homogeneously composed of poor peasants. Rather, a good proportion of the irregulars were landowners, clergy, or professionals.[123] This indicates 'that participation in *la guerrilla* of those with some stake in society—something to lose, indeed—was far higher than that of those on the margins'.[124] Those who joined the *partidas* often seem to have done so in order to defend traditional social and political structures and to fend off Napoleonic reform; this was hardly the same as greed or criminal energy. Seemingly criminal behaviour on the part of the irregulars often resulted from the need to survive in desperate and lawless circumstances, rather than from the attempt to make a fortune as a wartime profiteer, although the latter may have applied in some individual cases.

In this sense, the *partidas* were more motivated by local rather than national concerns: 'whilst the masses were frequently worked on by members of the establishment, they do not seem to have been much moved by anti-French propaganda per se, but rather by their experience of enemy occupation and the extent of their ability to feed their families and to carry on some semblance of normal life'.[125] Revenge for atrocities suffered at the hands of the French occupiers often seemed to be an important factor in the decision to join the *partidas*.[126]

[122] John Lawrence Tone, *The Fatal Knot: The Guerrilla War in Navarre and the Defeat of Napoleon in Spain* (Chapel Hill, NC: University of North Carolina Press, 1994), 128 ff.

[123] Charles J. Esdaile and Leonor Hernández Enviz, 'The Anatomy of a Research Project: The Sociology of the Guerrilla War in Spain, 1808–14', in Esdaile (ed.), *Popular Resistance in the French Wars*, 119.

[124] Esdaile and Enviz, 'The Anatomy of a Research Project', 124.

[125] Esdaile, 'Patriots, Partisans and Land Pirates', 22.

[126] Esdaile, *Fighting Napoleon*, 112. Cf. Rothenberg, 'The Age of Napoleon', 95: 'While invaders were commonly disliked, active resistance was most frequently provoked by excesses of requisitioning.'

Nonetheless, protest at the local or regional level against the new social and economic structures could be highly political. In some parts of Spain, for instance, in particular in the Serranía de Ronda, the irregulars' activities were more akin to a peasant uprising than to guerrilla warfare against foreign occupation.[127] The same was true for the irregulars in Italy, who often seemed to resemble 'social bandits' in Eric Hobsbawm's sense.[128] Yet a comparison of the regions in which irregular fighting broke out in the Napoleonic era shows that poverty was not the main driving force. The Spanish region of Navarre, for instance, was relatively wealthy at the beginning of the nineteenth century, and yet it saw fierce irregular activities. As John Tone showed in his seminal study, people in Navarre had nothing to gain from the Napoleonic reforms imposed on their region, but a lot to lose. The region had enjoyed considerable political and economic privileges before the arrival of Napoleon, and the feudal system had not manifested itself in great social disparities between its inhabitants. The exactions of the Napoleonic regime meant a degree of taxation that was unprecedented in Navarre. Hence resistance to Napoleon was fierce and persistent.[129]

The category of 'guerrilla' was blurry around the edges not only because of the fighters' mixed motives, but also because of the unclear relationship between apparent irregulars and regular fighters. Many resistance fighters were actually scattered regulars who had survived the French offensive in the summer of 1809, although their appearance and attitude resembled those of irregulars. In addition, allied British troops became involved in irregular activities as well. Finally, some of the irregular bands were actually composed of people's militias or home guards.[130] To complicate things further, it was often unclear whether the apparently scattered regulars were actually deserters from regular units, since unequivocal desertion from the Spanish army and the militia forces was a significant problem.

Also the relationship between the irregulars and the local populace was far from clear-cut. In Spain, the majority of the local population did not rush to arms in order to join the resistance movement against Napoleon. On the contrary, they 'wanted nothing to do with the war'.[131] In some regions, peasants collaborated with the irregulars and supported them, as seemed to have been the case in some segments of the rural society in northern Italy.[132] However, the relationship between the irregulars and the local population was

[127] Esdaile, *Fighting Napoleon*, 124.

[128] Grab, 'State Power', 44. Hobsbawm's 'social bandit' is an ambivalent figure: while he is the result of social inequalities, his behaviour does not necessarily aim at improving the lot of the lower strata of society. Rather, it is egoistic in the first place. Nevertheless, his actions may acquire political significance that encourages social change in the long run. Eric J. Hobsbawm, *Bandits* (Harmondsworth: Pelican Books, 1972), 83 ff.

[129] Tone, *The Fatal Knot*.

[130] Esdaile, *Fighting Napoleon*, 44 ff.

[131] Esdaile, *Fighting Napoleon*, 88.

[132] Grab, 'State Power', 59.

more often tense, not least because the former 'requisitioned' goods and money from the peasants and tried to force them to join their ranks.

Finally, when it comes to the military effectiveness of the irregulars, the picture is equally mixed. In Spain, many of the larger armed groups such as the *partidas* led by the famous *cabecillas* Espoz y Mina or El Empecinado proved to be quite effective, in particular during the later years of the conflict. Yet the assessment that the irregulars effectively 'won' the Peninsular War is not plausible. After all, they were supported by the British troops. To a certain extent, the people's war even undermined the Spanish state, hence stymieing a potentially more effective defence against the Napoleonic forces. The *partidas* sapped the strength of the Spanish regular army throughout the conflict.[133] In some regions, such as Navarre, guerrillas such as Espoz y Mina took over state functions such as tax collection and courts and provided a system of governance, which effectively competed with the central state at the end of the war.[134] Consequently, it should come as no surprise that the reassertion of Spanish state power involved forceful action against the irregular bands: 'in 1813 Patriot Spain was effectively at war with the guerrillas.'[135]

The empirical picture of the irregular during the Napoleonic Wars fits neither the French depiction of the 'brigand' and 'savage', nor the nationalist, liberal, or even socialist view of the irregular as the patriot, the freedom fighter, or the social rebel. But, if the phenomenon of the irregular is far more complex and nuanced than that, we also have to take a critical look at the dichotomy between the irregular and the regular that contributed to denouncing the irregular in the first place. Again, here we see more similarities than differences between the regular and the irregular.

At a very basic level, this is due to the fact that the French were likely to denounce all military resistance they encountered in Spain as 'guerrilla' action. This was to a certain extent understandable, as the French 1808 summer offensive had left many of the regular Spanish troops scattered around the country. But it begs the question what the real differences between the regulars and the 'irregulars' were in this case, apart from the fact that the alleged irregulars had suffered quite a decisive blow. Moreover, as mentioned above, many of those troops that the French regarded as irregular were actually locally recruited militias that had been tolerated by the state such as the Catalonian *miquelets*, or members of the home guard (*somatens*). According to Antonio Moliner Prada, 'there was little or no difference between the behaviour of the *somatens*, *miquelets* and guerrillas when they were in their ranks and that which they displayed when they were out of the control of

[133] Esdaile, *Fighting Napoleon*, 154.
[134] Tone, *The Fatal Knot*, 173.
[135] Tone, *The Fatal Knot*, 192.

the authorities'.[136] Similarly, the Cossacks that the Napoleonic forces encountered in Russia in 1812 were traditional 'irregular' units associated with the regular Russian forces, but they were not guerrillas in the sense of a national resistance movement.[137]

In terms of the tactics they used, the differentiation between regulars and irregulars also proved tricky at times. Well-established and successful bands such as the one led by Espoz y Mina in Navarre stretched their repertoire by the end of the conflict to include more conventional operations and tactics. In 1813 they even acquired siege weapons and conducted siege operations, although with limited success.[138] On the other hand, as already mentioned, atrocities and terror tactics were by far not the sole reserve of the irregular forces, although the French regulars tended to claim that their indiscriminate violence was a mere reaction to previous misdeeds by the irregulars.

But what about the linchpin of criticism towards the irregulars—namely, their denouncement as mere 'brigands' and the allegation that they were subject to no political authority and discipline? It is easy to forget that, in many cases, the guerrillas had actually been called to arms by the state. Spain was a case in point. In the Tyrol, Andreas Hofer claimed he had a commission from the Austrian emperor Francis II. Thus, while Prussia and its 1813 instigation of a popular resistance movement from above was the most extreme case of such a state sponsorship of 'irregular' warfare, it was by no means an isolated case.

Equally, it is easily forgotten that the besieged state authorities in occupied territory often made attempts at regulating the irregulars. This was the case in Spain. After receiving complaints about the *partidas'* autonomous activities, the Spanish *junta central* issued a *reglamento* on 28 December 1808, which was to become the first in a series of attempts to organize and discipline the irregulars. It contained rules on the activities of the irregular bands, their discipline, their pay, and their hierarchical structure.[139] Hence, at least theoretically, the irregulars were by no means without any authority or discipline, although regulating their activities proved difficult, and 'the majority of historians agree that compliance with the *reglamento* was almost non-existent'.[140] However, the lack of authority of the *junta central* was not necessarily due to the innate indiscipline of the irregulars. Rather, it was due to a conflict between the central and the local *juntas*: 'by clashing with one another, rival codes of conduct were at the heart of many of the competence

[136] Antonio Moliner Prada, 'Popular Resistance in Catalonia: *Somatens* and *Miquelets*, 1808–14', in Esdaile (ed.), *Popular Resistance in the French Wars*, 105.

[137] Hartley, 'The Patriotism of the Russian Army', 186.

[138] Tone, *The Fatal Knot*, 141 ff.

[139] Scotti-Douglas, 'Regulating the Irregulars', 142.

[140] Scotti-Douglas, 'Regulating the Irregulars', 143.

disputes that from the very beginning affected the relationship between the guerrillas and the civilian authorities'.[141]

Finally, many of the guerrilla leaders themselves were rather successful in establishing their own systems of discipline within their *partidas*. While part of the Spanish guerrilla movement displayed ideological leanings towards anarchism and called themselves 'Men of Idea', guerrilla leaders such as Espoz y Mina dismissed such a mindset.[142] Where guerrilla leaders proved successful in implementing discipline, the only marked difference between their authority and that of the regular commander was that the former relied more on charismatic leadership than on the impersonal authority of the office. This was reflected in the fact that the death or injury of a guerrilla leader always caused confusion and disruption among his men.[143]

In some cases the state authorities themselves seemed to free the irregulars from all constraints in the conduct of their operations. The statute of April 1809 issued by the *junta central*, for instance, insists that the irregulars are encouraged to 'inflict[] all the evil and damage possible' on their opponent.[144] Moreover, the question of whether and how to impose discipline was subject to political contestation. Many argued that 'any attempt to regulate the *partidas* would necessarily discourage the populace' and make the irregulars less effective.[145] The irregulars presented the political authorities in Spain with the question how to get the balance right between spontaneity and discipline, between bottom-up 'enthusiasm' and top-down political control.

As ineffective as the attempts at regulating the irregulars may have been, they proved that the concept of the 'lawless' irregular who is under no command is inaccurate. There was, after all, 'a corpus of legislation that, at least theoretically, ruled the guerrillas and governed their legal status'.[146] Furthermore, even if the irregulars frequently ignored the political authorities and their attempts to impose discipline among them, their French regular counterparts did not necessarily act in a more disciplined way. It is true that the French often did not see the need to apply the customs of war against an opponent that did not do so either. But at the heart of the problem was the fact that, although France could recruit and deploy an army of a size that was unprecedented in European history, it could not supply it with the resources it needed. Hence, the scale of requisitioning required by the Napoleonic forces was equally unprecedented. The French commanders clearly perceived this as a problem, since they knew that excessive requisitioning led to indiscipline:

[141] Scotti-Douglas, 'Regulating the Irregulars', 149.

[142] Tone, *The Fatal Knot*, 116 ff. Mina forced his men to wear their hair short, thus distinguishing themselves from the 'Men of Idea', whose characteristic feature was their long hair.

[143] Tone, *The Fatal Knot*, 104; Broers, *Napoleon's Other War*, 29.

[144] Tone, *The Fatal Knot*, 144.

[145] Esdaile, *Fighting Napoleon*, 168.

[146] Scotti-Douglas, 'Regulating the Irregulars', 147.

'All too often it [requisitioning] led to looting and indiscriminate violence, which eroded discipline, while taking supplies, especially in impoverished areas, threatened the population with starvation and provoked popular resistance.'[147] Thus, far from keeping within the limits of the customs and laws of war, the Napoleonic forces frequently exacerbated the problem of irregular warfare with their own undisciplined behaviour. After all, experience of the harsh behaviour of the occupying forces and economic despair were among the chief recruitment factors for the resistance movements. Other regular forces did not necessarily behave any better: discipline was a huge problem for both the Prussian and the Austrian armies, as well as for the British forces under Wellington's command.[148] It comes as no surprise, then, that the popular views on irregulars and regulars did not differ much. Arguably, the local populace in Spain often regarded irregulars as a cross between criminals and deserters, but this image was also similar to that of the regular armies as being 'drawn from the dregs of society, the human detritus from prisons and poorhouses who, when they no longer had enemy armies to attack, would turn against the civilian population'.[149]

Thus, narratives of differences between irregulars and regulars are often externalizations of the problematic character of the regular forces themselves. By attributing negative characteristics such as individualism, desire for personal gain, indiscipline, and a propensity towards indiscriminate violence to the irregular 'brigand', the image of the regular was purged from all negative traits. When successfully inculcated in the local populace, this externalization supported an influential, hierarchically dichotomous image of politically legitimate regular troops and illegitimate irregular troops: 'To that degree, the war against "brigands" was also a victory for political rhetoric, for one discourse over another, a victory by revolutionary and Napoleonic *spinmeisters* which effectively complemented that of the nation's military battalions in the field.'[150]

Conclusion

The period between 1740 and 1815 was a period in which established concepts of illegitimate war underwent a thorough re-evaluation. In an era in which the 'people in arms' emerged as the legitimate entity to generate and organize military force, the rebel and the peasant in arms could no longer be condemned as illegitimate combatants per se. At the same time, differences between regular and irregular warfare emerged. At first, this pertained only to a

147 Rothenberg, 'The Age of Napoleon', 95.
148 Rothenberg, 'The Age of Napoleon', 95.
149 Forrest, 'The Ubiquitous Brigand', 29.
150 Forrest, 'The Ubiquitous Brigand', 41.

tactical differentiation between increasingly drilled and heavily armed 'regular' formations, on the one hand, and 'irregular' light infantry and cavalry units, on the other. This differentiation did not indicate any moral or legal assessment initially. However, the perception of the irregular as inferior crept in to the extent that irregular auxiliary troops were recruited from the native inhabitants of the colonies, who were perceived as 'savages', even if they were occasionally useful ones.

After the French Revolution and with the experience of counter-revolutionary rebellions within France and in Europe under Napoleonic occupation, the irregular evolved as the illegitimate belligerent. His—or her, since the irregular defied the gendered notion of the soldier as a male citizen—illegitimacy was based on the fact that they were fighting not for political motives, but for personal gain and under no command or discipline. Denounced as 'brigand', the image of the irregular oscillated between the criminal and the savage. This narrative was not only a forceful political weapon, undermining the legitimacy of political opponents. Given that the empirical differences between the regular and the irregular were often no more than nominal, in particular with respect to military discipline, the irregular served as a smokescreen onto which all negative empirical traits of the regular could be projected and thus externalized. With the depiction of the irregular as the outlaw and the deviation from the norm, the regular constituted itself as its legitimate and lawful counterpart.

3

The Nineteenth Century

Rebels, Rifles, and the Laws of War

The nineteenth century saw three main trends in the evolution of the idea of the irregular fighter. First, there were attempts at making the definition of the irregular more precise. In particular, this took place in the framework of the early efforts to codify the law of armed conflict and in Francis Lieber's work. In his legal opinions for the Union government, Lieber distinguished between different categories of irregular fighters and accordingly recommended different approaches to their treatment. However, this degree of sophistication was lost on Union commanders on the battlefield, who equated all sorts of irregular fighters with 'guerrillas' and 'bushwhackers', sometimes even at the expense of tactical reason. Attempts to codify the laws of war following the Franco-Prussian War also failed to provide a clear definition of the irregular. The question of where to draw the line between the lawful belligerent and the unlawful irregular had become so entangled with conflicts of political ideology and balance-of-power issues within Europe that no agreement could be reached among European states.

Secondly, punishing civilians for irregular attacks and activities that occurred in the area where they lived was established as a legitimate measure justified by military necessity. However, it is important to keep in mind that the notion of the 'civilian' in today's legal terms was not known to contemporaries. Once again, Lieber's work was significant for the emerging legal approach to the treatment of what contemporaries called 'peaceful' or 'inoffensive citizens'. While he acknowledged that certain categories of people should not become the object of military attack, he also followed military commanders' practice in distinguishing between three types of civilians, depending on their loyalty or disloyalty to one's own side during the American Civil War. German officers and troops in the Franco-Prussian War reinforced this approach by emphasizing that French civilians would be spared from harm only if they showed firm loyalty to the occupying forces. This legal tradition is often forgotten in today's accounts of the legal history of the 'civilian', but it was highly influential throughout the late nineteenth and the first half of the twentieth centuries, and probably had its gravest repercussions in Nazi Germany's occupation policies during the Second World War.

Thirdly, technological innovations during the nineteenth century transformed not only battlefield tactics and strategic practice, but also the concept of the irregular. Chief among these developments were the spread of railways and their use for military logistics as well as the introduction of new and more precise rifles. The evolving military importance of railways had rather obvious implications for irregular warfare, as it introduced new and potentially significant vulnerabilities for regular armed forces: railroads were difficult to protect against irregular attacks, which could cut off regular armies from their supplies. Yet, it was the second innovation, the introduction of rifles with a longer range, that changed the image of the irregular more fundamentally, since it transformed the meaning of courage on the battlefield. This resulted in two parallel but contradictory developments. On the one hand, the conceptual opposite of courage—namely, cowardice—was projected onto the irregular and evolved into the new lead theme of the official discourse on irregular fighters. On the other, irregular fighters fostered an image of themselves that emphasized traditional notions of courage and nostalgic images of hand-to-hand fight with old-fashioned weaponry.

This chapter focuses on irregular fighters in the American Civil War and the Franco-Prussian War. These two wars were not the only ones in which irregulars played a role. Irregular fighters were also involved in the Austro-Prussian War, for instance. But the American Civil War and the Franco-Prussian War are probably the best-researched armed conflicts of the nineteenth century, in particular with respect to the involvement of irregulars. Moreover, they also precipitated a heightened awareness among contemporaries of the problem of irregular fighters and led to efforts at tackling this problem by codifying the laws of war.

This chapter is divided into four parts. The first part deals with the American Civil War. It provides an overview of the different types of irregular fighters involved in the war and depicts official responses to irregular warfare. It focuses in particular on legal approaches to the problem of irregular warfare, mainly Francis Lieber's writings. The second part is devoted to the Franco-Prussian War. It discusses the role played by *francs-tireurs* as well as German reactions to the *francs-tireurs* problem. The third section analyses the emergence of new technologies during the nineteenth century, in particular the railroad and the rifle, and their impact on the evolution of the concept of the irregular. The fourth part discusses the early attempts at codifying the laws of war, which were strongly influenced by the experience in the American Civil War and the Franco-Prussian War.

The American Civil War, 1861–5

The American Civil War developed out of longstanding disputes over the role of central government in the United States. Although the conflict over the abolition of slavery was not among the direct causes of the war, the issue of slavery was central to the growing North–South divide during the first half of the nineteenth century. It also led to political turmoil in the newly created states in the west. Many northern states had opted for the emancipation of slaves in the late eighteenth and early nineteenth centuries. By contrast, many southerners felt that northern calls for emancipation threatened the southern economic structure, in particular the South's large cotton farms. The first visible rift over the slavery issue was the 1819 Missouri Crisis. It concerned the question of whether the newly formed state of Missouri would be admitted to the Union. The admission of Missouri, where slavery was permitted, would have brought the number of slave-holding and free states in the Union to parity. In 1820 Congress reached a compromise that envisaged admitting Maine as a new free state and prohibiting slavery in the territory north of the parallel 36°30′ north, except within the boundaries of the proposed state of Missouri. However, the Missouri Compromise, as it became known, did not resolve the issue itself: westward expansion meant that the problem would recur. Predictably, the question of the admissibility of slavery in newly acquired territories flared up again after the Mexican War (1846–8), when the Union had to determine the political fate of Texas, New Mexico, and California. Once again, an uneasy compromise was reached in 1850, and the matter was resolved temporarily, only to be brought up again in the 1854 Kansas–Nebraska Crisis, during which Missouri Senator David Atchison challenged the slavery-limiting provisions of the earlier Missouri Compromise. The resulting 1854 Kansas–Nebraska Act established the principle that the issue of slavery was to be resolved by popular sovereignty within individual states. However, the institution of slavery divided not only the Union, but also settlers within the new states. Hence, to a certain extent the American Civil War started as a guerrilla war in those states that were struggling to find a solution to the question of slavery. In Kansas it soon led to fierce clashes between abolitionists and pro-slavery settlers, many of whom crossed the border from Missouri in order to vote in territorial elections in Kansas in 1856. The ensuing low-level guerrilla warfare that came to be known as 'Bleeding Kansas' set a precedent for what was to happen in many border states during the Civil War.

The immediate trigger for the southern states' secession, which started with South Carolina's decision to secede from the Union on 20 December 1860, was the election of Abraham Lincoln as President. Lincoln was a liberal who held a decidedly anti-slavery position. Although he did not propose any immediate and direct action against slavery and was anxious to strike a conciliatory note towards the South, pro-slavery factions perceived his election as a revolution.[1]

The first military action of the war was the newly formed Confederacy's shelling of Fort Sumter, a Union outpost on a small island off the coast of South Carolina, in April 1861. Lincoln reacted to this opening of hostilities by imposing a sea blockade on the South, thereby attempting to undercut its supply lines. The blockade immediately raised the issue of the nature of the conflict and revealed the Union's ambiguous position. On the one hand, Lincoln insisted that states did not have the right to secede from the Union and that the South was therefore engaging in a rebellion against the legitimate government. On the other, a blockade was an instrument typically used by sovereign nations against each other.[2] Regardless of that dilemma, southern secession confronted the North with the task of restoring the Union. The ensuing war was much longer, much bloodier, and much less restrained than anyone had anticipated in 1861. As we shall see, it was irregular warfare in particular that prompted a loss of restraint on both sides.

Irregulars in the American Civil War

The involvement of irregular fighters in the American Civil War was considerable. Irregulars numbered presumably in the tens of thousands, which appears to be a rather small number when compared to the 3.5 to 4 million soldiers who fought in total in the Union and Confederate armies.[3] However, guerrillas had a decisive impact on the war. Their existence and activities did not affect so much the outcome of war, but rather the operations, tactics, policies, and, finally, public narratives of the war.[4] The role of irregulars was slightly more important on the Confederate side, for various and partly interrelated reasons. First, occupation by Union forces often fostered irregular activities. Second, irregulars thrived in a southern (and western) cultural context that included a strong vigilante and ranger tradition.[5] In these

[1] James M. McPherson, *Battle Cry of Freedom: The American Civil War* (Oxford: Oxford University Press, 1988), 233.

[2] McPherson, *Battle Cry of Freedom*, 316; Louis P. Masur, *The Civil War: A Concise History* (Oxford: Oxford University Press, 2011), 27.

[3] Daniel E. Sutherland, *A Savage Conflict: The Decisive Role of Guerrillas in the American Civil War* (Chapel Hill, NC: University of North Carolina Press, 2009), p. xii.

[4] Sutherland, *Savage Conflict*, p. xii.

[5] Sutherland, *Savage Conflict*, 43 ff.

regions, where central government and administration had developed only slowly, many settlers were still used to policing their own communities and to defending them against external threats by forming self-constituted armed groups. Third, the Confederate government maintained a highly ambivalent stance towards irregular fighters throughout the war. On the one hand, the government in Richmond and many Confederate generals loathed irregulars, because they knew that they were almost impossible to control. On the other hand, on 28 April 1862 the Confederate Congress adopted the Partisan Ranger Act, which stipulated that the President could authorize bands of partisan rangers.[6] Hence, some of those men whom the Union denounced as irregulars had, in fact, a presidential commission, and many more seem to have interpreted this act as a sort of carte blanche for the establishment of irregular armed groups. The Confederate district commander of Arkansas, Thomas C. Hindman, took this even further, issuing an order on 17 July 1862 that called for the formation of small bands of irregular fighters in order to fight back the Union advance west of the Mississippi.[7] In the end, the suspicion towards the partisan system and its impact on discipline and morale among regular Confederate soldiers prevailed: the Partisan Ranger Act was repealed on 17 February 1864. However, the Confederate leadership proved irresolute in this decision as well, since Mosby's Rangers and McNeill's Rangers were allowed to persist, and the other armed groups continued to operate as well.

At the tactical and operational levels, though not at the legal level, the Union showed a comparable ambivalence towards northern irregulars. As commander of the district of Missouri, General Henry W. Halleck tolerated the presence of Kansas 'Jayhawkers' in Missouri and did not expel them until it became obvious that they had plundered local Confederate sympathizers and Unionists alike.[8] Officially, however, Halleck was one of the most adamant opponents of irregular warfare and accused the Confederacy repeatedly of violating the laws of war by sanctioning armed bands of irregulars. Halleck was, himself, a lawyer and had written a treatise on international law before the war. It was his experience of irregular warfare during his time in Missouri that prompted Halleck to ask Francis Lieber to write a legal opinion for the Union on the treatment of irregulars.[9]

Irregular fighters were most active in the border states, the populations of which were split between Union and Confederacy supporters, mainly

[6] For the full text of the Partisan Ranger Act, see Robert R. Mackey, *The Uncivil War: Irregular Warfare in the Upper South, 1861–1865* (Norman, OK: University of Oklahoma Press, 2004), 205 ff.

[7] Richard S. Brownlee, *Gray Ghosts of the Confederacy: Guerrilla Warfare in the West, 1861–1865* (Baton Rouge, LA: Louisiana State University Press, 1984), 76 ff.

[8] Sutherland, *Savage Conflict*, 60.

[9] Lieber, 'Guerrilla Parties', printed in Richard Shelly Hartigan (ed.), *Lieber's Code and the Laws of War* (Chicago: Precedent, 1983), 31 ff.

Missouri, Kansas, Arkansas, Kentucky, and Tennessee. Virginia was geographically a border state as well, albeit with a solidly southern population. However, irregular fighters were also active in the Deep South: Alabama and Florida saw the emergence of armed bands in 1862, and Texas was used as a refuge by Missouri guerrillas in the winter of 1863. North Carolina and Georgia saw activities by irregulars in 1864.[10] The roles, constitution, and activities of irregulars varied, depending on the political and military situation in which they operated. In states such as Tennessee, which neither side actually controlled for most of the war, irregular warfare often resembled a small-scale civil war, in which neighbour fought against neighbour. In border states under Union occupation, such as Missouri, it amounted to a people's war in which parts of the population fought against Union forces. Finally, in the Deep South, irregular fighters mostly served the purpose of local defence against Union invasion. Partly as a result of this variation in circumstances, the notion of the 'irregular' in the American Civil War covers a broad spectrum of different fighters, ranging from commissioned partisan cavalry detachments to bandits who used the war as a smokescreen for their criminal activities.

The structure of the armed forces in pre-Civil War America compounded the difficulties of defining and identifying the irregular. The United States had a tradition of a 'dual army'.[11] The 1792 Uniform Militia Act stipulated universal military obligation to service in the state militias. However, these tended to be weak and poorly trained. Within the state militias, there were also elite volunteer companies. At the same time, there was a tradition of raising ad hoc volunteer forces at the national level, such as George Washington's Continental army during the American War of Independence.[12] During peacetime, there was a small federal standing army, as envisaged by the Constitution of 1789. This resulted in potential confusion as to which authorities could raise what kinds of troops. In addition, the long-held emphasis on local defence often led to engrained reservations towards service in the central armies.

During the American Civil War, both sides introduced conscription, the Confederacy in 1862 and the Union in 1863. However, conscription was never the main source of troop numbers on either side: only 21 per cent of Confederate troops and a mere 6 per cent of Union troops were drafted.[13]

[10] Sutherland, *Savage Conflict*, 103, 114, 250.

[11] Peter Maslowski, 'To the Edge of Greatness: The United States, 1783–1865', in Williamson Murray, MacGregor Knox, and Alvin Bernstein (eds), *The Making of Strategy: Rulers, States and War* (Cambridge: Cambridge University Press, 1994), 213.

[12] See Chapter 2.

[13] John Whiteclay Chambers II, 'American Views of Conscription and the German Nation in Arms in the Franco-Prussian War', in Daniel Moran and Arthur Waldron (eds), *The People in Arms: Military Myth and the National Mobilization since the French Revolution* (Cambridge: Cambridge University Press, 2003), 81.

The imposition of conscription upon split and/or occupied states such as Tennessee, for instance, fostered the formation of bands of irregular fighters, as Unionists understandably resisted service in the Confederate army and vice versa.[14] In addition, many conscripted soldiers returned to their home states, where they felt they could more directly defend their own families, property, and local communities. Technically they were deserters, but they did not conceive of themselves as such. Rather, they felt justified by the Federalist tradition of using conscription for local defence. To complicate things even further, in regions ridden by guerrilla warfare the local authorities formed local anti-guerrilla forces called 'home guards'—a measure that was rooted in the same tradition. They were often poorly controlled and commanded, thereby disintegrating into irregular bands themselves.[15]

Proliferation of irregular groups occurred not only at the local but also at the national level. Both the Union and the Confederate leadership established specific partisan ranger corps, tasked with fighting guerrillas and other irregular bands. The Union Fifth Tennessee Cavalry and the Confederate Fifth Tennessee Cavalry were examples of such anti-guerrilla forces.[16] In 1863 Union Major General William Rosecrans raised new cavalry regiments for the Army of the Ohio and the Army of the Cumberland, which were intended to operate as counter-raiding forces against Confederate raiding cavalry.[17] One of the most notorious among these various anti-guerrilla forces was Alfred Ellet's Marine Brigade (Union), which was charged with patrolling the Mississippi river and protecting Union steamboats from irregular attacks. However, owing to poor recruitment and training, Ellet's men operated often in precisely the same way as the guerrillas they were intended to combat.[18]

In sum, contemporaries of the American Civil War had considerable difficulties in distinguishing between regular and irregular forces, because the pre-war United States did not have a clear template of the regular soldier to start with. This difficulty was exacerbated by the fact that it was often impossible to differentiate visually between regular and irregular forces. Both guerrillas and militias often wore plain clothes. Confederate guerrillas sometimes wore Union uniforms as a disguise. Conversely, Union soldiers dressed up as guerrillas at times, in particular when they went on looting sprees, so that they could blame the attacks on irregulars.[19]

[14] Noel C. Fisher, *War at Every Door: Partisan Politics and Guerrilla Violence in East Tennessee, 1860–69* (Chapel Hill, NC: University of North Carolina Press, 1997), 42; cf. Sutherland, *Savage Conflict*, 216.

[15] Sutherland, *Savage Conflict*, 230.

[16] Sutherland, *Savage Conflict*, 59 ff.

[17] Mackey, *Uncivil War*, 166.

[18] Mackey, *Uncivil War*, 57.

[19] Michael Fellman, *Inside War: The Guerrilla Conflict in Missouri during the American Civil War* (Oxford: Oxford University Press, 1989), 28 ff.

Not only were the boundaries between the regular and the irregular blurred; there also existed a number of different types of irregular fighters. Robert Mackey differentiates between three main categories of irregular warfare: people's war, partisan warfare, and raiding war.[20] This approach is a useful starting point, although the categories should be seen as ideal types rather than as actual descriptions of the empirical reality, since the boundaries also between these three types were often not clear.

Occupied states with a split population saw irregular warfare that came close to a people's war in the nineteenth-century sense of the term, meaning a popular uprising of the citizens against invasion and occupation. Depending on the political circumstances and the military situation, it could also take the form of a small-scale civil war between Union and Confederate sympathizers. Prominent examples in this respect were Confederate resistance against Union occupation in Arkansas and Missouri, and Unionist resistance against Confederate attempts at gaining control over East Tennessee. This kind of guerrilla warfare was often sanctioned by individual military leaders on both the Confederate and the Union side. In Missouri, Confederate General Sterling Price called for the formation of guerrilla bands to harass Union occupation forces early on in the war. In Arkansas, Confederate General Hindman's 'Bands of Ten' decree sanctioned guerrilla groups that harassed Union General Samuel R. Curtis's troops marching to Helena after the Confederate defeat at Pea Ridge.[21] Irregular Kansas Union regiments, especially General Jim Lane's brigade, consisting of the Fourth Kansas Volunteer infantry and Fifth Kansas cavalry regiments, entered Missouri and terrorized its inhabitants. In September 1861, they were joined by 'Doc' Charles Jennison's notorious 'Jayhawkers', the Seventh Kansas Volunteer cavalry regiment.[22] Irregular warfare along the Missouri–Kansas border often had its roots in the pre-war cross-border hostilities.

But irregular groups also formed spontaneously, without instigation by military commanders. One of the most notorious groups was William Clarke Quantrill's guerrillas. They operated in both Missouri and Arkansas. Their prime purpose was to strike at Jennison's 'Jayhawkers' and disrupt Union communications.[23] Quantrill's group also achieved some small-scale 'regular' successes, such as the capture of a Union outpost in Independence, Missouri, in August 1862. Their greatest triumph was the sacking of Lawrence, Kansas, in August 1863, which started as a revenge raid for the Union's banishment and arrest policy, but quickly slipped into slaughter and indiscriminate

[20] Mackey, *Uncivil War*, 10.
[21] Sutherland, *Savage Conflict*, 67.
[22] Brownlee, *Gray Ghosts*, 38 ff.
[23] Brownlee, *Gray Ghosts*, 58 ff.

looting.[24] Quantrill's band declined and split in the winter of 1863, with 'Bloody' Bill Anderson assuming command of part of the group.

Irregular fighters also included Confederate partisan rangers, which were legitimized by the Partisan Ranger Act, at least during its implementation between 1862 and 1864. Activities by such groups were especially prevalent in states such as Virginia, Kentucky, and Tennessee, which were strategically important for the Union, largely for logistic reasons. Partisan rangers conducted small cavalry raids against Union pickets and railroad guards. Under the Partisan Ranger Act, they had permission to seize and sell captured Federal property, but the rangers often interpreted this permission liberally and plundered civilian passengers on the trains.[25] The most well known among these partisan ranger groups were Mosby's Rangers. John S. Mosby was a Confederate commissioned officer, and his group included many former regular officers and West Point graduates.

The third form of irregular warfare was raiding war conducted by regular cavalry operating independently. This form of irregular warfare was rooted in eighteenth-century notions of the irregular as described in the previous chapter. Prominent examples of this type of irregular fighter were John Hunt Morgan and Nathan Bedford Forrest. They sometimes conducted supporting missions for regular forces such as scouting, for instance, in the spring of 1862, when both Morgan's and Hunt's cavalry were scouting for General Albert S. Johnston's Army of the Tennessee.[26] However, their main activity was conducting raids on supply lines deep into enemy territory, such as Morgan's 'Christmas Raid' in 1862. During this raid Morgan led a considerable force of 2,100 cavalry into Union-controlled Kentucky in order to destroy supply depots, railroads, and bridges.[27] Raiding cavalry operated not only with considerable troop numbers, but also within a large geographical range deep into enemy territory, as demonstrated by Morgan's 'Great Raid' through Kentucky, Indiana, and Ohio in July 1863.

The composition of the irregular bands and the motivation of individual fighters varied greatly as well. In both the North and the South, the spectrum ranged from petty criminals to West Point graduates and members of the upper stratum of society. The respective opponent's public denunciation emphasized that irregulars sought private gain rather than fighting for a political cause and that they engaged in violence for the sake of violence. However, the irregulars' real motivations were more complex. Don Bowen showed in a detailed study of guerrillas in Jackson County, Missouri, that irregulars tended to come from wealthy southern families who had a great stake in maintaining the pre-war social order, in particular the institution

[24] Brownlee, *Gray Ghosts*, 122 ff.

[25] Mackey, *Uncivil War*, 87.

[26] Mackey, *Uncivil War*, 129.

[27] Mackey, *Uncivil War*, 154 ff.

of slavery.[28] Biographies of prominent irregular fighters from across the social spectrum also support the impression that irregular fighters did not come from the dregs of society: William Quantrill had been a teacher before the war, Mosby originated from the Virginia plantation aristocracy and had been a lawyer before the war, Morgan came from a similar social background as Mosby and had had a pre-war career as an officer, and Forrest, although not himself from a wealthy family, had made a fortune with cotton plantations in Tennessee before the war.[29] Nonetheless, there were lawless elements among the guerrillas, and their number increased towards the end of the war.[30] 'Bloody' Bill Anderson and George Todd, for instance, both active in Quantrill's group before they became guerrilla leaders themselves, represented the rather more violent and adventure-seeking part of the spectrum.[31]

Motivations varied among the different groups as well as between individual irregular fighters. According to Daniel Sutherland, the defence of local communities was the prime motive for irregular activities, but the picture was probably more complex.[32] Revenge played an important part as well. Bill Anderson, for instance, justified his actions by pointing out that the Federals had killed his father and sister: 'I have fully glutted my vengeance... I have tried to war with the Federals honourably, but for retaliation I have done things, and am fearful will have to do that which I would shrink from if possible to avoid.'[33] Another strongly motivating factor was to avoid the unpleasant service in the regular armed forces while at the same time being able to operate closer to home.[34] Finally, to the extent that irregulars were able to retain the spoils of war (and Confederate partisans were even entitled to this by law), personal gain certainly played a role as well. Some irregulars transformed themselves into outright criminals and continued to operate as criminal bands after the war, such as the James brothers and Younger brothers. However, the presence of criminal elements should not be mistaken as proof that all irregulars were bandits. Most had mixed motives, but, as Noel Fisher wrote with respect to guerrilla warfare in East Tennessee, 'the political motives, though sometimes blurred and smudged, were always visible'.[35] Even in the case of outright banditry, political nuances were present. In a context in which the 'normal' legal fabric of a society had unravelled, murder and theft, even if only for the purpose of protecting and revenging oneself, had a political meaning. The James and the Younger brothers were often cheered (and supported) by local communities, because they were regarded

[28] Don R. Bowen, 'Guerrilla War in Western Missouri, 1861–1865: Historical Extensions of the Relative Deprivation Hypothesis', *Comparative Studies in Society and History*, 19/1 (1977), 30–51.

[29] Brownlee, *Gray Ghosts*, 54; Mackey, *Uncivil War*, 73, 126.

[30] Sutherland, *Savage Conflict*, 211.

[31] Sutherland, *Savage Conflict*, 65.

[32] Sutherland, *Savage Conflict*, p. xi.

[33] Quoted in Fellman, *Inside War*, 139.

[34] Sutherland, *Savage Conflict*, 43.

[35] Fisher, *War at Every Door*, 63.

as strongmen who could defend themselves against their enemies' corruption. They became folk heroes although they did not fight for any ostensible political cause.[36]

Regarding the degree of violence used by irregular fighters, the picture is equally mixed. Many irregular groups were rather well disciplined and did not commit random acts of violence. Mosby's Rangers, for instance, mostly behaved in a disciplined fashion and did not target local Virginia citizens, with the exception of civilian train passengers falling victim to their raids on railways. They also treated captured Union troops well and mostly paroled them immediately after capture.[37] Quantrill, too, paroled Union soldiers. His men apparently started to execute Union prisoners only after they had learned of Halleck's 'no-quarter' policy against the guerrillas.[38] Figures such as Bill Anderson and his second in command, Archie Clement, seemed to resemble more closely the image of the bloodthirsty killer. They were notorious for decorating the saddles and bridles of their horses with the scalps of their victims. But what may seem as gratuitous violence has to be understood within the particular cultural context in which they were operating: scalping was a reference to earlier forms of irregular warfare in America, as described in the previous chapter. It was a calculated symbol of a particular cultural self-image, depicting the irregular as the noble savage, and it also acted as a form of deterrence by instilling fear in their enemies.[39] Moreover, violent acts perpetrated by irregulars should not be measured by today's legal standards. Rather, they must be seen in the context of the American Civil War, whereby also regular troops terrorized civilians and engaged in large-scale looting, in particular where they were harassed by irregulars.[40] While it is true that notorious guerrilla operations such as the sacking of Lawrence slipped into a massacre, the irregulars felt this was vindicated by the harsh treatment meted out by Union General Jim Lane and district commander Thomas Ewing towards the pro-Confederate population of Missouri.[41]

Dealing with Irregulars: Ambiguity or Dichotomy?

In 1862, General Halleck commissioned Francis Lieber to write a legal memorandum on the problem of guerrilla warfare. In the resulting text entitled 'Guerrilla Parties Considered with Reference to the Laws and Usages of War', Lieber distinguishes between different types of irregular fighters. On the first

[36] Richard White, 'Outlaw Gangs of the Middle Border: American Social Bandits', *Western Historical Quarterly*, 12/4 (1981), 387–408.

[37] Mackey, *Uncivil War*, 75.

[38] 'Order No. Two' of 13 March 1863, in Brownlee, *Gray Ghosts*, 64 ff.

[39] Sutherland, *Savage Conflict*, 28.

[40] Fellman, *Inside War*, 29 ff.

[41] Brownlee, *Gray Ghosts*, 115 ff.

type, the 'freebooter', Lieber writes that 'freebooters are nothing less than armed robbers of the most dangerous and criminal type'.[42] The second type is the 'brigand', 'the soldier who detaches himself from his troop and commits robbery, naturally accompanied in many cases with murder and other crimes of violence'.[43] Thirdly, 'partisans and freecorps' are 'bodies detached from the main army', hence regular units that operate independently.[44] The fourth and the fifth categories, 'the spy, the rebel, and conspirator' and the 'war rebel', designate individuals and groups operating covertly or rising up overtly against occupation forces.[45] The last category was the 'rising en masse, or "the arming of peasants"', which designated the spontaneous constitution of armed groups 'to repel invasions'.[46] Of these six types of irregulars, only the partisans, free corps, and the rising en masse, even without uniforms, were lawful belligerents in Lieber's eyes. The guerrilla oscillated between the brigand, the partisan, and the rising en masse, which, according to Lieber, made it particularly difficult to determine his or her legal status.[47] Lieber's treatment of irregular fighters in the Lieber Code, written a year later, was less nuanced. He retained the distinction between the lawful partisan and unlawful other types of irregular fighters, hence insisting on this difference in opinion with Halleck, who considered partisans to be unlawful belligerents.[48] However, this time he did not mention the rising en masse, nor did he discuss the problematic question of the legality of the guerrilla.[49] On the one hand, this may have been due to the different character of the text, which was intended to be a military manual; on the other hand, he may have intended to meet Halleck halfway. In any case, Lieber's discussion of irregular fighters was remarkably detailed and complex. However, Lieber himself did not perceive of his writings as legal guidance or even as law as such. He conceded that it was for the government to decide on the actual treatment of irregulars. He was also aware that the usefulness of his legal memoranda would be 'much diminished by the fact that the soldier generally decides these cases for himself'.[50]

Indeed, the Union government and its military leaders did not follow Lieber's opinion on the treatment of irregular fighters. Many nuances of Lieber's analysis were lost on Union decision-makers and troops in the field. The prevailing opinion was that all irregular fighters were unlawful belligerents and hence to be punished by death. Union officers did not acknowledge

[42] Lieber, 'Guerrilla Parties', printed in Hartigan (ed.), *Lieber's Code*, 34.
[43] Lieber, 'Guerrilla Parties', printed in Hartigan (ed.), *Lieber's Code*, 34.
[44] Lieber, 'Guerrilla Parties', printed in Hartigan (ed.), *Lieber's Code*, 35.
[45] Lieber, 'Guerrilla Parties', printed in Hartigan (ed.), *Lieber's Code*, 36 ff.
[46] Lieber, 'Guerrilla Parties', printed in Hartigan (ed.), *Lieber's Code*, 38 ff.
[47] Lieber, 'Guerrilla Parties', printed in Hartigan (ed.), *Lieber's Code*, 40 ff.
[48] Sutherland, *Savage Conflict*, 127.
[49] Lieber, 'General Orders No. 100' [Lieber Code], printed in Hartigan (ed.), *Lieber's Code*, 60 ff.
[50] Quoted in Sutherland, *Savage Conflict*, 128.

the legitimacy of armed local citizens fighting against invading Union troops, Lieber's 'rising en masse'. Nor did they recognize the authorization of irregular soldiers by Confederate General Hindman's 'Bands of Ten' order—for instance, when Union troops did not give quarter to authorized irregular groups during the invasion of Arkansas by General Curtis's Army of the Southwest in June 1862. Hindman responded by threatening to kill Union POWs should Union Colonel Graham N. Fitch continue his 'no-quarter' policy towards Confederate irregular forces. In the autumn of 1863, Fitch was relieved from his post and succeeded by Colonel Frederick Steele. In an attempt to undercut local support for the guerrillas and to prevent them from being perceived as 'martyrs' by their local communities, Steele introduced a new programme of leniency towards the guerrillas and commuted death sentences to hard labour sentences.[51]

The Union also regarded partisan rangers as unlawful belligerents, despite the fact that they were often led by officers and authorized by the Confederate Partisan Ranger Act. Mosby's Rangers, for instance, were mostly well disciplined, and Mosby himself was a commissioned officer. Yet the Union regarded them as bushwhackers; Halleck, for instance, spoke of 'the robber bands of Mosby'.[52] This not only implied that they would be executed when captured. The Union's equation of partisan rangers with guerrillas also led Union commanders to use anti-guerrilla tactics on opponents who differed from guerrilla forces in terms of their character, skills, and tactics. The dominant Union method in fighting against partisan rangers was cavalry sweeps and search-and-destroy missions by counter-guerrilla forces. Operating as detached cavalry units, these counter-guerrilla forces often closely resembled the partisan rangers they were intended to fight. In addition, Union commanders took passive defensive actions, such as protecting railroads and other strategically important points. They also implemented retributive measures intended to punish the local population for the partisan rangers' activities. The most notorious of these retributive operations was General Philip Sheridan's 1864 campaign in the Shenandoah Valley in western Virginia, which involved the burning of entire villages and counties. In the Shenandoah Valley campaign, which partly served the purpose of destroying Confederate supplies in an area that was considered the Confederacy's 'breadbasket', Union commanders perceived the severity of that destruction to be justified as a retributive measure against irregular warfare in the region, in particular the activities of Mosby and his men.[53] But the Union's retributive stance in the Shenandoah Valley backfired greatly.

[51] Mackey, *Uncivil War*, 61. [52] Mackey, *Uncivil War*, 119.

[53] Clay Mountcastle, *Punitive War: Confederate Guerrillas and Union Reprisals* (Lawrence, KS: University Press of Kansas, 2009), 103 ff.

> In effect, Mosby the *partisan*, because of Federal retribution on the citizens of the Shenandoah, became the semblance of legal authority in the region, fulfilling the role that the Union Army in Arkansas played in providing law and order.... In a partisan war, burning towns and plundering the countryside aided the irregulars, who gained ample support from the civilian populace, and, since they were seen as legitimate members of the Confederate military, were seen as representatives of a legal government with all the authority therein.[54]

Some Union commanders and Federal newspapers even denounced raiding cavalry, such as John Hunt Morgan and Nathan Bedford Forrest's cavalry, as guerrillas. In 1862 a special commission was set up in Nashville, Tennessee, to investigate Union General Don Carlos Buell's failure to advance to Chattanooga in the summer of 1862, when his supply lines were disrupted by Morgan and Forrest's raiders. The commission pressed Buell to explain why he had not conducted retributive operations against civilians in order to suppress the 'guerrilla' activities that had stalled his advance. Buell insisted that cavalry raiders were not guerrillas, but this position was evidently not very widespread among Union commanders.[55] The Union's main countermeasure against the cavalry raids was building blockhouses around strategically important points. This defensive measure had worked reasonably well against guerrillas and partisan rangers, but it achieved little success against raiding cavalry with artillery support.[56] It could also not prevent raiding parties from destroying lines of communication in Kentucky and Tennessee. Hence, the Union again committed tactical errors owing to their misperception of cavalry raiders as guerrillas. In short, while the denunciation of all sorts of irregulars as mere brigands and robbers was intended to win the war of ideas, it had a negative effect on the war on the ground: when commanders took the propaganda seriously, they committed preventable tactical blunders, which sometimes had operational and strategic repercussions.

Irregulars and the Punishment of Civilians

The Lieber Code has often been hailed as the first step in the legal codification of the distinction between combatants and civilians in war.[57] But this is only partly true. Lieber was ambivalent when it came to protecting civilians from the effects of war. The Lieber Code granted military commanders a high level of discretion in the treatment of civilians. Large-scale destruction of property was permissible as long as it was warranted by military necessity, being

[54] Mackey, *Uncivil War*, 120.
[55] Mackey, *Uncivil War*, 139.
[56] Mackey, *Uncivil War*, 134.
[57] See, e.g., Theodor Meron, 'Lieber's Code and the Principles of Humanity', *Columbia Journal of Transnational Law*, 36/2 (1997), 269–81.

defined fairly generously as 'the necessity of those measures which are indispensable for securing the ends of the war'.[58] Only wanton destruction and cruelty were prohibited. The Lieber Code permitted civilian reprisals, the starving of the enemy both armed and unarmed, and hostage-taking.[59] Most importantly, however, the Lieber Code contained two contrasting approaches to the treatment of civilians. On the one hand, Lieber stated that the 'unarmed citizen is to be spared in person, property, and honour as much as the exigencies of war will admit'.[60] On the other, in the section on 'Insurrection—Civil war—Rebellion', Lieber wrote:

> The military commander of the legitimate government, in a war of rebellion, distinguishes between the loyal citizen in the revolted portion of the country and the disloyal citizens. The disloyal citizen may further be classified into those citizens known to sympathize with the rebellion without positively aiding it, and those who, without taking up arms, give positive aid and comfort to the rebellious enemy without being bodily forced thereto. . . . The commander will throw the burden of war, as much as lies within his power, on the disloyal citizens, of the revolted portion or province, subjecting them to a stricter police than the non-combatant enemies have to suffer in regular war . . . he may expel, transfer, imprison, or fine the revolted citizens who refuse to pledge themselves anew as citizens obedient to the law and loyal to the government.[61]

Lieber's second approach made protections for civilians dependent upon their behaviour and even their political *Weltanschauung*. It effectively put all civilians in enemy territory under a general suspicion. Hence, the Lieber Code 'erected very few strong barriers against severe treatment' of civilians.[62] This was even more significant, as it was almost impossible for citizens in a split border state such as Missouri, for instance, to show themselves as loyal supporters of either side, as they were terrorized by both Union troops and Confederate guerrillas alike. According to Michael Fellman, in these circumstances many civilians resorted to 'survival lying' and 'the creation of a wartime alter-self' that could respond to contacts with either side in a self-preserving way.[63]

Lieber's second approach to the treatment of civilians was firmly rooted in the practice of Union commanders. However, its significance went far beyond the American Civil War. Although Lieber had intended to restrict his second approach to situations of civil war and rebellion, it would set a precedent for

58 Lieber, 'General Orders No. 100' [Lieber Code], printed in Hartigan (ed.), *Lieber's Code*, 48.

59 Lieber, 'General Orders No. 100' [Lieber Code], printed in Hartigan (ed.), *Lieber's Code*, 48 ff.

60 Lieber, 'General Orders No. 100' [Lieber Code], printed in Hartigan (ed.), *Lieber's Code*, 49.

61 Lieber, 'General Orders No. 100' [Lieber Code], printed in Hartigan (ed.), *Lieber's Code*, 71.

62 Mark Grimsley, *The Hard Hand of War: Union Military Policy towards Southern Civilians 1861–1865* (Cambridge: Cambridge University Press, 1995), 150.

63 Fellman, *Inside War*, 49.

the treatment of civilians for decades to come, in particular in areas where irregular fighters were present. For almost a century after the American Civil War, the legal definition of the civilian and the protections available to this category were derived from the fundamental dichotomy between the regular and the irregular. In short, the 'unarmed citizen' was worthy of protection only if he or she showed absolute loyalty to the regular forces and had no connections with irregular forces or resistance movements, let alone take up arms him or herself to fight against the regular forces.

How did the Union practice of treating civilians harshly, often referred to as 'hard war', develop? Initially Union military policy towards civilians emphasized the ideas of restraint and conciliation. This was based upon the conviction that the southern 'cause' had never been a popular one and that it had been imposed by the minority plantation aristocracy on the majority of 'ordinary people' in the South.[64] It was mainly the experience of guerrilla warfare that changed the Union commanders' stance towards civilians in the border states and in the South.[65] Many of the 'hard war' practices developed from the bottom up: punishment policies were adapted by individual commanders in a decentralized way before they were sanctioned and often adopted by their superiors. These practices suited the disposition of the troops.[66] For instance, in his role as district commander of North and Central Missouri, Brigadier General John Pope announced in June 1861 'that he would hold citizens living along the North Missouri Railroad responsible for "wanton destruction" of bridges, culverts, and track. Residents within a five-mile radius of guerrilla outrages would pay a "levy of money or property sufficient to cover the whole damage"'.[67] Other commanders in Missouri copied this approach,[68] and a few days later Major General George B. McClellan, commander of the Department of the Ohio, adopted it as official policy.[69] One of the reasons for the rapid spread of the punitive approach was most likely the vulnerability of Union logistics: the need to requisition food, fuel, and fodder from the local population was higher wherever Union supply lines were disrupted, whether due to irregular attacks or to innately inadequate conditions (which were often blamed on irregulars). In other words, punishment, requisitioning, and looting often went hand in hand.

Official Union policy towards civilians encompassed several measures: assessments, arrest and banishment, hostage-taking, and retributive burning.

[64] Grimsley, *Hard Hand*, 9.

[65] Mountcastle, *Punitive War*, 141. See Grimsley, *Hard Hand*, for a critique of this explanation. According to Grimsley, if the experience of irregular warfare were the main reason for adopting a 'hard war' approach, it should have started at the very beginning of the war. Yet, Grimsley himself gives examples of individual commanders adopting a punitive approach early on in the war. This approach was later endorsed and copied at higher command levels.

[66] Grimsley, *Hard Hand*, 41.

[67] Sutherland, *Savage Conflict*, 20.

[68] Grimsley, *Hard Hand*, 39.

[69] Sutherland, *Savage Conflict*, 33.

On 16 July 1861, the Union adopted the Confiscation Bill, which permitted the confiscation of property being used to support the Confederate cause, including slaves. This measure was obviously aimed at weakening the southern elite, hence still being in line with the conciliatory approach toward the wider population.[70] But harsher measures were soon to follow. In July 1862, Brigadier General John Pope issued General Orders Numbers Five, Six, and Seven, stipulating that the Army of Virginia should live off the land and that only loyal citizens could claim indemnities for requisitioned goods. Local inhabitants were to be held responsible for any damage done by guerrillas.[71] General Order Number Eleven, which soon followed, envisaged the arrest of all 'disloyal male citizens'.[72] Major General Samuel R. Curtis and Major General John M. Schofield soon copied the assessment system and applied it along the Missouri–Arkansas border and in Missouri.[73] But the assessment system rapidly extended beyond its punitive purpose and became a general additional source of income for resource-starved Union commanders. In August 1862, Schofield used the system to raise $500,000 to clothe and arm the Missouri militia, which was to augment the local defence against guerrillas.[74] In November 1862, Major General William S. Rosecrans, who succeeded Don Carlos Buell as commander of the Department of the Cumberland, issued directives on foraging, in which he introduced a three-tier approach to treatment of civilians; this approach was picked up later by Lieber in the Lieber Code. In March 1863, Halleck, now General in Chief of all Union armies, approved and reinforced the three-tier system.[75]

Arrests and banishments were other measures that soon became a staple of the Union's treatment of 'disloyal' civilians. The institutionalization of these measures occurred in a decentralized way. In June 1862 Brigadier General Benjamin Loan ordered the banishment of disloyal citizens in and around Lexington, Kansas. On 24 December 1862, Major General Curtis, commander of the Army of the Southwest, extended the banishment policy in his Order Number Thirty-Five, in which he stipulated the banishment of individuals 'though no specific act of disloyalty can be proven against them', if required on grounds of security.[76] On 18 August 1863, Major General Ewing introduced banishment in the District of the Border, targeting the families of known guerrillas and those families who were suspected of having aided guerrillas. After Quantrill's raid on Lawrence on 21 August 1863, Ewing extended the banishment policy in his notorious General Order Number

[70] Grimsley, *Hard Hand*, 78.
[71] Grimsley, *Hard Hand*, 86.
[72] Grimsley, *Hard Hand*, 88.
[73] Grimsley, *Hard Hand*, 98, Brownlee, *Gray Ghosts*, 165 ff.
[74] Brownlee, *Gray Ghosts*, 167.
[75] Grimsley, *Hard Hand*, 102, 145.
[76] Quoted in Brownlee, *Gray Ghosts*, 163.

Eleven, which envisaged the banishment of *all* inhabitants from Jackson, Cass, Bates, and parts of Vernon County.[77]

Hostage-taking was practised in regions where neither side had firm control over the territory, such as East Tennessee. It was used as a retaliatory measure for guerrilla attacks: sympathizers of the opposite side were arrested and threatened with banishment, should the guerrilla attacks not stop. In East Tennessee, it was practised by both Union and Confederate forces.[78]

The final element in the Union's policy towards civilians was the deliberate destruction of property, including retributive burnings. Brigadier General John McNeil authorized his troops in the District of North-East Missouri to burn the homes of known irregulars.[79] But the destruction soon became less discriminate, and whole villages fell victim to the torch of Union soldiers. Burnings often happened spontaneously and without orders, conducted out of revenge by troops who had come under irregular attack. One of those cases was the burning of Winton, North Carolina, on 20 February 1862, which occurred as retaliation for a guerrilla ambush on the previous day.[80] However, the indiscriminate destruction of entire villages was increasingly the result of commanders' orders. An early instance of such authorized indiscriminate destruction was Colonel Graham Fitch's White River expedition in the summer of 1862.[81] However, the most notorious operation that involved large-scale indiscriminate destruction was General Philip Sheridan's 1864 Shenandoah Valley Campaign. When Lieutenant John R. Meigs, Sheridan's chief engineer and son of the Union quartermaster-general Montgomery C. Meigs, was killed by Confederate scouts on 3 October 1864, Sheridan ordered his troops to burn every building in a 3-mile radius of the attack, including the town of Dayton.[82] In late November 1864, Sheridan ordered his subordinate Major General Wesley Merritt to devastate Loudon County, where Mosby's Rangers had operated. He formulated his orders in a way that clearly put them into the context of punitive war rather than that of the denial campaign against the Confederacy that he conducted in the Valley at the same time.[83]

It is important to note that the Lieber Code did not set standards for the treatment of what would later be referred to as civilians, but rather mainly codified existing conduct of Union forces during the Civil War into a military manual. By the time Lieber wrote the code in 1863, the Union's 'hard war' policy had fully evolved, and Lieber made some evident references to battlefield practices and military directives that were in place at the time, most importantly the three-tier approach. Initially, the three-tier model was rooted

[77] Brownlee, *Gray Ghosts*, 126.
[78] Fisher, *War at Every Door*, 148 ff.
[79] Brownlee, *Gray Ghosts*, 171.
[80] Grimsley, *Hard Hand*, 60.
[81] Mackey, *Uncivil War*, 53.
[82] Mountcastle, *Punitive War*, 128 ff.
[83] Mountcastle, *Punitive War*, 133.

in the idea that secession was a cause that was pursued only by the minority elite of the southern population. It targeted mainly wealthy slave-owners. Many aspects of the Union policy towards civilians were initially based on the three-tier model—for instance, the confiscation of property of disloyal southerners and the burning of the homes of known guerrillas. However, it is doubtful whether Union policy towards southern civilians bore the mark of 'surgical precision' and 'directed severity', as Mark Grimsley insisted.[84] For one, it was almost impossible to know who was 'disloyal'. More importantly, however, punitive measures against civilians had an inherent tendency to become increasingly indiscriminate over time: assessments were eventually raised among the entire population of a district, the family members of guerrillas were arrested and banished, and villages were burnt indiscriminately. This went far beyond the limits of the three-tier system and indicated that, rather than limiting violence and destruction, the system was a slippery slope that soon became a convenient excuse for wholesale persecution and destruction.

The Franco-Prussian War

The Franco-Prussian War had its roots in an escalating power rivalry between France and Germany. Prussia's victory in the 1866 Austro-Prussian War had significantly changed the balance of power in Europe. In the aftermath of the war, Prussia annexed Schleswig, Holstein, Hanover, Hessia-Kassel, Nassau, and Frankfurt am Main. It also amalgamated the rest of North Germany's states (Saxony, Hessia-Darmstadt, Mecklenburg, the Thuringian duchies, and the free cities of Hamburg, Bremen, and Lübeck) into a North German Confederation, which was effectively under Prussian control with respect to its foreign and military affairs. Moreover, Prussia's rapid industrialization had considerably increased its economic weight. By the second half of the 1860s, Prussia had transformed itself from an unimportant, small country into one of Europe's foremost powers. France felt threatened by this rapidly emerging neighbour. To counterbalance Prussia's rise after its victory in the Austro-Prussian War, Napoleon III demanded that Prussia cede the left bank of the Rhine as a guarantee of France's power position; Bismarck refused. Directly after the Austro-Prussian War, both France and Prussia contemplated an armed attack on one another. German chief of staff General Helmuth von Moltke urged Bismarck to attack France, while Prussian forces where still mobilized.[85]

[84] Grimsley, *Hard Hand*, 116, 175.

[85] Michael Howard, *The Franco-Prussian War: The German Invasion of France, 1870–1871* (London: Routledge, 1981), 41.

On the French side, prominent political voices demanded an immediate attack on Prussia to halt its rise to power.[86]

After 1866, both France and Germany considered war inevitable. All that was required was an immediate provocation. The Hohenzollern candidature for the vacant Spanish throne in 1870 provided this trigger. Bismarck had pressured Leopold, Hereditary Prince of Hohenzollern-Sigmaringen, to accept the throne in order to ensure political influence in Spain. When the news of the Hohenzollern candidature was leaked prematurely, France's political leadership was outraged. Leopold withdrew his candidature, but on 13 July 1870 the French ambassador to Prussia submitted Napoleon III's demands for a Prussian guarantee to refrain from supporting a renewed candidature of the Hohenzollerns in the future; William I refused to give such a guarantee. Bismarck was informed of the outcome of the meeting in a telegram from Ems, where the meeting had taken place, on the eve of 13 July, the famous 'Ems Despatch'. Bismarck made the content of the despatch public, editing it slightly so as to emphasize France's 'outrageous' behaviour.[87] On 19 July, France declared war on Germany. In both countries, mobilization was already underway. However, Prussia's mobilization was much better organized than France's, which was confused and delayed. Hostilities finally started on 2 August with a French attack on Saarbrücken. After a series of German battlefield successes in August 1870, the French Army of Châlons was defeated in the battle of Sedan on 1 September and surrendered subsequently. Napoleon III was captured. German troops advanced further into France and started to besiege Paris on 19 September. After Bazaine's Army of the Rhine capitulated at Metz on 27 October, France fought on with its remaining forces. On 26 January 1871, France signed an armistice in Versailles.

The francs-tireurs *and the Myth of the* levée en masse

After the Army of Châlons had surrendered at Sedan, France's regular forces were severely decimated, and it became clear that it would be difficult for them to repel the German invasion. Small bands of irregular fighters, called *francs-tireurs* (literally: free-shooters), started to emerge on a local basis. Their emergence was encouraged soon after the capture of Napoleon III by the new Government of National Defence, which had formed under General Louis Jules Trochu and the new Minister of the Interior and Minister of War, Leon Gambetta. Gambetta also made efforts to regulate the *francs-tireurs*. In a decree of 29 September, he included them in the same disciplinary regime as the

[86] Geoffrey Wawro, *The Franco-Prussian War: The German Conquest of France, 1870–1871* (Cambridge: Cambridge University Press, 2003), 17.
[87] Howard, *Franco-Prussian War*, 48 ff.

Garde Nationale Mobile, the French militia force. In a decree of 4 November, he gave regular military commanders authority over the *francs-tireurs* who operated in their area.[88]

French regular forces at the time consisted of the regular army and the *Garde Nationale Mobile*. The regular army was in principle raised by conscription, but service with the colours was far from universal. Draftees were selected by a lottery, and wealthy families could buy an exemption for their offspring if they had drawn a *mauvais numéro*, a bad number. Soldiers served five years with the colours, plus four years in the reserve forces. The *Garde Nationale Mobile* had been introduced, or rather revived, in a bill of 1868 with the aim of increasing troop numbers in order to match the large Prussian army in the event of war. Although a *levée en masse* had not been part of French military planning before the war, the emergence of *francs-tireurs* fit very well into the political landscape of the newly formed Third Republic, which had emerged after the capture of Napoleon III. The Government of National Defence was republican, and the republican political discourse had long emphasized the idea of popular war and the role of the people for the defence of the country.[89] The reform of the French army had been the subject of fierce political debates before the war. Traditionalists had wanted to strengthen the regular army, whereas republicans had intended to place national defence entirely into the hands of a militia, following the Swiss example. The *Garde Nationale Mobile* had been an uneasy compromise between those two camps.[90]

For the republicans, the emergence of *francs-tireurs* units was seen as a sign of the renewed republican spirit of the French people in the face of foreign invasion. Gambetta made explicit reference to the *levée* of 1792 when he called his fellow citizens to arms in October 1870: 'The republic appeals for the support of everyone; this government will make it its duty to utilize everyone's courage, to employ all capacities... *Levons-nous en masse* [let us rise *en masse*] and die rather than submit to the shame of [national] dismemberment.'[91] Yet, at the same time he also equated the efforts of the *francs-tireurs* with 'chouannerie', thereby placing them more clearly into the operational spectrum of guerrilla warfare.[92]

If official figures are accurate, the French people followed the call to arms quite enthusiastically. According to official calculations, approximately

[88] Howard, *Franco-Prussian War*, 253.

[89] On the different strands of the republican discourse in France during the war, see Sudhir Hazareesingh, 'Republicanism, War and Democracy: The Ligue du Midi in France's War against Prussia, 1870–1871', *French History*, 17/1 (2003), 48–78.

[90] Howard, *Franco-Prussian War*, 32 ff.

[91] Quoted in John Horne, 'Defining the Enemy: War, Law, and the Levée en masse from 1870 to 1945', in Daniel Moran and Arthur Waldron (eds), *The People in Arms: Military Myth and National Mobilization since the French Revolution* (Cambridge: Cambridge University Press, 2003), 108.

[92] Quoted in Howard, *Franco-Prussia War*, 249.

57,600 *francs-tireurs* operated on French territory during the Franco-Prussian War.[93] Other sources indicated even higher numbers.[94] Yet, the official records of 33 departments mentioned only 143 casualties among the *francs-tireurs* throughout the war, which seems to indicate that the official numbers were inflated.[95] The *francs-tireurs* units were composed of a mixture of individuals from different political and social backgrounds, 'from the violently radical *francs-tireurs* of the Seine to Colonel de Cathelineau's legitimists from Vendée, from the wealthy bourgeois *tirailleurs volontaires* from the Gironde to the companies of papal Zouaves which Colonel de Charette brought back to France after the Italian occupation of Rome'.[96] They were complemented by a number of foreign fighters, including one unit under the command of Giuseppe Garibaldi, who had been an ally of Prussia on the Italian side in the Austro-Prussian War, but then switched sides and fought against the Germans.

The activities of the *francs-tireurs* included organizing defensive measures against German troops in those areas that were not yet occupied, as well as destroying crops and livestock in regions that were threatened by invasion. They conducted strikes against German lines of communication in occupied territory and sometimes acted in support of regular forces—for instance, as scouts or screens.[97] If they acted alone, they essentially operated as guerrilla forces; if found together with regular forces, the *francs-tireurs* more closely resembled the eighteenth-century model of irregular fighters as auxiliary forces. Occasionally, *francs-tireurs* fought German forces in the open. On 26 November, for instance, units under the command of Garibaldi's son Ricciotti attacked the Prussia XIV Army Corps under General August von Werder in Dijon. However, such attacks usually ended in failure for the irregulars, as did this one.[98]

Francs-tireurs units soon became notorious for their lack of discipline. Gambetta's decrees of 29 September and 4 November intended to put a stop to that, but imposing discipline on the *francs-tireurs* proved almost impossible. According to some accounts, local communities feared the *francs-tireurs* even more than the German occupying forces and requested the latter to provide

[93] Howard, *Franco-Prussia War*, 252.

[94] Cf. Katja Mitze, '"Seit der babylonischen Gefangenschaft hat die Welt nichts derart erlebt": Französische Kriegsgefangene und Franctireurs im Deutsch-Französischen Krieg 1870/71', in Rüdiger Overmans (ed.), *In der Hand des Feindes: Kriegsgefangenschaft von der Antike zum Zweiten Weltkrieg* (Cologne: Böhlau Verlag, 1999), 237.

[95] These official records document that 53 *franc-tireurs* were killed, 39 died of illness, 44 were wounded, and 7 were missing: Sanford Kanter, 'Exposing the Myth of the Franco-Prussia War', *War and Society*, 4/1 (1986), 15.

[96] Howard, *Franco-Prussian War*, 252.

[97] Howard, *Franco-Prussian War*, 255, 296, 304, 374 ff.

[98] Howard, *Franco-Prussian War*, 409.

protection against the former.[99] The effectiveness of the *francs-tireurs* varied. They were a nuisance to the German forces, but never developed into an existential threat or even a serious operational challenge. It also seems to be clear that the *francs-tireurs'* mobilization never reached the threshold of a true *levée en masse*. Even if the official figure of 57,600 were correct, it is not clear how many of these actually engaged in military activities and if so, for how long. Especially in those areas that were under tight German control, *francs-tireurs* activities seemed to die down quickly as a result of German repression. Sometimes efforts to suppress *francs-tireurs* were initiated by local French authorities, in order to spare their communities from German reprisals.[100]

German Counter-Measures

Despite the French government's insistence that *francs-tireurs* were lawful belligerents, as early as 22 August the German Headquarters issued an order according to which *francs-tireurs* were not to be treated as POWs, but to be executed upon their capture. This order was later amended, and the new decree stipulated ten years of forced labour rather than immediate execution. Despite such orders, it seems that German forces treated most *francs-tireurs* as POWs when they captured them.[101] Nonetheless, civilians had to endure harsh treatment by the German forces, in particular in those areas where *francs-tireurs* were active. Like the Union forces in the American Civil War, the Germans chose a punitive approach, based on the assumption that it was ultimately the local population who was responsible for the *francs-tireurs* problem. The catalogue of counter-measures comprised implementing assessments and fines against the local population when irregular attacks had taken place, taking local authorities hostage in retaliation for irregular activities, and punitive burnings and summary executions of civilians.[102] After I Bavarian Corps had come under fire in the village of Bazeilles near Sedan on 1 September and had been shot at by civilian snipers hiding in the village's buildings, the Bavarian troops took revenge on the village. They burned it down and killed a number of civilians.[103] In the night of 7–8 October, German forces razed the town of Ablis near Orléans and killed all male inhabitants in retaliation for *francs-tireurs* activities. The village of Fontenoy-sur-Moselle was completely destroyed, and all inhabitants were killed by soldiers from the

[99] Kanter, 'Exposing the Myth', 15; Mitze, '"Seit der babylonischen Gefangenschaft"', 241.
[100] Kanter, 'Exposing the Myth', 15.
[101] Mitze, '"Seit der babylonischen Gefangenschaft"', 238 ff.
[102] Mitze, '"Seit der babylonischen Gefangenschaft"', 240, Mark R. Stoneman, 'The Bavarian Army and French Civilians in the War of 1870–1871: A Cultural Interpretation', *War in History*, 8/3 (2001), 275.
[103] Stoneman, 'The Bavarian Army', 275 ff.

Fifty-Seventh Infantry after *francs-tireurs* had managed to blow up a viaduct that connected the train line between Frouard and Blasne on 22 January 1871.

Just as in the American Civil War, the official justification for the punishment of civilians was military necessity. Whereas the Union's retributive measures during the American Civil War had often backfired, the Germans' 'hard war' seemed to work, and irregular activities died down. However, operational requirements were only one part of the picture. The harsh treatment of civilians suspected of aiding the *francs-tireurs* was also an expression of an ideological conflict between the opponents. According to John Horne, the Germans regarded the emergence of the *francs-tireurs* and the French leadership's framing of the phenomenon as a *levée en masse* as 'trebly illegitimate':

> First, it refused to accept the verdict of formal French defeat in conventional battle. Second, it generated a 'treacherous' [*heimtückisch*] style of combat, including ambush, streetfighting and barricades, and hence a new and totally negative image of the enemy as an insurrectionary civilian and possibly (by a complete inversion of the gendered codes of combat) as a woman. Third, such warfare seemed to be directly inspired by the principles of republicanism and emanated from the Paris revolution.[104]

In short, the double meaning of the *levée en masse*, which had been so significant for the emergence of the irregular during the French Revolutionary Wars and the Napoleonic Wars, once again became relevant in the Franco-Prussian War. While France claimed that the *francs-tireurs* were legitimate and patriotic soldiers, the Germans regarded them as an unlawful popular uprising against their legitimate authority as occupying forces.

Technology and the Irregular: Logistics, Weapons, and Courage on the Battlefield

The American Civil War and the Franco-Prussian War shattered the hopes of nineteenth-century rulers and militaries that irregular warfare was an aberration confined to the period of revolutions around 1800. However, besides this realization, external events, mainly changes in military technology, conveyed additional importance upon the problem of irregular fighting. Technological advances in the nineteenth century had an impact on the conduct of irregular fighters as well as on the ways in which irregular warfare was perceived. The main technological developments in the nineteenth century were the spread

[104] Horne, 'Defining the Enemy', 109 ff.

of railways and the introduction of long-range rifles. Of these two developments, the latter had a more decisive impact on irregular warfare.

Railroads: The Achilles' Heel of Regular Armies?

The American Civil War was the first major war in which railroads were extensively used for logistics purposes. Some historians claimed that the Union's superiority regarding railroads was one of the factors that allowed the Union to prevail against the Confederacy.[105] Conversely, others argued that the Confederacy's inadequate management of their railways contributed to southern defeat in the war.[106] In a similar vein, Geoffrey Wawro pointed out that one of the decisive strengths of the Prussians during the Franco-Prussian War was the superiority of their railway system.[107]

If railways did indeed have this decisive impact on the conduct of the war, then one would assume that they gave the threat of irregular warfare an entirely new meaning, given that railways were particularly vulnerable to irregular attacks. In fact, commanders in both the American Civil War and the Franco-Prussian War were acutely aware of the irregular threat to their supply lines.[108] According to Union General William T. Sherman: 'Railroads are the weakest things in war; a single man with a match can destroy and cut off communications.' Sherman went on to emphasize that especially the irregulars threatened the railroads.[109] However, attacks on lines of communication were not new. They had been a preferred method among many of the more successful and professional irregulars in the Peninsular War.[110] Moreover, the importance of railways for both the Union and the Prussians should not be overestimated. In the American Civil War, horse-drawn wagons were still an important mode of transportation for supplies.[111] Irregular attacks on railways often had less of an impact on the supply system than the structural problems built into the system itself, such as the failure to ensure transport from railheads to the troops in the field, or poor planning, which resulted in blocked lines and shortages of rolling stock. More fundamentally, the strategic importance of the railways for the victorious parties in both the American Civil War and the Franco-Prussian War has also been disputed. Herman Hattaway and Archer Jones, for instance, argue that the Union's success in the

[105] John Keegan, *A History of Warfare* (New York: Vintage, 1993), 305.

[106] Robert G. Angevine, *The Railroad and the State: War, Politics and Technology in Nineteenth-Century America* (Stanford: Stanford University Press, 2004), 151.

[107] Wawro, *Franco-Prussian War*, 48 ff., 74, 80.

[108] For more information on the role of irregulars in the Franco-Prussian War, see Howard, *Franco-Prussian War*, 377.

[109] Quoted in McPherson, *Battle Cry of Freedom*, 515.

[110] e.g. Tone, *The Fatal Knot*, 100.

[111] David Kirkpatrick, 'Logistics of the American Civil War', *RUSI Journal*, 152/5 (2007), 78.

American Civil War came about only once the Union had stopped relying on railroads for their logistic requirements.[112] With respect to the role of railroads for the Franco-Prussian War, Martin van Creveld conceded that they were significant for the deployment of forces at the opening of the campaign and for the siege of Paris, but not for the rest of the war:

> Between these two phases of the struggle, however, the railways do not seem to have played a very important role, partly because of difficulties with the lines themselves and partly because of the impossibility of keeping the railheads within a reasonable distance of the advancing troops. Most surprising, however, is the fact that none of this had much influence on the course of operations, or indeed caused Moltke any great concern.[113]

According to van Creveld, contemporary military strategists, mainly Moltke himself, were aware of the fact that railways were not the strategic silver bullet that later historians saw in them, and hence they did not conceive of them as a strategic priority.[114] To be sure, strategists tried to protect railways against irregular attacks, but they were well aware that railways would always be vulnerable to such attacks and that their disruption would not change the course of the war in any fundamental way. Strategic planners in the nineteenth century simply did not rely on railways sufficiently to build their campaigns upon them. When the railway system failed, either owing to the structural faults of the system itself or owing to its vulnerability to irregular attack, they simply reverted back to the older system of requisitioning from the local population.[115]

Weapons and Courage on the Battlefield

The second technological innovation of the nineteenth century was the invention of long-range firearms. They were quickly introduced into Western armies. At the outset of the American Civil War, the infantry on both sides was equipped with smooth-bore muskets, which had an effective range of 50 yards. These were soon replaced with more accurate muzzle-loading rifles, mainly the US-produced Springfield and the British Enfield models, both with an effective range of 300–400 yards. Breech-loading rifles that additionally had a greater rate of fire were introduced later in the war as well, but the muzzle-loading rifle remained the standard weapon of the Civil

[112] Herman Hattaway and Archer Jones, *How the North Won: A Military History of the American Civil War* (Champaign, IL: University of Illinois Press, 1983), 192.

[113] Martin van Creveld, *Supplying War: Logistics from Wallenstein to Patton* (Cambridge: Cambridge University Press, 1977). 96.

[114] Creveld, *Supplying War*, 106.

[115] Creveld, *Supplying War*, 107; McPherson, *Battle Cry of Freedom*, 515.

War soldier.[116] The Prussian army adopted the breech-loading Dreyse needle rifle in the late 1840s. The French army had used the muzzle-loading Minié rifle in North Africa in the mid-nineteenth century, but in 1866 replaced it with the breech-loading Chassepot, which was superior to the German Dreyse in terms of its range.

The introduction of these new weapons precipitated changes in battlefield tactics. The weapons and, in particular, their greater range had two main effects: they produced more casualties, and they strengthened the tactical defensive.[117] However, in the nineteenth century, the established military wisdom emphasized that wars were won by decisive battles and that battles were mainly won by large, battalion-size, close-order formations, closing up with the opponent, and trying to decimate the opponent's forces with bayonet charges. Yet, with the introduction of long-range rifles, advancing columns were often severely reduced before they could lead a bayonet charge. Hence, the new weapons confronted commanders with a problem. On the one hand, offensive action on the battlefield was costly and often unsuccessful. On the other, battles could not be won by defensive tactics alone. This dilemma was not entirely new, as both Clausewitz and Jomini had already known that the tactical defensive tended to be stronger than the offensive.[118] However, the introduction of long-range rifles exacerbated the problem by increasing the costs of offensive battlefield tactics dramatically.

This context challenged the meaning of courage on the battlefield in several ways. In the face of the new battlefield situation, some commanders and military thinkers adopted a decidedly conservative stance and held that the new challenges facing attacking forces had to be overcome by strengthening morale and training of the troops. In short, they wanted soldiers to move forward on the battlefield in spite of the greater risk of death and injury.[119] According to this approach, the side that showed more courage would prevail in battle.

During the American Civil War, popular culture initially underpinned this central importance of courage on the battlefield. The press on both sides was obsessed with bayonet charges, although they hardly played a role in the new battlefield conditions.[120] Courage was perceived as having protective qualities. It was thought that the soldier who did not dodge bullets would survive in battle—a cultural image that had often been confirmed by battlefield

[116] McPherson, *Battle Cry of Freedom*, 475.

[117] McPherson, *Battle Cry of Freedom*, 475; Dennis Showalter, 'Infantry Weapons, Infantry Tactics, and the Armies of Germany, 1849–64', *European Studies Review*, 4/2 (1974), 119–40.

[118] McPherson, *Battle Cry of Freedom*, 125.

[119] McPherson, *Battle Cry of Freedom*, 132.

[120] Adam I. P. Smith, '"On the Altar of the Nation": Narratives of Heroic Sacrifice in the American Civil War', in Sibylle Scheipers (ed.), *Heroism and the Changing Character of War: Towards Post-Heroic Warfare?* (Basingstoke: Palgrave, 2014), 33–46.

experience before the introduction of the muzzle-loading rifle.[121] However, the new weapons technology dispelled that idea very quickly, and 'convictions about the potency and protectiveness of courage yielded to the suspicion that special courage had become the mark of death'.[122] Officers now found it difficult to lead their men into battle and to motivate them by displaying particular courage, as they were often singled out by sharp-shooters, who could now hit them at long ranges.[123] A more realistic approach to the new battlefield reality eventually developed, one that acknowledged the high costs of the offensive and stressed the adoption of defensive measures. However, such adaptations profoundly challenged the meaning of courage on the battlefield. At the outset of the American Civil War, trench-digging and the construction of breastworks were ridiculed. Soldiers thought that only cowards used the shovel.[124] However, as soon as it became obvious that the new weaponry could inflict casualties on a hitherto unimaginable scale, the ridicule stopped, and trench-digging became a widespread practice.

Finally, to the extent that offensive tactics were revised, these revisions also challenged established cultural images of the courageous soldier. Where tactical formations were loosened in order to become more flexible, they also became harder to control and command. The result was often confusion among advancing troops. A further innovation was a greater emphasis on sharp-shooters. Yet, the role of the sharp-shooter stood in stark contrast to the prevailing idea of courage on the battlefield: killing the enemy at a long range and from the protection of covered positions was deemed a cowardly way of fighting.[125]

But the introduction of long-range infantry weapons, which undoubtedly shook the popular image of the regular soldier, in particular that of the infantryman, in the second half of the nineteenth century, also had repercussions for the perception of the irregular fighter. The dichotomy between the regular and the irregular mainly works by highlighting the most ambiguous and problematic feature of the regular soldier and then projecting the regulars' negative traits on the irregular. In the French Revolutionary and Napoleonic armies, discipline had been the most problematic aspect. While often aspired to, it was rarely achieved, especially off the battlefield. But this shortcoming was glossed over by highlighting the indiscipline of the irregular. In the second half of the nineteenth century, the cultural underpinnings of the regular–irregular dichotomy had slightly shifted. To be sure, indiscipline was still a huge problem for the American Civil War armies. However, courage also

[121] Gerald F. Linderman, *Embattled Courage: The Experience of Combat in the American Civil War* (New York: Free Press, 1987), 20.

[122] Linderman, *Embattled Courage*, 59.

[123] Linderman, *Embattled Courage*, 140; McPherson, *Battle Cry of Freedom*, 475.

[124] Linderman, *Embattled Courage*, 143.

[125] Linderman, *Embattled Courage*, 148.

held a central position within the cultural universe of nineteenth-century America, precisely because it could partly make up for a lack of discipline, at least on the battlefield.[126] But, as the concept of courage slowly unravelled as well, cowardice was increasingly attributed to the irregular fighter. This partly stemmed from the pragmatic frustration among regular troops about the fact that irregulars did not give battle. Hence, one Illinois Private wrote about irregular fighters in Missouri:

> They are a pack of cowards. They are like a thief in the night.... If they are pursued by troops they cut & run for their houses. If you go to their houses to question them & they are good union men & unless you catch them in the very act you don't know who they are & most of them are smart enough not to get catched at all.[127]

But the concept was also rooted in cultural expectations and hierarchies. In his study of the behaviour of the Bavarian Army in occupied France during the Franco-Prussian War, Mark Stoneman explained with respect to the soldiers' perception of the *francs-tireurs*:

> The Bavarians had expected a "fair" fight, which to them meant fighting a manly enemy whom they could see and recognize as such, an enemy who would not shoot and then run, hide or pretend to be an innocent civilian.... Officers were trained to perceive war in such ritualized terms. Civilian snipers had no place in this mental universe.[128]

At the same time, a counter-narrative of the irregular fighter as the true hero emerged. This picked up on the challenges to the image of the courageous soldier emerging from the new battlefield tactics. While front-line soldiers dug trenches and ostensibly fought pointless battles with no discernible impact on the course of war, irregulars 'saw themselves as noble knights, cunning and daring, rescuing damsels in distress, saving desperate civilians just in the nick of time, slaying the foe, avenging wrong, defending the southern cause'.[129] Where the regular armies mostly fell short—that is, in local defence—irregulars saw themselves as the true defenders of home and hearth, especially in those regions invaded by the opponent where no regular army was present to protect the population—for instance, in Tennessee or Florida. Some irregulars prided themselves on their discipline and restraint. William Quantrill, for instance, regularly paroled captured Union prisoners in the early years of the war.[130] Others insisted on fanciful military ranks and titles. Increasing levels of violence and brutality were usually blamed on the regular opponent, and

[126] Linderman, *Embattled Courage*, 43.
[127] Private E. C. Sackett, quoted in Fellman, *Inside War*, 164.
[128] Stoneman, 'The Bavarian Army', 276–7.
[129] Fellman, *Inside War*, 139. [130] Brownlee, *Gray Ghosts*, 64.

irregulars insisted that they were simply replying in kind to the regular armies' illegitimate behaviour.[131]

Alternatively, guerrilla leaders linked irregular warfare with the image of 'Indian warfare' and the theme of the noble savage: 'Born of the woods and fields, they acted as "natural man", imbued with inherent virtue and pastoral innocence, as much a product of the natural world as Indian peoples.'[132] Gestures such as scalping, as practised by 'Bloody' Bill Anderson and Archie Clement, underpinned this cultural reference.

Interestingly, many irregular fighters also had a preference for close-range weapons and did not use the new weaponry, thereby reversing the prevailing battlefield trend. Mounted partisans and raiding cavalry preferred the Bowie knife to the cavalry sabre.[133] When it came to firearms, irregulars generally used revolvers, in particular the Colt 0.44-calibre 'Army' model, rather than rifles. Its effective range was between 75 and 100 yards, meaning that the standard irregular tactic—if there was such a thing—was charging at the enemy on horseback and shooting from close range.[134] Hence, while regular soldiers were struggling to bridge the gap between cultural expectations regarding courage on the battlefield and the new technological battlefield realities, irregular fighters fostered a nostalgic self-image that emphasized their warrior qualities and their daring style of combat.

Legal Codification: Establishing the Battlefield of Legal Definitions

The early attempts at codifying the laws of war were to a large extent a reaction to the challenge of irregular fighters. The Lieber Code, or General Order No. 100, was written by Francis Lieber to broaden his original legal opinion paper on guerrilla fighters. Both papers were commissioned by Union General Halleck, who had made his original request as a reaction to his frustration with guerrilla fighters during the American Civil War. Both this war and the Franco-Prussian War were characterized by divergent ideas among the opponents concerning who was a lawful belligerent. The Confederate government in Richmond insisted that partisan rangers were lawful belligerents, just as the French Government of National Defence decreed the *francs-tireurs* to be part of the official defence forces. Both the Union government and the Germans had a much more restricted view of who was a lawful belligerent. These conflicting views arose partly out of the disparity of military power between the opponents: it was only logical for the invading forces to insist that popular resistance

[131] Fellman, *Inside War*, 137.
[132] Sutherland, *Savage Conflict*, 29.
[133] Sutherland, *Savage Conflict*, 29.
[134] Mackey, *Uncivil War*, 83.

against invasion and occupation were unlawful, as popular resistance obviously favoured the defensive. One's own military advantage and political expediency hence played a large role in legal debates. However, a second important factor was that in both the American Civil War and the Franco-Prussian War two different military ideologies clashed. For the US-American South, the suspicion against an ever-increasing centralization of political power meant that defence was still very much understood in terms of local defence. In the states of the Deep South, this notion was underpinned by a long tradition of ranger forces and vigilantes. In the Franco-Prussian War, it was the republican tradition of popular defence and the *levée en masse* that clashed with the Prussian idea that it was the army that formed the (male) population into national citizens and not the other way around.

It was no wonder, then, that contemporaries felt the need to clarify the laws of war and in particular its definition of irregular fighters. After the Franco-Prussian War, in 1874 European delegations and international lawyers met in Brussels in order to negotiate and codify the laws of war. That project failed to produce anything that was acceptable to all parties involved, and hence no law was adopted. Thus, the early attempts at codifying the laws of war did not do much to clarify, let alone define and regulate, the status of irregular fighters. The question of the lawfulness of irregular fighters had become so entangled with issues of individual national military and political power position as well as with ideological differences, that no agreement was possible among the major European powers. However, the attempts at legal codification did succeed in opening up the definitional battlefield and clarifying the different positions within it.

The Brussels Conference was influenced by the Lieber Code, which formed the basis for several draft proposals that were negotiated in Brussels.[135] The conference had convened on the initiative of the Russian emperor Alexander, but without a clearly stated agenda or purpose. Once the delegations had met in Brussels, the political cleavage lines soon became clear: Germany and Russia, which both possessed large regular armies, intended to keep the definition of lawful belligerency highly restrictive, hence not allowing for a defensive *levée en masse* or even militia forces. Germany preferred to have no codified law at all and would have rather relied on the precedent its forces had set during the Franco-Prussian War. Russia submitted a draft proposal that envisaged tilting the legal balance so much in favour of states with large standing armies that the British delegate to the conference, Savile Lumley, referred to it as a 'Code of Conquest'.[136] In contrast, smaller states that relied

[135] Karma Nabulsi, *Traditions of War: Occupation, Resistance, and the Law* (Oxford: Oxford University Press, 1999). 5.

[136] Quoted in Nabulsi, *Traditions of War*, 6.

on militias for their national defence were opposed to this restrictive proposal. They were supported by Britain and France. The Brussels Conference broke up after one month of negotiations, and no legal text was adopted. Only a declaration was issued. However, international lawyers were apparently not discouraged by this outcome. In 1880 the Oxford Institute of International Law published its own manual entitled 'The Laws of War on Land'.

The Brussels declaration and the Oxford Manual were rather similar with respect to their structure and content. Both texts were more inclusive than the Lieber Code with respect to the question of the lawfulness of irregulars. Articles 9 and 10 of the Brussels declaration stipulated that not just regular armies, but also militia forces, volunteer corps, and a spontaneous rising en masse against invasion, were lawful forms of national defence.[137] Article 2 of the Oxford Manual read:

> The armed forces of a State includes [*sic*]:
> 1. The army properly so called, including the militia;
> 2. The national guards, *landsturm*, free corps, and other bodies which fulfil the three following conditions:
>   (a) That they are under the direction of a responsible chief;
>   (b) That they must have a uniform, or a fixed distinctive emblem recognizable at a distance, and worn by individuals composing such corps;
>   (c) That they carry arms openly;
> 3. The crews of men-of-war and other military boats;
> 4. The inhabitants of non-occupied territory, who, on the approach of the enemy, take up arms spontaneously and openly to resist the invading troops, even if they have not had time to organize themselves.[138]

The fact that neither the Brussels declaration nor the Oxford Manual garnered widespread acceptance reflected the contestedness that still surrounded the definition and treatment of irregular fighters. Consensus on the definition of lawful belligerency was not reached until the 1907 Hague Convention on the Rules and Customs of War on Land. The Hague Convention retained the provisions on the lawfulness of the rising en masse, but it did not mention the various other military organizations included in the Oxford Manual, such as national guards, *landsturm*, and free corps. This established the legal definition of the irregular *ex negativo*: all those forces that do not meet the included criteria of lawful belligerency are irregular forces. In fact, the legal approach to irregular fighters was characterized by a double negativity: not only did the Hague Convention not define the irregular in positive terms; it also failed to include provisions on their treatment. Instead, the Martens Clause provided

[137] International Committee of the Red Cross, 'The Laws of War on Land', Oxford, 9 September 1880 <http://icrc.org/ihl.nsf/140-FULL?OpenDocument>.

[138] International Committee of the Red Cross, 'The Laws of War'.

minimal and weak protections that were intended to prevent the worst excesses towards persons who were not explicitly protected under the terms of the Hague Convention:

> Until a more complete code of the laws of war has been issued, the high contracting Parties deem it expedient to declare that, in cases not included in the Regulations adopted by them, the inhabitants and the belligerents remain under the protection and the rule of the principles of the law of nations, as they result from the usages established among civilized peoples, from the laws of humanity, and the dictates of public conscience.[139]

While both the Brussels declaration and the Oxford Manual were more inclusive than the Lieber Code with respect to the question of who qualified as a lawful belligerent, they reaffirmed the legality of punishing civilians for the activities of irregular fighters. However, it should be kept in mind that these texts were written at a time when other legal texts still neither used nor defined the category 'civilian' explicitly, much less protected them. To be sure, both the Brussels declaration and the Oxford Manual tried to limit the means of punishment available to the belligerent parties, thereby evidently drawing lessons from both the American Civil War and the Franco-Prussian War. Article 37 of the Brussels declaration and article 47 of the Oxford Manual, for instance, stipulated that the local population in occupied territory must not be forced to swear an oath of allegiance to the occupying regime.[140] This had been common practice of the Union occupation forces during the American Civil War. Moreover, both texts insisted that occupying forces must respect private property and that they must issue receipts for contributions and goods requisitioned from private persons. However, they stipulated that occupying powers could still levy fines from the population.[141] While the Oxford Manual was more explicit concerning protections for civilians, stating that 'it is forbidden to maltreat inoffensive populations',[142] it also made clear that 'the inhabitants of an occupied territory who do not submit to the orders of the occupant may be compelled to do so'.[143] Hence, the upshot of the early attempts at codifying the laws of armed conflict was that the civilian population warranted protection only if it fully subjected itself to the occupying forces. This approach to the laws of war was as much about disciplining

[139] Preamble to the Hague Conventions, printed in Adam Roberts and Richard Guelff (eds), *Documents on the Laws of War* (Oxford: Oxford University Press, 2000), 70.

[140] Preamble to the Hague Conventions, printed in Roberts and Guelff (eds), *Documents*, 70; International Committee of the Red Cross, 'Project of an International Declaration concerning the Laws and Customs of War', Brussels, 27 August 1874 <http://www.icrc.org/ihl.nsf/135-FULL?OpenDocument>.

[141] International Committee of the Red Cross, 'Project of an International Declaration', articles 36–48; International Committee of the Red Cross, 'The Laws of War', articles 43–60.

[142] International Committee of the Red Cross, 'The Laws of War', article 7.

[143] International Committee of the Red Cross, 'The Laws of War', article 48.

civilians as it was about 'humanizing war' and protecting local populations against wanton cruelty. Its outlook on civilians was always potentially punitive, and the reason for this punitive posture was the civilians' potential links with irregular fighters: they could aid them, or they could even take up arms themselves. So, if the line between the regular and the irregular was unclear at the end of the nineteenth century, the boundaries between the irregular and the civilian were even more questionable. This tendency was to be carried forth in the Hague law.[144]

However, with the increasing industrialization of war, a second ambiguity regarding the image of the civilian complemented the irregular-civilian ambivalence: the civilians' relevance for the war economy. According to Amanda Alexander: 'Governments and population alike acknowledged the vital role of non-combatants in the modern, industrialized war machine, while military strategists described them as a key military target. It was this paradoxical reconstruction of non-combatants as both weak and critically important, as both pitiful victims and primary targets, that constituted the new idea of the civilian population.'[145]

Nonetheless, the older ambiguity regarding the distinction between civilians and irregulars persisted. For instance, this was a significant factor escalating the German forces' fear of *francs-tireurs* during the first months of the First World War to collective paranoia and then to atrocities committed in Belgium and northern France.[146] Substantial legal clarifications in the treatment of civilians would come only with the 1949 Geneva Conventions, which provided more thorough protections for civilians in occupied territory. As we shall see in the next chapter, this change was once again precipitated by a transformation in the perception of the irregular.

Conclusion

In the American Civil War and the Franco-Prussian War irregular fighters played a variety of different roles. Some resembled the eighteenth-century model of irregular fighters as light cavalry units operating independently. Others stood in the guerrilla tradition that had been established during the Napoleonic Wars, in particular the Peninsular War. Some irregulars fighting in

[144] Karma Nabulsi, 'Evolving Concepts of Civilians and Belligerents: One Hundred Years after the Hague Peace Conferences', in Simon Chesterman (ed.), *Civilians in War* (Boulder, CO: Lynne Rienner, 2001), 9–24.

[145] Amanda Alexander, 'The Genesis of the Civilian', *Leiden Journal of International Law*, 20/2 (2007), 360.

[146] John Horne and Alan Kramer, *German Atrocities, 1914: A History of Denial* (New Haven: Yale University Press, 2002).

the American Civil War had their roots in particular traditions of local defence, such as the ranger and vigilante tradition, and some made conscious reference to the history of Indian warfare. In the Franco-Prussian War, the republican discourse placed *francs-tireurs* in the framework of the historical tradition of the *levée en masse*. A distinctive development of the nineteenth century was the increased political and military awareness of the irregular as a problematic figure in contemporary wars. The opposing sides in the American Civil War and the Franco-Prussian War had different ideas regarding the question of who qualified as a lawful belligerent. As a result, the mere suppression of irregulars by force was not seen as sufficient. Legal clarification was needed. The Lieber Code and the Brussels Conference arose from this perceived need for legal clarification. However, the end result of these legal efforts was confusion rather than clarification. The reasons for this outcome were twofold. First, there was a tension between nuanced legal reasoning, as provided, for instance, by Lieber in his legal memo on 'Guerrilla Parties', and the requirements of military commanders on the field, who were neither inclined to try to understand these nuanced legal distinctions nor able to apply them amid the fog of war. More importantly, however, the issue of irregular fighters also became increasingly entangled with questions of political and military power. Smaller countries preferred to allow a large margin of discretion for organizing defensive forces, as this would strengthen their position vis-à-vis larger countries. Conversely, great powers such as Germany and Russia favoured, of course, restricting lawful belligerency to standing regular armies, as this reinforced their own position. By the time these long political struggles were resolved in a widely accepted agreement, the 1907 Hague Convention on the Customs and Rules of War on Land, these definitions had changed considerably from Lieber's memo on 'Guerrilla Parties'. Whereas Lieber's memo had introduced positive definitions of a variety of different irregular fighters and proposed differentiated treatment for those different categories, the Hague Convention defined the irregular *ex negativo* by outlining the character of regular armed forces. This had two consequences. First, all different sorts of irregulars were lumped together in the category of the unlawful belligerent. Second, the legal definition *ex negativo* made it almost impossible to stipulate the treatment of irregulars in positive—if, of course, punitive—terms. Irregular fighters stood outside the law, both with respect to legal definitions and in the sense that they were excluded from legal protections. The only protections available to them were the weak and vague principles mentioned in the Martens Clause. This approach persisted throughout the legal developments of the twentieth century. As we shall see, it is a legacy that political decision-makers, military commanders, and lawyers are still grappling with in the wars of the twenty-first century.

A second, crucial feature of the early legal approaches to irregular fighters was that they did not clearly differentiate between the irregular and the civilian. The notion of the civilian as such was not defined. Instead, contemporary legal texts made protections for civilians dependent on their attitude towards the occupying forces and on their potential connections with irregular fighters. The responsibility for irregular warfare was often completely imposed upon the local population, be it in the form of fines and assessments to cover the damage caused by irregulars or in the form of civilian reprisals. Under these circumstances, it was literally impossible to define the civilian as a separate category of individuals and to stipulate protections for them. This would change only after the experience of the Second World War and with the codification of the 1949 Geneva Conventions.

The history of irregular fighters in nineteenth-century wars also illustrates that the concept of the irregular evolved in a close relationship with that of the regular. The perception of an actor as an irregular often depended upon existing military traditions in a given country. Part of the confusion and political contestation about lawful belligerency in the American Civil War stemmed from the fact that the pre-Civil War United States had a heterogeneous military tradition, in which states could raise armed forces just as well as the central government and in which volunteer forces coexisted alongside conscripted troops. In a similar vein, the *francs-tireurs* issue in the Franco-Prussian War partly grew out of a cultural and ideological clash between the republican tradition in France, according to which the army consisted of citizens who were eager to defend 'la patrie en danger', and the Prussian approach that regarded the army as the school of the nation, and not as its outgrowth.

Changes in the cultural image of the soldier also precipitated changes in the notion and the self-image of the irregular. Whereas the irregular had featured as the undisciplined brigand in the French Revolutionary and the Napoleonic Wars, in the nineteenth century the irregular became increasingly scorned for his or her cowardice. This slight cultural shift was a result of technological changes, mainly the introduction of long-range rifles, which had undermined traditional notions of courage among regular troops. Courage was still the central part of the cultural script for the regular soldier, but it was difficult to uphold in a situation where offensive tactics had become increasingly costly and defensive measures such as trench-digging increasingly widespread. In this situation of technological transformation and cultural uncertainty, cowardice was blamed on the irregular. At the same time, irregular fighters fostered an image of themselves as the true war heroes, who managed to preserve their warrior qualities in spite of the modernization of the battlefield.

4

The Second World War

Anti-Partisan Warfare, Genocide, and the Rebirth of the Auxiliary Fighter

The Second World War is not usually primarily associated with irregular warfare. While armed resistance and partisan warfare may not have had a large influence on the overall strategic outcome of the war, they did involve large numbers of mobilized activists and affected the lives of many contemporaries.[1] The experience of armed resistance and its repression by German and Japanese occupying forces and their auxiliaries also had a decisive impact on the development of the law of armed conflict after the war in the Geneva Conventions.

The historiography of irregular warfare in the Second World War poses several challenges. On the one hand, as is usually the case with irregular fighting, mainstream military history tends to treat it as a deviation from the more 'decisive' conventional operations. Hence there is not much information on how regular and irregular warfare interacted and influenced one another in strategic and operational terms. On the other hand, resistance against Nazi occupation was a central foundational myth of many countries in post-war Europe. This often placed an undue burden on historical facts and the actors involved and led to a lopsided historiography that emphasized heroic sacrifice and national unity.

[1] By 1944 there were an estimated 1.5 million people in arms participating in irregular warfare in the framework of the Second World War. Jonathan E. Gumz even makes the case for reconceptualizing the Second World War 'as a series of unbounded insurgencies in that they were not solely confined to professional militaries alone, but rather involved various ideological and nationalized groups' ('Reframing the Historical Problematic of Insurgency: How the Professional Military Literature Created a New History and Missed the Past', *Journal of Strategic Studies*, 32/4 (2009), 553–88).

A further problem is the close factual linkage between the German suppression of irregular warfare in the occupied territories and the Holocaust. This generated a considerable interest in German occupation policies and the question of who was involved in the large-scale killing of civilians, Jewish civilians in particular, as reflected, for instance, in the German debate surrounding the role of the Wehrmacht or the Goldhagen debate.[2] These discussions moved the issue of irregular warfare and its repression into the centre of the historical debate, but they often framed the issue rather rigidly in terms of the ideological dispositions at play. Nazi ideology and its components, in particular anti-Semitism, racism, anti-bolshevism, and the notion of *Lebensraum* were the focus of this research. However, this obscured the fact that there is a large degree of continuity between earlier wars and the Second World War regarding the concept of the irregular, as well as the institutional responses to irregular warfare and certain outcomes of its repression. In other words, by approaching the issue of irregular warfare in the Second World War from the perspective of the Holocaust, there is a risk of projecting the singularity of the Holocaust on to the broader issue of anti-partisan warfare. However, while the Holocaust was undoubtedly singular, the fact that during the Second World War the notion of the irregular became charged with racial, ideological, and cultural stereotypes was not.

This chapter is divided into four parts. The first part gives a general overview of armed resistance movements across Europe. The second part addresses the German response to armed resistance and irregular fighting. The focus of this part is a critical discussion of mainstream arguments regarding the German response to armed resistance, in particular the unique relevance of Nazi ideology and the so-called criminal orders for the ruthless anti-partisan campaign on the eastern front. It will be shown how both race and political *Weltanschauung* were attached to the already existing stigma of the irregular fighter as the bandit and outlaw. The third part discusses the return of the idea of the irregular as the auxiliary to the regular armed forces during the Second World War. It addresses the strategic concepts behind the support and use of irregular fighters by Allied powers such as Britain and the Soviet Union, but also Nazi Germany's use of *Volksdeutsche* or native auxiliaries in the occupied territories. The final part is devoted to the discourse of armed resistance itself, and in particular to the irregular forces' inversion of the notions of political legitimacy, social participation, and political representation.

[2] Hannes Heer and Klaus Naumann (eds), *Vernichtungskrieg—Verbrechen der Wehrmacht 1941–1944* (Frankfurt: Hamburger Edition, 1995); Daniel J. Goldhagen, *Hitler's Willing Executioners—Ordinary Germans and the Holocaust* (London: Abacus, 1997). For a critique of Goldhagen's book, see, e.g., Norman G. Finkelstein and Ruth Bettina Birn, *A Nation on Trial: The Goldhagen Thesis and Historical Truth* (New York: Henry Holt and Company, 1998); and Robert R. Shandley (ed.), *Unwilling Germans? The Goldhagen Debate.* (Minneapolis, MN: University of Minnesota Press, 1998).

This chapter will focus on irregular fighting in Europe, the Soviet Union, and the Balkans. Given the extraordinarily comprehensive spectrum of military theatres and operations in the Second World War, it is indispensable to be selective. Central to this text is the European experience of armed resistance against Nazi occupation, since this was the most important factor impacting on the post-war legal development of the law of armed conflict in the Geneva Conventions.[3] It should also be mentioned that by highlighting *armed* resistance against Nazi occupation, the chapter does not intend to belittle other forms of resistance. Nonetheless, while non-violent forms of resistance may have been equally effective and legitimate, they did not contribute to the development of the concept of the irregular fighter.

A Historical Overview of European Resistance Movements in the Second World War

1939–1940: Poland, Scandinavia, and the Benelux countries

Military operations in the Second World War began on 1 September 1939 with the German attack on Poland. Irregular fighting ensued almost immediately, with the Polish government's call for general armed resistance on 9 September.[4] Anticipating armed resistance by irregular fighters, already on 3 September Hitler issued 'an order calling for "insurgents" to be "shot on the spot"'.[5]

In Poland, as elsewhere, the initial organization of armed resistance against German occupation emerged mainly from the defected regular armed forces in conjunction with political parties. On 13 November, Polish Army General Władisłav Sikorski, who had fled to Paris and formed the Polish Government in exile, ordered the creation of the Związek Walki Zbrojnej (ZWZ, Union of Armed Struggle). Initially, its main task was sabotage. After the German invasion of the Soviet Union in June 1941, the ZWZ was renamed Armja Krajowa (AK, Home Army). The AK grew rapidly, and in 1943 it had an estimated 200,000 fighters in arms. With its growth and with the change in the strategic environment, the AK also revised its approach from sabotage to armed resistance. At the end of 1942, it began to engage in partisan warfare, but it also increasingly conducted actions against *Volksdeutsche* (ethnic Germans) who

[3] Nazi German occupation policies developed in close connection with the genocide of Europe's Jewish population. After the war the Holocaust was perceived as the worst instance of violence against civilians, notwithstanding the fact that high levels of violence against civilians also characterized Japanese occupation policies. Cf. Geoffrey Best, *War and Law since 1945* (Oxford: Oxford University Press, 1994), 64.

[4] Mark Mazower, *Hitler's Empire: Nazi Rule in Occupied Europe* (London: Allen Lane 2008), 67.

[5] Mazower, *Hitler's Empire*, 67.

had been resettled on Polish territory. AK operations culminated in the unsuccessful Warsaw uprising in the summer and autumn of 1944.

In 1940, Germany occupied Denmark, Norway, the Netherlands, and Belgium.[6] Denmark remained formally independent; hence resistance was slow to develop. The first sabotage actions occurred in 1942. In 1943 Danish resistance became more organized with the emergence of the Freedom Council, which acted as an umbrella organization for the various resistance groups. At the same time, the British Special Operations Executive (SOE)[7] put increasing efforts into sabotage actions in Denmark. The SOE's plan was to provoke German repression, and thereby to spark off resistance in Denmark on a larger scale. The plan succeeded: in August 1943 Nazi Germany imposed direct rule on Denmark,[8] and by the end of 1943 an underground army had developed there. It was soon joined by the Danish regular armed forces, which had been briefly interned after the imposition of direct rule but then released.[9] Yet, despite its strong organizational basis, sabotage—not partisan warfare—remained the central activity of the Danish underground army. Since German occupation forces left Denmark without fighting, the underground army played no role in the liberation.

After the German occupation of Norway, King Hakoon fled to London, where he established a government in exile. Vidkun Quisling, head of the Norwegian national socialist party Nasjonal Samlin, served a brief term as self-declared prime minister, but he was soon marginalized, and Norway was put under civilian administration. An armed resistance organization, Milorg, had already developed in 1940, but its internal organization suffered owing to enduring negotiations with the Norwegian government in exile and the British SOE over the best approach to resistance. In the end, Milorg abandoned the idea of a general uprising and concentrated on sabotage.[10]

The Netherlands was placed under a civilian administration headed by Arthur Seyss-Inquart immediately after German occupation. Resistance efforts emerged partly from the demobilized military and from defected officers. The SOE also made efforts to support Dutch resistance groups, but had to face the greatest debacle of its history with the *Englandspiel* episode: when the first agents were dropped into the Netherlands in 1942, they failed to realize that the Dutch networks with which they linked up had already been penetrated

[6] Both Denmark and Norway were occupied on 9 April 1940. The Wehrmacht invaded both the Netherlands and Belgium on 10 May 1945. The Dutch armed forces surrendered on 15 May 1940 and the Belgian armed forces on 28 May 1940.

[7] See also the section 'Armed Resistance and the Revival of the Irregular as the Auxiliary'.

[8] David Stafford, *Britain and European Resistance, 1940–1945: A Survey of the Special Operations Executive, with Documents* (London: Macmillan, 1980), 13.

[9] Hans Kirchhoff, 'Denmark', in Bob Moore (ed.), *Resistance in Western Europe* (Oxford: Berg, 2000), 108.

[10] Arnfin Moland, 'Norway', in Moore (ed.), *Resistance in Western Europe*, 229.

by the German security police (Sicherheitspolizei, SIPO). As a result, all supplies provided by SOE fell into German hands, and forty-three British agents were killed.[11] Partly as a result of this incident, organized armed resistance in the Netherlands developed late and under firm coordination of the Allies. In 1944 all individual Dutch resistance groups merged into the Dutch Forces of the Interior. As resistance operations picked up, German occupation became more repressive in turn. Still, the Dutch Forces of the Interior engaged mainly in sabotage actions and played no military part in the liberation of the Netherlands.[12]

After the occupation of Belgium, the Belgian King Leopold remained in the country, hoping that it would remain formally independent. However, because of Belgium's strategic importance, it was placed under military administration, even if many administrative tasks remained in Belgian hands.[13] In 1941 a Belgian underground army, the Belgian Legion, was formed. Its members were mainly Belgian officers and reserve officers, and it was later renamed into the Secret Army of Belgium. It was supported by the SOE, and its main tasks were large-scale sabotage acts.[14] However, in general Belgian resistance remained highly fragmented throughout the war and remained split over the question of King Leopold's failure to challenge Nazi Germany's occupation.

1940: France

The history of French resistance to German occupation is complex. On 10 May 1940 Nazi Germany attacked France. On 22 June 1940 France signed an armistice with Nazi Germany after Wehrmacht forces had pushed far into French territory and had captured a large part of the French armed forces. The terms of the armistice divided France into five parts: the north-western-most part of France came under the military administration of Belgium and Northern France. Alsace and Lorraine were directly annexed to Germany. The two major parts of the French territory were the occupied zone in the north stretching down the entire Atlantic coastline, and the Free Zone under the French authority of the Pétain government, which had its seat in Vichy. Finally, Italian forces occupied small chunks of territory on France's south-eastern border.

The terms of the armistice immediately caused a conflict within the French military leadership; some 'dissidents' refused to follow Pétain and went into exile in Britain. Among those was Charles de Gaulle, who was soon to rise to

11 Stafford, *Britain and European Resistance*, 94.
12 Dick van Galen Last, 'The Netherlands', in Moore (ed.), *Resistance in Western Europe*, 206.
13 Mazower, *Hitler's Empire*, 107.
14 Pierre Lagrou, 'Belgium', in Moore (ed.), *Resistance in Western Europe*, 37.

the central figure of the French resistance.[15] De Gaulle fled to London and formed the Free French Forces, which consisted mostly of other dissident officers. Later they would fight alongside the British in the Battle of Britain, in Africa, and in the Middle East, and finally in the battle of the Atlantic.[16]

Resistance within France was slow to develop. Armed resistance groups emerged both in the occupied territory and in Vichy France. Among the groups that pursued military resistance were, for instance, Ceux de la Libération (CDLL) and Ceux de la Résistance (CDLR) in the occupied territory and the Organisation Métropolitaine de l'Armée in Vichy France.[17] These armed groups increasingly had linkages with the Free French Forces under de Gaulle. With the German attack on the Soviet Union in June 1941, communist resistance also gained momentum in France. As long as the Hitler–Stalin pact had been in place, communist resistance had been stymied by the Comintern doctrine that the Soviet Union and Nazi Germany were formally allies.[18] After the signing of the Hitler–Stalin pact in 1939, the Parti Communiste Français (PCF) had been declared illegal in France. In 1940 the underground PCF formed an armed wing, the Organisation Spécial de Combat (OS). In spite of the official PCF line of not resisting German occupation, the OS soon engaged in acts of sabotage. After the German attack on the Soviet Union, the PCF renamed and reorganized the OS into the Francs-Tireurs et Partisans (FTP) and launched the Front Nationale as a putative umbrella organization for all French resistance movements. Despite this claim to national representation, the PCF was mainly active in the occupied zone.[19]

The communist milieu in France had the advantage that it could muster some knowledge and experience with irregular fighting: many members of the PCF and many non-French communists who were stranded in France in 1940 had fought in the International Brigades during the Spanish Civil War. Especially those communists who had fled from Spain to France after the end of the Spanish Civil War felt abandoned by Moscow and showed no allegiance to the Hitler–Stalin pact. Hence they were among the first to engage in armed resistance.[20]

Armed resistance in France received a further impetus with the German occupation of the free zone that followed the Allied landings in North Africa in 1942. Especially conservative circles and military officers now realized that the Vichy regime would not be able to restore French sovereignty against the

[15] Olivier Wieviorka, 'France', in Moore (ed.), *Resistance in Western Europe*, 130.

[16] Henri Michel, *The Shadow War: Resistance in Europe 1939–1945* (London: Corgi, 1972).

[17] Wieviorka, 'France', 129.

[18] Lynne Taylor, 'The Parti Communiste Français and the French Resistance in the Second World War', in Tony Judt (ed.), *Resistance and Revolution in Mediterranean Europe, 1939–1948* (London: Routledge, 1989), 53.

[19] Taylor, 'The Parti Communiste Français', 71.

[20] Tony Judt, 'Introduction', in Judt (ed.), *Resistance and Revolution*, 1–28.

German occupiers. From these circles a further armed resistance organization emerged, the Organisation Métropolitaine de l'Armée, later renamed Organisation de Résistance de l'Armée (ORA).[21]

An even more important factor pushing the French resistance to a new level was the German decision in February 1943 to implement a foreign labour scheme named *service du travail obligatoire* (*STO*). As in other countries, it was the German attempt to deport young men to Germany in order to exploit their labour that fuelled armed resistance more than any other measure of the German occupation forces. Many of those who were about to be drafted into the *STO* tried to avoid it and went into hiding; they became *réfractaires*. Many of these *réfractaires* took up arms as a way of surviving. These included the many urban resistance groups of the years 1940–2, as well as rural resistance as part of the so-called *maquis*. The structure and the activities of the *maquis* bore the traits of its origin, the *réfractaire* situation: they were based on escape and refuge.[22]

The '*maquis* discourse', as H. R. Kedward called the *maquisards'* more or less coherent identity narratives, harked back to two contradictory historical roots. On the one hand, their rural and decentralized structures resembled those of the counter-revolution at the end of the eighteenth century. Like the counter-revolutionaries, the *maquis* adopted certain functions of social regulation within the local communities. In Kedward's words, 'the causes were profoundly different, but the mechanisms remarkably similar'.[23] On the other hand, contemporaries who had fought in the *maquis* pointed out the parallels between the *maquis* and the *levée en masse* of 1792. Confronted with the charge that most *maquisards* were driven by delinquency or mental disorders, a young student of medicine who had fought with a *maquis* battalion in the Haute-Lozère insisted that the *maquis* had acted on behalf of the political collective rather than for self-seeking motives and compared the *maquis* movement with the *levée en masse*.[24]

As with the urban resistance movements, the *maquis* was split into communist and non-communist groups. Among the non-communist groups, the Gaullist Armée Secrète (AS) had the greatest influence. As elsewhere, both in the urban resistance and in other occupied countries, there was a fierce conflict between communist and non-communist armed groups over the

[21] Wieviorka, 'France', 133.

[22] H. R. Kedward, *In Search of the Maquis: Rural Resistance in Southern France, 1942–1944* (Oxford: Oxford University Press, 1993), 45.

[23] Kedward, *In Search of the Maquis*, 99 ff. On the counter-revolutionary uprisings in eighteenth-century France, see also Chapter 2.

[24] Kedward, *In Search of the Maquis*, 110. This contradictory historical reference to both the *levée en masse* and the counter-revolution may seem paradoxical, but it was by no means unprecedented. Leon Gambetta had used the same contradictory historical references when calling for a general mobilization during the Franco-Prussian War; see Chapter 3.

best approach to resistance. AS accused the communist FTP of wasting their resources on 'direct action' that achieved little, but provoked the occupying forces into severe, repressive actions against the civilian population. FTP, in turn, charged AS of 'hoarding resources which could have been used directly against the enemy, of keeping *maquisards* in a state of inactivity and frustration waiting for Jour-J [D-Day], and of surrendering initiatives to the British, Gaullist, and American agents, who were not necessarily accepted as working for the interests of the ordinary French people'.[25]

Some *maquis* groups, especially those who were oriented towards the Gaullist AS, were led by army officers, whose regular army mentality sometimes had an undeniable impact on *maquis* operations. In Manigod in Haute-Savoie, a group of army officers established military training facilities, an *école de cadres*, for the *maquis*. The curriculum at the *école de cadres* was inspired by two ideas: first, the officers who ran the *école* elevated irregular warfare to a scientific issue and referred to it as '*la science militaire et maquisarde*'. Secondly, they thought of irregular warfare as a sub-form or sub-discipline of regular warfare.[26] Essentially, they harked back to the mid-eighteenth-century view of irregular warfare as a semi-independent auxiliary tool to conventional operations. Of course, this perspective fit perfectly into the Gaullist wait-and-see approach and its emphasis on close integration with Allied strategic preferences.[27]

However, when translated into practical operations, this concept soon proved its limits or, rather, showed how easily such a regular warfare mindset could lead to a complete misjudgement of the strategic situation. It was arguably their regular warfare mindset that tempted the officers of the AS *maquis* of Haute-Savoie into engaging in more conventional operations such as the entrenchment of Glières in the first months of 1944. The operation envisaged the fortification and armed defence of a plateau that was supposed to serve as a drop-off site for military supplies from London. It ended in a complete disaster, in which several hundred *maquisards* and civilians were killed.[28] The *maquis* had been neither equipped nor sufficiently trained to conduct operations on such a scale.

In spite of the internal differences among the various resistance groups within France, in February 1944 Charles de Gaulle in conjunction with Jean Moulin, a resistance leader who was betrayed to the Germans and died in Gestapo hands,[29] achieved an at least formally united resistance organization under the umbrella of the Forces Françaises de l'Intérieure (FFI) commanded

25 Kedward, *In Search of the Maquis*, 126; cf. Michel, *Shadow War*, 214.
26 Kedward, *In Search of the Maquis*, 134.
27 See also the section 'Inversions: Discipline, Legitimacy, and the *levée en masse*'.
28 Kedward, *In Search of the Maquis*, 137 ff.
29 While Moulin worked to establish this unification, he did not live to see it, as he had already died in 1943. De Gaulle and Koenig continued these efforts and realized the formal unification.

by General Marie Pierre Koenig. However, even within this unified framework, internal differences over the most suitable approach to armed resistance persisted. The operation at Mont Mouchet illustrated though that pragmatic considerations sometimes overrode seemingly rigid positions in this internal conflict. This operation to instigate an uprising had been planned and initiated by de Gaulle and Koenig, although both normally insisted that armed resistance should focus on sabotage and cooperate with the Allies. The Mont Mouchet episode was a localized uprising in early June 1944 in the Auvergne, with Mont Mouchet as its centre. Militarily the operation was not particularly effective, but it engendered among the *maquis* the sense that the *maquisards* themselves were able to contribute substantially to the liberation of France.[30] This enthusiasm among the French resistance and the sense of self-liberation persisted during the Allied advance through France and led to the attempted 'self-liberation' of Paris.[31] The idea of self-liberation became one of the founding myths of post-war France.

1941: The Soviet Union

What historians refer to as 'partisan warfare' in the narrower sense of the term ensued after the German invasion of the Soviet Union, which started on 22 June 1941. Wehrmacht forces, along with some units from Slovakia, Hungary, Romania, and Finland, advanced rapidly during the first few months of Operation 'Barbarossa', as the Germans had codenamed the invasion. However, the operation ground to a halt before reaching Moscow at the beginning of December 1941.[32]

The Soviet leadership was taken by surprise by this rapid advance.[33] Soviet pre-war military planning had not anticipated the German forces' 'in depth' occupation nor the necessity of defending Soviet territory with irregular forces. Even so, in the 1920s and 1930s the Red Army had accumulated some expertise on irregular warfare. Mikail Frunze had been the founding father of the Soviet theoretical and doctrinal approach to irregular warfare. After his death in 1925, Red Army officers such as Ilya Starinov, Vasily Bliukher, Ieronim Uborevich, Ian Berzin, and Vitaly Primakov built on Frunze's work and organized military training for irregular warfare within the Red Army. However, most of these officers fell victim to the Stalinist

[30] Kedward, *In Search of the Maquis*, 166 ff. [31] Michel, *Shadow War*, 324.

[32] On Operation 'Barbarossa', see David Stahel, *Operation Barbarossa and Germany's Defeat in the East* (Cambridge: Cambridge University Press, 2009). Stahel argues that Operation 'Barbarossa' was flawed from the beginning; rather than conforming to the *blitzkrieg* template, the summer 1941 campaign lacked the momentum necessary for a decisive German victory.

[33] On Soviet war preparations, or, rather, the lack thereof, see David E. Murphy, *What Stalin Knew: The Enigma of Barbarossa* (New Haven: Yale University Press, 2005).

purges of the Red Army officer corps in the late 1930s. Stalin himself was deeply suspicious of irregular warfare. First, he regarded the notion of defence with irregular forces in case of a German attack as defeatist. Stalinist grandeur demanded that it would be the Red Army that would repel any invasion attempt at the Soviet border. Second, Stalin and the Soviet leadership feared that a call for popular armed resistance could unleash nationalist forces, especially in the territories in the west of the Soviet Union that had come under Soviet control only in 1939 as a result of the Hitler–Stalin pact. The result of uncontrolled popular resistance, it was thought, could be the break-up of the Soviet Union.[34]

The need to react quickly to the rapid German advance into Soviet territory caused the Soviet leadership to reconsider its stance on irregular warfare. In a radio address on 3 July 1941 Stalin called upon all citizens to defend the Soviet Union against the invaders. However, as Kenneth Slepyan pointed out, there was a discrepancy between Soviet propaganda and policy:

> Propaganda was one thing; state policy quite another. In 1941 and the first nine months of 1942, actual Soviet policy discounted and even discouraged popular initiative and participation. On 29 June and 18 July 1941, the first national directives calling for partisan warfare envisioned a much more limited struggle, to be based on the Party and state security apparatus.[35]

Yet, by stressing the guiding role of the party and the security apparatus, Slepyan downplayed the role of Red Army officers and soldiers within the partisan movement. They played a major role in the formation of the first partisan groups.[36] The main reason for this massive irregularization of regular Red Army troops was political: on 16 August 1941, the Red Army High Command (Stavka) issued order 270, which stated, 'if a commander or unit of Red Army soldiers instead of the organization of resistance to the enemy prefers to surrender, they are to be destroyed by all means, both on the ground and from the air, and the families of Red Army soldiers who have given themselves up into captivity will be deprived of support and aid'.[37] Essentially, this meant that surrender was a bad option for Red Army soldiers and officers, and this drove many Red Army troops to continue fighting as partisans if they could avoid capture. But pressure to take to the forests and swamps not only

[34] A. A. Maslov, 'Concerning the Role of Partisan Warfare in Soviet Military Doctrine of the 1920s and 1930s', *Journal of Slavic Military Studies*, 9/4 (1996), 885–94. Cf. Kenneth Slepyan, *Stalin's Guerrillas: Soviet Partisans in World War II* (Lawrence, KS: University of Kansas Press, 2006), 17 ff.

[35] Slepyan, *Stalin's Guerrillas*, 24.

[36] John Armstrong, 'Introduction', in John Armstrong (ed.), *The Soviet Partisans in World War II* (Madison: University of Wisconsin Press, 1964), 21 ff.

[37] Quoted in Robert W. Thurston, 'Cauldrons of Loyalty and Betrayal: Soviet Soldiers' Behavior, 1941 and 1945', in Robert W. Thurston and Bernd Bonwetsch (eds), *The People's War: Responses to World War II in the Soviet Union* (Urbana, IL: University of Illinois Press, 2000), 247.

came from the Soviet leadership: if Red Army members had been captured by the Germans, the horrible conditions of German captivity provided a good incentive to escape. Nazi Germany denied that the 1929 Geneva Convention on Prisoners of War applied on the eastern front on the grounds that the Soviet Union had not ratified it. German provisions for the treatment of Soviet POWs precluded independent third-party controls. Moreover, given Stalin's denunciation of Soviet POWs as traitors, they could not expect any interest in their fate on the part of the Soviet leadership either. The conditions for Soviet POWs in German captivity were so harsh that, of the 3.3 million Soviet POWs captured in the first few months of the war, the majority had died by March 1942. If the bleak conditions of German captivity provided a major incentive to escape, the chaos of mass detention and the lack of suitable transportation for Soviet POWs offered ample opportunities to do so.[38]

Partisan warfare in the Soviet Union developed unevenly. In the Baltic states, which had just come under Soviet authority in 1939, parts of the population initially regarded the German occupation forces as liberators rather than oppressors. Especially the rural population had resented the pre-war Soviet land reforms and hoped that the German occupiers would revert to the pre-Soviet system of rural agriculture. German occupation authorities recruited many local natives into the auxiliary police formations and even raised two Latvian SS units. Throughout the war, no well-organized partisan movement developed in the Baltic states.

The picture was somewhat similar in Ukraine. A plethora of irregular movements had already existed before the war, but their agenda was nationalist rather than communist, and they had been formed to resist the Soviet authorities rather than the German occupiers. The Ukranian Insurgent Army (Ukrajinska Powstanska Armija (UPA)), for instance, claimed to represent the armed wing of the émigré government of the Ukranian People's Republic. It fought against both the German occupation forces and the Soviet partisans trickling into Ukraine from Belarus.[39] The Polish Home Army (Armja Krajowa) was also active in the western districts of Ukraine, as these were former Polish territory with a large Polish minority population.

Finally, Belarus was the site of much partisan activity throughout the war. Large numbers of Red Army soldiers had been cut off from their command by advancing German troops in the summer of 1941; those not captured by

[38] Rüdiger Overmans, 'Die Kriegsgefangenenpolitik des Deutschen Reiches 1939–1945', in Jörg Echternkamp (ed.), *Das Deutsche Reich und der Zweite Weltkrieg, Band 9/2: Die Deutsch Kriegsgesellschaft: Ausbeutung, Deutungen, Ausgrenzung* (Munich: Deutsche Verlags-Anstalt, 2005), 805 ff.

[39] Alexander Statiev, 'Was *Smuglianka* a Lunatic or a Siguranţa's Agent-Provocateur? Peculiarities of the Soviet Partisan Struggle in the Western Borderlands', *Journal of Strategic Studies*, 31/5 (2008), 762.

Wehrmacht forces survived in the forests and swamps and continued to fight as partisans. Because of the German agenda of 'Belarussification' of the local administration, many Belorussian nationalists supported the occupying forces initially and joined the native auxiliary police forces. But the Germans' harsh and repressive occupation policies soon fuelled support for the Soviet partisans in Belarus. The Soviet partisans, in turn, targeted many of their operations at local collaborators and their families rather than at the German occupation forces. As in Ukraine, the Polish Armja Krajowa was active in the western provinces of Belarus, which had been Polish territory before the break-up of Poland in 1939.[40]

By the spring of 1942, the emerging Soviet partisan movement had almost been destroyed by German occupying forces, but it recovered and grew over the second half of 1942. In 1943 two factors contributed to the increasing growth of armed resistance. First, as the German defeats at Stalingrad and Kursk made it clear that the Germans could not win the war in the Soviet Union,[41] the retreating German authorities increased the recruitment of forced labour in the still occupied territories in the east. While the labour scheme had been largely voluntary before January 1943, the authorities now exerted pressure on the local population and operated with fixed quotas. As they steadily retreated westwards, the Germans had no incentive to keep the labour force in the occupied territories intact in order to maintain a viable economic structure.[42] At the same time, after a brief interlude of more benign measures towards the civilian population, the military leadership switched its approach to a dead zones policy. This included deporting the inhabitants of entire villages where partisan resistance was suspected into forced labour service.[43] As in other occupied countries, the option of avoiding such a fate provided a compelling incentive for the local population to join the partisans.

The second factor that strengthened the partisan movement in 1943 was the Soviet leadership's decision to broaden the base of partisan warfare and to include elements that had been previously deemed too unreliable: women, ethnic minorities (such as Tatars, Jews, and Ukrainians), and former collaborators who had re-defected to the Soviet side. The People's Commissariat of Defence (Narodny Kommissariat Oborony (NKO)) order 189 'On the tasks of

[40] Hans-Heinrich Nolte, 'Partisan War in Belorussia, 1941–1944', in Roger Chickering, Stig Förster, and Bernd Greiner (eds), *A World at Total War: Global Conflict and the Politics of Destruction* (Cambridge: Cambridge University Press, 2005), 261–76; Bernhard Chiari, *Alltag hinter der Front: Besatzung, Kollaboration und Widerstand in Weißrussland, 1941–1944* (Düsseldorf: Droste Verlag, 1998).

[41] Slepyan, *Stalin's Guerrillas*, 190.

[42] Nolte, 'Partisan War', 274; Alexander Hill, *The War behind the Eastern Front: The Soviet Partisan Movement in North-West Russia 1941–44* (London: Frank Cass, 2005), 151.

[43] Ben Shepherd, *War in the Wild East: The German Army and Soviet Partisans* (Cambridge, MA: Harvard University Press, 2004), 177.

the partisan movements' of 5 September 1942 encapsulated this broadening of the partisan base: 'In fact, this order, for the first time, really put "the people" into the regime's formula all-people's war and gave state sanction and support to a mass-based movement.'[44] And 'the people' responded to this call to show allegiance to the partisan cause, if partly for opportunistic reasons: the clearer it became that the German occupying forces would eventually be evicted from the Soviet Union, the greater the fear of Soviet reprisals after liberation.[45]

The most suitable approach to armed resistance was not a subject of major conflict among Soviet partisans, as it had been in the French resistance movement. Partisan activities in occupied Soviet territory centred on sabotage, attacks on the railway network, and assassinations. The most high-level victim of partisan assassinations was Wilhelm Kube, the German general commissioner of Belarus who was killed by a bomb planted by partisans on 22 September 1943 in Minsk. However, the most common victims of partisan assassinations, especially in Belarus, were local collaborators and their families rather than the German occupiers.[46] As in France, the irregulars also conducted more 'conventional' operations aimed at liberating specific areas or villages. The village of Klichev in Belarus was a case in point: after three partisan brigades with heavy weaponry had attacked and liberated it in March 1942, it served as an operational centre for partisan activities throughout the war.[47] In particular towards the end of the war, partisan activities were also integrated into Red Army operations. The Operation 'Rail War' in the summer of 1943, for instance, was a large railway sabotage operation conducted by partisans with the aim of assisting the Red Army counter-offensive at Kursk. 'Rail War' was followed by Operation 'Concert', which was designed to disrupt German reinforcements and supplies during the Battle of the Dnieper.[48] On occasion partisans also conducted long-range raids. Ukrainian partisans under the leadership of Sydir Kovpak, who had already participated in partisan operations during the Russian Civil War, for instance, conducted several raids into western Ukraine. Militarily, long-range raids tended to have limited effectiveness, but they proved to be a valuable recruitment tool for the partisan movement.[49]

1941: Yugoslavia and Greece

In both Yugoslavia and Greece, occupation by the Axis powers and resistance to it sparked a dynamic that led to civil war in both countries. Germany's

[44] Slepyan, *Stalin's Guerrillas*, 49.
[45] Hill, *The War behind the Eastern Front*, 154.
[46] Nolte, 'Partisan War', 270.
[47] Nolte, 'Partisan War', 271.
[48] Slepyan, *Stalin's Guerrillas*, 98 ff.
[49] Slepyan, *Stalin's Guerrillas*, 99 ff.

occupation of Greece and Yugoslavia was not so much inspired by the expansionist aim to gain additional *Lebensraum* or to exploit these countries economically. Rather, Germany was more or less sucked into the Balkans by the hapless actions of its foremost ally Italy. Mussolini had tried to invade Greece so as further to realize Italy's ambitions of dominating the Balkans. The Italian invasion had started in October 1940, but in April 1941 Mussolini's forces were pushed back by the Greek military. On 6 April 1941, Nazi Germany came to Mussolini's aid and achieved a decisive push into Greece. On 27 April 1941 Athens fell, and on 3 May German forces finally occupied Crete. Subsequently, Greece was jointly occupied by German, Italian, and Bulgarian forces.

Resistance in Greece started almost immediately with the occupation by Axis powers. Initial resistance acts were uncoordinated. The perpetrators were individuals, such as the two students who tore down the swastika banner from the Acropolis on 30 May 1941. Organized resistance soon emerged from the communist milieu. On 27 September 1941 the National Liberation Front (Ethniko Apeleftherotiko Metopo (EAM)) was created. In December 1941, EAM formed an armed wing, Ellinikós Laïkós Apeleftherotikós Stratós (ELAS; Greek People's Liberation Army). Republican resistance was slower to organize, but at the end of 1941, republican circles created their resistance organization named Ethnikos Dimokratikos Ellinikos Syndesmos (EDES; National Republican Greek League). Also from a republican background emerged Ethniki kai Koinoniki Apeleftherosis (EKKA; National and Social Liberation). However, EKKA was much smaller than EDES and remained restricted to the mountainous region of central Greece and the region of Fokida. Throughout the occupation of Greece, ELAS asserted itself as the biggest of those three armed resistance organizations. Greek irregular fighters were referred to as *andartes*, a term that went back to Greek resistance against Ottoman domination in the nineteenth century.

Some of the irregular attacks in Greece were conducted by surprisingly large forces.[50] The most dramatic action, which took place on 25 November 1942, was the destruction of the strategically important Gorgopotamos viaduct on the rail line from Athens heading north. This operation, codenamed 'Harling', was conducted jointly by ELAS and EDES and supported by the SOE. However, it was to be the last joint operation of ELAS and EDES, as the political rift between the two organizations grew and turned them into opponents. The SOE contributed to this rift. It grew increasingly wary of working with the communist-inspired EAM/ELAS, as London feared that a communist-dominated post-war Greece would threaten British interests in

[50] Mark Mazower, *Inside Hitler's Greece: The Experience of Occupation* (New Haven: Yale University Press, 1993), 135.

the Middle East.[51] In July 1943 Brigadier Eddie Myers, head of the SOE mission in Greece, had helped negotiate the so-called National Bands Agreement that stipulated that the three resistance organizations, EAM/ELAS, EDES, and EKKA, were to cooperate.[52] But the Foreign Office sidelined the SOE and dispatched Sir Orme Sargent as its representative to Greece. Eventually the more EAM/ELAS-sceptic line of the Foreign Office prevailed, and EAM/ELAS was cut off from British support.[53]

Fighting between EAM/ELAS, on the one hand, and EDES and EKKA, on the other, broke out in late 1943. Despite the Allies' lack of support for EAM/ELAS, the latter succeeded in liberating large areas of mainland Greece by mid-1944. In these areas, referred to as 'Free Greece', EAM/ELAS often targeted and terrorized EDES sympathizers.[54] After the liberation of Greece in October 1944, EAM initially participated in the Papandreou government, but in December 1944 clashes ensued between Greek government forces, British forces, and ELAS. In March 1946 the continued political polarization between republican forces and ELAS led to the outbreak of the Greek Civil War.

Nazi Germany's involvement in Yugoslavia developed in a similarly accidental fashion as its occupation of Greece. In 1941 Germany had pushed Prince Paul of Yugoslavia to join the Axis and to support its Balkans policy. As a result of this perceived betrayal of Yugoslavian interests, Prince Paul fell victim to a *coup d'état* and was replaced by Prince Peter II.[55] The Axis powers started to invade Yugoslavia on 6 April 1941 and occupied it subsequently. Yugoslavia was carved up into five parts: the Independent State of Croatia (Nezavisna Država Hrvatska (NDH) under Ante Pavelić, Serbia and the Banat under German occupation, the Dalmatian coast, which was annexed by Italy, Montenegro under Italian occupation, and, finally, the eastern border regions, which were annexed by Hungary and Bulgaria.

As in Greece, in Yugoslavia armed resistance to the Axis occupation evolved out of a dynamic of political polarization. The Wehrmacht and other German occupation forces maintained only a light footprint in Yugoslavia and left the main burden of the occupation to their Italian allies and their native collaborators.[56] In the NDH, this was Ante Pavelić and his *Ustasha* movement. Pavelic was an extremist Croatian nationalist whose aim was to create an ethnically pure and independent Croatia. Between 1941 and 1944 the *Ustasha* killed at least 200,000 Croatian Serbs. *Ustasha* persecution of the Serbs fuelled

[51] Stafford, *Britain and European Resistance*, 126.
[52] Mazower, *Inside Hitler's Greece*, 142.
[53] Stafford, *Britain and European Resistance*, 126.
[54] Mazower, *Inside Hitler's Greece*, 289.
[55] Ben Shepherd, *Terror in the Balkans: German Armies and Partisan Warfare* (Cambridge, MA: Harvard University Press, 2012), 75.
[56] Matteo J. Milazzo, *The Chetnik Movement and the Yugoslav Resistance* (Baltimore, MD: Johns Hopkins University Press, 1975), 5.

the development of a Serbian resistance movement and, in particular, the growth of Tito's Partisans.[57] To a lesser extent, it also encouraged the formation of Draža Mihailović's Serb-nationalist, monarchist *Chetnik* movement.[58]

Mihailović was a Serbian officer who had defied the Yugoslav High Command's decision to capitulate. With a few loyal officers and troops he withdrew to Ravna Gora in the Serbian Surobor mountains. The *Chetnik* movement grew as more dispersed officers and soldiers joined, mainly owing to the German failure to capture and intern the remnants of the—officially surrendered—Yugoslav army.[59] The *Chetniks* pursued a wait-and-see approach to armed resistance. Initially the Yugoslav government in exile recognized the *Chetniks* as a legitimate resistance organization, and the *Chetniks* tried to gain support from the Allies. However, as early as November 1941, Mihailović attempted to negotiate with the German occupation authorities.[60] Having identified Tito's Partisans as the *Chetniks'* main opponent, Mihailović asked the Germans for weapons and ammunition for their anti-Partisan operations. Throughout the war the *Chetniks* continued to collaborate both with the Italian and German occupation forces and even the anti-Serb *Ustasha*.[61] In 1943 *Chetnik* collaboration with the Axis powers caused Britain to stop all support to the *Chetniks* and to focus their aid efforts on Tito's Partisans, despite its communist orientation.[62] British policy was hence markedly different in Yugoslavia from in Greece, where its efforts had followed a fiercely anti-communist stance.

Tito's Partisan movement emerged only after the start of Operation 'Barbarossa'. On 4 July 1941 the Yugoslav communist party launched an armed uprising. In the weeks after this event, an organizational structure evolved under the leadership of Josip Broz Tito. In contrast to the *Chetniks*, the Partisans chose a strategy of direct action against the occupation forces and the *Ustasha*.[63] In the autumn of 1941, the Partisans attempted to liberate some territory around the town of Užice in western Serbia, but the Germans ejected them. Nevertheless the Partisan movement continued to grow, and over the next two years Tito regularized it to a certain extent, with the result that, in 1943, a semi-regular Partisan army had emerged. While German repressive measures towards the civilian population were severe, local German commanders did not necessarily execute all Partisan captives. Starting in the

[57] In the case of Yugoslavia, the term 'Partisans' with a capital P refers only to the armed resistance movement led by Tito.

[58] Jonathan Gumz, 'German Counterinsurgency Policy in Independent Croatia, 1941–1944', *Historian*, 61/1 (1998), 38. The term 'Chetnik' went back to the Serb irregular formations that had fought against Turkish domination in the nineteenth century and to irregular fighters in the Balkan wars of 1912 and 1915.

[59] Milazzo, *The Chetnik Movement*, 15.

[60] Milazzo, *The Chetnik Movement*, 38.

[61] Milazzo, *The Chetnik Movement*, 90.

[62] Stafford, *Britain and European Resistance*, 167 ff.

[63] Michel, *Shadow War*, 277.

summer of 1942, the Germans negotiated prisoner exchanges with the Partisans. After the battle of Neretva in the spring of 1943, in which Tito's Partisans prevailed against their German, Italian, and *Chetnik* opponents, the Partisan leadership offered the Germans a ceasefire and a collaboration arrangement that went beyond the mere exchange of prisoners. Lieutenant General Edmund Glaise von Horstenau, Hitler's military delegate to the government of the NDH, and Siegfried Kasche, German ambassador to the NDH and *SS-Obergruppenführer*, had a favourable disposition towards the proposal, but their hopes were soon quashed by a negative response from Berlin. Upon hearing of the proposal, Hitler reportedly said: 'We don't negotiate with rebels, we shoot them!'[64]

In spite of the failed agreement between the Partisans and the Germans, the battle of Neretva became the turning point for the Partisans. Tito successfully convinced large parts of the population that the Partisan movement had a certain degree of political legitimacy. This was unsurprising as his local competitors (Ante Pavelić, Draža Mihailović, the German and the Italian occupation forces) offered less attractive political options. By the end of the war, the Partisans were the dominant military force in Yugoslavia. They eventually were the only resistance movement that succeeded in liberating their own country with their own forces.[65]

1943: Italy

In Italy resistance to fascism was largely dormant during the Mussolini years. The first signs of resistance ensued in 1943 with a wave of factory strikes in which the Partido Comunista Italiano (PCI) played a central organizational role.[66] Italy's situation changed dramatically when Allied forces landed in Sicily on 9 and 10 July 1943 and pushed northwards into mainland Italy up to Naples. On 25 July the Italian King Victor Emmanuel III imposed Mussolini's resignation and installed Marshal Pietro Badoglio as the new head of the Italian government. Badoglio secretly tried to negotiate an armistice with the Allies while at the same time declaring that Italy would continue to fight alongside Germany. Anticipating an Italian 'betrayal', German forces occupied the northern part of the country in early September 1943 and subsequently pushed southwards past Naples. The king and Badoglio fled to Brindisi, while the German occupation authorities reinstated Mussolini as head of the government of the Repubblica Sociale Italiana (RSI).

64 Klaus Schmider, *Partisanenkrieg in Yugoslawien 1941–1944* (Hamburg: E. S. Mittler & Sohn, 2002), 250.

65 Schmider, *Partisanenkrieg in Yugoslawien*, 337 ff.

66 David Travis, 'Communism and Resistance in Italy, 1943–8', in Judt (ed.), *Resistance and Revolution*, 86.

This change in Italy's situation gave a major impetus to the development of armed resistance. Resistance emerged along party lines: the PCI formed the Garibaldi brigades, the Socialist party created the Matteoti formations, and the Action Party established the Justice and Liberty formations. Alongside these major organizations, some minor Catholic and monarchist formations developed as well. Although individual irregular fighters in Italy may not initially have been motivated by political considerations, armed resistance soon became politicized.[67] As in other occupied countries, in Italy there was a conflict among the partisan organizations over whether to choose a wait-and-see approach or whether to engage in direct action. The communist Garibaldi brigades again opted for direct action.[68]

The partisan movement grew rapidly at the beginning of 1944, mainly as a result of the RSI's introduction of conscription and of Nazi Germany's launch of a forced labour scheme in Italy. Partisans succeeded in establishing 'free zones' in various parts of the RSI. However, most of these were destroyed in the autumn of 1944. Amid this crisis, in November 1944 a proclamation signed by the British general Harold Alexander was broadcast by the Allies, announcing the immediate end of the Allied offensive in Italy and the withdrawal of all Allied support for the partisans. This decision was informed by the Allies' view of armed resistance organizations as auxiliaries to conventional forces.[69] According to this perspective, when no conventional attack was imminent, support for irregular fighters would be meaningless. However, as the Allies realized that the Italian partisan movement had become a force in its own right, they overturned the decision, and the partisan movement grew exponentially during the last few months of the war. Its activities culminated in several uprisings and insurrections in April 1945 and hence contributed, if largely symbolically, to the liberation of Italy.

German Repression of Partisan Warfare and Armed Resistance

The general assumption about Nazi Germany's occupation policy, both in academia and in popular historiography, is that German repression was much more severe in the east than in the occupied territories in western Europe.[70] The academic debate discussed several causal factors for this, in particular the interlinked notions of the *Lebensraum* concept, racism, and Nazi ideology.

[67] Gustavo Corni, 'Italy', in Moore (ed.), *Resistance in Western Europe*, 165.
[68] Corni, 'Italy', 167 ff.
[69] See the section 'Armed Resistance and the Revival of the Irregular as Auxiliary'.
[70] Cf. Mazower, *Hitler's Empire*, 139.

The debate oscillated between the 'rational' strategic aims of Nazi Germany's *Lebensraum* policy and the 'irrational' ideological underpinnings of the war in the east. When explaining the alleged qualitative differences between the western and the eastern front, Mark Mazower, for instance, attributed central importance to Hitler's statements about the war with the Soviet Union as a '"war of annihilation"' and a clash '"between two ideologies"', as opposed to Germany's war against its western European neighbours, which was supposedly conducted in a gentlemanly way.[71] According to Jürgen Förster, Hitler's war aims in the war with the Soviet Union were fourfold:

1. procuring colonial territory for German settlers;
2. annihilation of the 'Jewish–Bolshevik' leadership stratum, including its so-called biological root—the millions of Jews in East Central Europe;
3. decimation of the Slav masses and their domination by four German 'viceroys'—the Reich Commissars;
4. exploitation of the economic resources in the conquered countries to safeguard the planned Germanic Empire against any form of blockade.[72]

Racism and anti-bolshevism were regarded as the central pillars in this strategic concept. They determined, in the common historical interpretation, that German behaviour in the east was much more ruthless than in western Europe.

At the same time, racism and anti-bolshevism were perceived as the main link between the leadership's strategic concept and the mindset of those who had to implement it on the ground: the military leadership, the officers, and the troops, plus members of other security organizations such as SS and police. In his seminal work on Nazi Germany's war in the east, Omer Bartov traced the ideological indoctrination of Wehrmacht soldiers:

> During the war in Russia the process of dehumanisation of the enemy was probably more successful than in any other war in modern history—the Russians, Slavs, Jews, Mongols, all had lost any relationship to the human race, and were nothing more than satanic monsters trying in vain to appear human, imposters whose identity had to be exposed and whose existence endangered everything which civilised men held dear.
>
> Thus the Russians and Jews were not to be treated according to any accepted rules of military conduct, as they had lost their right to such a treatment both by their racial and cultural inferiority and by the historical role they had played against Germany.[73]

[71] Mazower, *Hitler's Empire*, 139.

[72] Jürgen Förster, 'The Wehrmacht and the War of Extermination against the Soviet Union', *Yad Vashem Studies*, 14 (1981), 10.

[73] Omer Bartov, *The Eastern Front, 1941–45: German Troops and the Barbarisation of Warfare* (Basingstoke: Palgrave, 2001), 83, 85. For a similar argument on the police force, see Christopher

According to Christian Streit, anti-bolshevism was also the strongest link between committed Nazis and conservative circles, which tended to find the Nazi leadership's virulent anti-Semitism repugnant.[74]

Many historians cited the so-called criminal orders as further proof of the exceptional character of the war with the Soviet Union. The first of these 'criminal orders' was the so-called Barbarossa decree of 13 May 1941. This decree stipulated the suspension of military jurisdiction in the occupied territories in the Soviet Union. Specifically, it ruled that *francs-tireurs* were to be eliminated by the Wehrmacht and other security forces and that reprisals and other collective punishment actions against the civilian population in the occupied territories were permissible. Furthermore, it stated that civilians in the occupied territories were no longer protected under the military jurisdiction of the occupying forces; crimes committed against the civilian population by the occupying forces would not be prosecuted unless the morale of the troops was at stake.[75] The so-called commissar order of 6 June 1941 stipulated the suspension of POW status for political commissars. It declared that 'in the struggle against bolshevism it cannot be expected that the enemy will show respect for the principles of humanity or the laws of war.... The instigators of barbarian Asiatic methods of fighting are the commissars.'[76] Finally the 'Guidelines for the behaviour of the troops in Russia' of 23 May 1941 reiterated the ideological framework of the war against the Soviet Union and listed 'bolshevist agents-provocateurs, francs-tireurs, saboteurs, and Jews' as Nazi Germany's main enemies in the east.[77]

The 'criminal orders', in conjunction with the ideological indoctrination of Wehrmacht officers and soldiers and other occupying forces regarding the inferiority of 'Jews', 'Slavs', and 'Bolsheviks', are regarded as the central transmission belt between the leadership's strategic–ideological outlook on the war in the east and its practical implementation on the ground. The mainstream historiography of the Second World War interprets them as indicators of the exceptional position of the Soviet Union among Nazi Germany's enemies. As such, they serve as the most common causal explanation for the ruthlessness of German repression on the eastern front.

However, the importance of the 'criminal orders' has to be put into perspective. Most of the provisions encapsulated in the criminal orders as they

R. Browning, *Ordinary Men: Reserve Battalion 101 and the Final Solution in Poland* (London: Penguin, 2001).

[74] Christian Streit, *Kein Kameraden: Die Wehrmacht und die sowjetischen Kriegsgefangenen, 1941–1945* (Bonn: Verlag J. H. W. Dietz, 1991), 20.

[75] Streit, *Kein Kameraden*, 41 ff.

[76] Quoted in Streit, *Kein Kameraden*, 48.

[77] Quoted in Streit, *Kein Kameraden*, 49. The so-called *Aussonderungsbefehl* (17 July 1941) is also part of this complex of 'criminal orders'. However, as it applied mainly to Soviet POWs in German POW camps, it was not directly relevant for the conduct of anti-partisan warfare.

applied to the repression of armed resistance in particular had already been in place in occupied Poland. In October 1939, Hitler had ordered a general amnesty for all crimes that German forces had committed against Polish soldiers or civilians.[78] Moreover, the German occupation forces had kept occupied Poland in a legal vacuum between the end of the military administration on 25 October 1939 and the extension of German criminal law to occupied Poland on 6 June 1940.[79]

More importantly, the 'criminal orders' *as far as they concerned the conduct of anti-partisan warfare* were not a breach of the contemporary law of armed conflict.[80] Denying irregular fighters the status of legal combatants and shooting them on the spot had long been common practice. Both the 1907 Hague Convention on the Laws and Customs of War on Land and the 1929 Geneva Convention on Prisoners of War did not prohibit these practices. The 1907 Hague Convention also remained silent on civilian reprisals and collective punishments. The main charge in the framework of the *SS-Einsatzgruppen* trial (1945–58) was thus based not on war crimes, but on the concept of crimes against humanity. These had not been explicitly defined until the Nuremberg Charter of 1945.[81] The provisions on the repression of armed resistance in the 'criminal orders' applying to Operation 'Barbarossa' were thus not manifestly illegal. In contrast, Hitler's so-called night and fog decree of 7 December 1941, which envisaged the secret deportation of those suspected of resisting German occupation, was clearly illegal. However, this applied to *all* territories under German occupation; in fact, most of the victims of the 'night and fog decree' were from the occupied territories in western Europe rather than from the Soviet Union. Thus, what set the 'criminal orders' apart was not so much their manifestly criminal content, but the way in which they intertwined and fused the legal perspective on anti-partisan warfare with racial and political stereotypes: by placing the genocide of the Jews and other political opponents into the framework of anti-partisan warfare, the 'criminal orders' provided a thin legal veneer, which may have sufficed as a justification for those perpetrators who had not been committed Nazis and racists at the start of the war. In fact, however flimsy and opportunistic their statements in defence of their actions

[78] Mazower, *Hitler's Empire*, 70.

[79] Streit, *Keine Kameraden*, 36.

[80] Those aspects of the criminal orders that were illegal pertained to the treatment of Soviet POWs and, in particular, the provision that 'political commissars', in spite of being members of the regular armed forces, were to be denied POW status owing to their political role or their racial identity. However, the eastern theatre of war was not the only territory where illegal orders were given with respect to the treatment of captured enemy soldiers. According to Overmans, the order to execute all Italian soldiers fighting on behalf of the Badoglio government after Italy's surrender was an even more blatant breach of the laws of war than the 'commissar order' (Overmans, 'Die Kriegsgefangenenpolitik des Deutschen Reiches', 812 ff.).

[81] Hilary Earl, *The Nuremberg* SS-Einsatzgruppen *Trial, 1945–1958: Atrocity, Law and History* (Cambridge: Cambridge University Press, 2009), 92.

might have been, some defendants at the *SS-Einsatzgruppen* trial in Nuremberg argued that the executions they had carried out had been lawful, as 'those who were killed had been found guilty of partisan warfare and robbery'.[82]

Some historians, especially those working on German occupation and armed resistance in western Europe, have questioned whether there was actually a qualitative difference between the German occupation policies in the eastern and in the western territories. Peter Lieb, for instance, has argued that 'towards the end of the war the situations in the West and in the East seem to have leveled out to a considerable extent'.[83] Indeed, there are no discernible differences between the German operational and tactical repertoires of anti-partisan warfare in western and eastern occupied territories. Both in the western and eastern territories, German occupation forces attempted to 'mop up' partisans through *Einkesselungsaktionen* (closing in from the periphery to the centre of a circular territory) or, when there were not sufficient troops available for *Einkesselungsaktionen*, through the use of mobile columns. They killed and arrested suspected irregular fighters and their supporters in all territories. Finally, in both the western and the eastern territories they conducted collective punishment actions such as the destruction of entire villages and the taking and killing of hostages as a reprisal for armed resistance or suspected armed resistance. Even the most extreme measures, such as the creation of 'dead zones', were not unique to German anti-partisan warfare in the occupied territories of the Soviet Union.[84] German occupation forces had also established dead zones in both Greece and Yugoslavia.[85] Massacres occurred not only in the eastern, but also in the western and the south-eastern occupied territories.[86]

Not only was the operational and tactical repertoire of the German occupiers similar across occupied Europe; the institutional structures and agents in charge of the repression of armed resistance were also largely identical. German occupation forces in Poland had developed a pattern of occupation

[82] Earl, *The Nuremberg* SS-Einsatzgruppen *Trial*, 144.

[83] Peter Lieb, 'Repercussions of Eastern Front Experiences on Anti-Partisan Warfare in France 1943–1944', *Journal of Strategic Studies*, 31/5 (2008), 799; cf. Galen Last, 'Netherlands', 215.

[84] Shepherd, *War in the Wild East*, 166 ff.

[85] Mazower, *Hitler's Empire*, 497; Klaus Schmider, 'Auf Umwegen zum Vernichtungskrieg? Der Partisanenkrieg in Jugoslawien, 1941–1944', in Rolf-Dieter Müller and Hans-Erich Volkmann (eds), *Die Wehrmacht: Mythos und Realität* (Munich: Oldenbourg, 1999), 907.

[86] The following list of massacres is only illustrative, not conclusive: 4 September 1939 Częstochowa (Poland): several hundred killed; 20–21 October 1941 Kragujevac (Yugoslavia): 2,778 executed; the Babi Yar massacres in 1941 (Soviet Union): up to 150,000 civilians, POWs, and suspected partisans shot; the Komeno massacre on 16 August 1943 (Greece): 317 inhabitants of the village killed; the massacre in Oradour-sur-Glane (France): 647 inhabitants of the village killed; the Ardeatine massacre on 24 March 1944 (Italy): 335 civilians killed in reprisal for a bomb attack on the German occupiers.

akin to standard operating procedures, which tended to be reproduced as Nazi Germany occupied ever-larger parts of Europe.

German terminology depicted irregular fighters in Poland as *francs-tireurs* (Reichsführer-SS Heinrich Himmler) and as *Banden* or *Banditen*.[87] Both the term *francs-tireurs* and the criminalization of irregular fighters as bandits show a clear continuity with previous episodes of irregular fighting and anti-irregular measures. The notion of the *franc-tireur* in particular had entered the German discourse during the Franco-Prussian War.[88] It had also been relevant during the First World War, especially during the German invasion of Belgium and Northern France; associated with the German troops' fears of armed attack by irregulars, the ensuing paranoia of *francs-tireurs* had led to widespread civilian killings during the first months of the First World War.[89]

Poland was not only the stage for establishing a discursive frame for irregular fighters more generally; it was also the context out of which the German occupying forces and the political leadership developed their template on an institutional approach to the problem of irregular fighting. Responsibility for the repression of armed resistance rested mainly with uniformed police battalions and the so-called *Einsatzgruppen*, which included both SS and police forces. The Sicherheitsdienst (SD) prepared lists of putative anti-German elements that were to be targeted as core initiators of armed resistance: 'Polish nationalists, Jews, communists and members of the Catholic clergy, the intelligentsia, and the Polish nobility.'[90] Hans Frank, governor general of occupied Poland, in particular, believed that resistance in Poland was inspired and orchestrated by the elites, and that these needed to be eliminated. He even assumed that the Polish workers and peasants would ally themselves with the German occupiers once they had been 'liberated' from their Polish 'oppressors'.[91]

This pattern of criminalization of armed resistance, along with the close integration of SS and police forces in the repression of partisan warfare, became Nazi Germany's standard institutional approach to occupation across Europe, in the eastern territories, and in the Balkans. Contrary to the common perception that *Einsatzgruppen* operated only in the eastern occupied territories, there were, in fact, also *Einsatzgruppen* in France, Belgium, and the

[87] Edward B. Westermann, *Hitler's Police Battalions: Enforcing Racial War in the East* (Lawrence, KS: University Press of Kansas, 2005), 26 ff.

[88] See Chapter 3. [89] Horne and Kramer, *German Atrocities*.

[90] Westermann, *Hitler's Police Battalions*, 128.

[91] Mazower, *Hitler's Empire*, 90 ff. As far-fetched as this seemed, even to some contemporaries, the idea that the elites are the source of irregular resistance was not completely uncommon: in a similar vein, Union commanders and political leaders in the American Civil War believed that irregular armed resistance in the South had been instigated by the rich plantation aristocracy. See Chapter 3.

Netherlands, Luxembourg, and Italy.[92] However, the criminalization of resistance did not imply a 'juridification' of repression, but rather the suspension of most legal protections for civilians and POWs in the occupied territories. This was sometimes established explicitly for an entire territory, as in the case of Hitler's retrospective amnesty for the security forces in Poland or in the case of the 'Barbarossa decree' or the 'night and fog decree'. In other cases, jurisdictions seemingly overlapped between the occupiers and their puppet regimes: after the entrenchment and capture of Glières in France, for instance, some *maquisards* were shot on the spot by German security forces, while others were tried by Vichy courts martial.[93] More generally, in a climate in which German occupying forces and their auxiliaries did not have to fear legal or disciplinary consequences for committing war crimes, the question of the legal treatment of partisans, suspects, and supporters did not play a large role. Yet, despite this or, rather, because of this de-juridified character of anti-partisan warfare, the German occupying forces regarded themselves as the custodians of 'law and order' in the occupied territories and saw their actions as policing rather than war.[94]

The number of civilians killed in the course of the German anti-partisan operations in the occupied territory of the Soviet Union was certainly larger than in any other occupied country. And yet German occupation policy in the Soviet Union was not unique in its ruthlessness. Rather, there were similarities with two other occupied countries: Yugoslavia and Greece. According to Ben Shepherd, German occupation forces reacted to irregular warfare in Yugoslavia 'with a campaign of hostage-taking and reprisals that was exceptional, even by Nazi standards, in the scale of indiscriminate butchery that it inflicted'.[95] And Mazower argued that, with the start of the German occupation, 'Greece became the arena for a sustained anti-guerrilla campaign, suffering the sort of destruction that had already devastated parts of Bosnia, Serbia and the Ukraine'.[96]

If the difference between German occupation in western Europe, in the east, and the south-east concerned the level of violence and the number of victims rather than the occupiers' mindset and institutional and operational repertoire,[97] how can we explain this difference? Most importantly, the examples

[92] Unfortunately, no systematic study of the history of the *Einsatzgruppen* has been published to date. On the emergence of *Einsatzgruppen* and *Einsatzkommandos* in western Europe, see Carsten Dams and Michael Stolle, *Die Gestapo: Herrschaft und Terror im Dritten Reich* (Munich: Beck, 2009), 151, 198, 207.

[93] Kedward, *In Search of the Maquis*, 137.

[94] Cf. Mazower, *Inside Hitler's Greece*, 160; Jonathan Gumz, 'Wehrmacht Perceptions of Mass Violence in Croatia, 1941–1942', *Historical Journal*, 44/4 (2001), 1031.

[95] Shepherd, *Terror in the Balkans*, 1.

[96] Mazower, *Inside Hitler's Greece*, 155.

[97] A further indication of this is that reprisal shootings were conducted in all occupied territories, but the ratio varied: after the bomb attack of via Rasella in Rome, German forces

of Yugoslavia and Greece undermine the importance of the strategic notion of *Lebensraum* as a factor for the escalation of violence. Neither Yugoslavia nor Greece had ever been the subject of German colonial ambitions. Rather, in both cases Nazi Germany had been sucked into the occupation by Italy's unsuccessful campaigns in the Balkans.

Nor are racism and Nazi ideology entirely convincing factors in the cases of Yugoslavia and Greece. In Nazi Germany's racial hierarchy, the Greeks were regarded as superior to the Slavs.[98] The Serbs did 'rank lower' in this hierarchy. Yet, it is not quite clear whether Serbophobia, if it played a part in the occupation of Yugoslavia, had its roots primarily in Nazi racial ideology or in (Austro-Hungarian) imperial prejudice against the 'Balkan peoples' and the Serbian 'traitors' in particular, stemming from the First World War. In fact, Ben Shepherd showed that Austrian-born officers were more likely to engage in campaigns of ruthless violence in Yugoslavia than their German-born colleagues.[99] Jonathan Gumz's work on Wehrmacht perceptions of *Ustasha* anti-Serb violence in Croatia buttressed this doubt that racism was the central factor for the escalation of Nazi German violence in Yugoslavia. Gumz outlined how Wehrmacht commanders took issue and criticized the *Ustasha*'s racially targeted violence against the Serb minority as acts of senseless barbarity and as counter-productive for their own campaign of 'pacification' in the Balkans.[100] He further argued that although the Wehrmacht's anti-partisan campaigns in Croatia tied in with 'National Socialist ideology that prescribed the brutal treatment of the Slavic *Untermenschen*', Wehrmacht commanders in Croatia did not tend to justify their actions in racist terms. Rather, they perceived their approach to anti-partisan warfare as evolving from a strategic concept which emphasized not only the need to control a given territorial space but also the belief that this control could be achieved by maximizing the use of force.[101] Thus, although there is no doubt that German occupation policy in Yugoslavia was characterized by utmost brutality, it is questionable whether racism was a primary causal factor for this behaviour.

The three cases of extreme escalation in the use of force in anti-partisan operations—in the Soviet Union, Yugoslavia, and Greece—had two factors in common: first, the rapid development of a partisan movement that included large numbers of officers and troops from the regular army who had not been

intended to kill ten Italians for one dead German. In Yugoslavia and Greece, the German occupation forces applied much higher ratios of 1:50 or 1:100; cf. Schmider, 'Auf Umwegen zum Vernichtungskrieg?', 919; Mazower, *Inside Hitler's Greece*, 177.

[98] Cf. Mazower, *Inside Hitler's Greece*, 158 ff., who concedes that the Greeks 'stood well above the Serbs in the Nazi racial hierarchy', but still maintains that German brutality towards Greece was fuelled by racial prejudice against the 'Balkan peoples', under which the Greeks were subsumed.

[99] Shepherd, *Terror in the Balkans*, 253.

[100] Gumz, 'Wehrmacht Perceptions of Mass Violence', 1028 ff.

[101] Gumz, 'Wehrmacht Perceptions of Mass Violence', 1019, 1037.

demobilized or detained and, second, a politically polarized climate in which anti-German partisan warfare was but one component of an overarching local civil war.

All three cases featured a large number of regular officers and troops that served as an initial recruitment pool for the partisan movement or, rather, were forced into irregularity for lack of viable alternatives. As outlined above in the Soviet case, the twin 'push factors' for this development were Stalin's stigmatization of Red Army members who surrendered to the Germans and the bleak and potentially lethal perspective of Germany captivity. In Yugoslavia, only Serb troops remained in German captivity throughout the war, while Croats were released soon after the start of the occupation.[102] However, even in the case of the Serbs, the German occupiers did not make sufficient efforts to track down officers and soldiers who had not been demobilized, at least not before the outbreak of the Partisan revolt.[103] Greek soldiers were also soon released from captivity after Greece had been occupied in its entirety.[104] This created a pool of military professionals whose best chance of survival was often to join the partisan movements. Some of them decided to fight on in irregular formations, because they disagreed with the terms of surrender negotiated by their high command, as was the case with the Serbian *Chetniks*.

Both in Greece and Yugoslavia, former members of the regular armed forces tended to join republican (EDES and EKKA in Greece) or monarchist (*Chetnik* movement in Yugoslavia) groups rather than the communist partisans. In the case of the Soviet Union, former Red Army personnel tended to fight in Soviet partisan formations, but their opponents were not necessarily solely the German occupation forces. Rather, in the western borderlands, they often fought against nationalist partisans such as the Ukrainian UPA. This climate of civil war influenced the occupation forces' perceptions and their measures to suppress partisan war:

> As such, anti-partisan war overlapped with the various civil wars that raged throughout Eastern Europe and in turn the prosecution of the anti-partisan war, though undertaken by common agents in the form of the Wehrmacht, SS and police, became a war that was both governed by certain shared principles on the part of Reich forces and fundamentally shaped by these local civil wars.[105]

[102] Overmans, 'Die Kriegsgefangenenpolitik des Deutschen Reiches', 868.

[103] Milazzo, *The Chetnik Movement*, 15.

[104] Overmans, 'Die Kriegsgefangenenpolitik des Deutschen Reiches', 868. The same was true for soldiers from the Netherlands, Norway, Denmark, and the Flemish part of Belgium. However, these countries were not as politically polarized as Greece, Yugoslavia, or the western border regions of the Soviet Union. Moreover, they either had strong collaborationist governments (Denmark) or governments in exile who collaborated closely with the Allies, as in the cases of the Netherlands, Belgium, and Norway.

[105] Gumz, 'Wehrmacht Perceptions of Mass Violence', 1036; cf. Alfred J. Rieber, 'Civil Wars in the Soviet Union', *Kritika: Explorations in Russian and Eurasian History*, 4/1 (2003), 129–62.

Highlighting the participation of regular officers and soldiers in the partisan movement and local civil wars should not be understood as an attempt to excuse the German escalation of violence in anti-partisan operations. Rather, it aims to show the close interconnectedness between regular war and irregular warfare. Regular soldiers can become irregular fighters depending on the circumstances in which they find themselves, just as they can turn back into regulars at the end of the occupation, as happened in the Soviet Union; or they can become the new regular forces, as in the case of Tito's Partisans in Yugoslavia.

Finally, we have to consider the argument that German military culture could be an explanatory factor for the particularly ruthless repression of armed resistance to German occupation during the Second World War.[106] Shepherd describes this factor more precisely as an 'institutionally conditioned abhorrence of irregular warfare'.[107] Rather than explaining the German escalation of violence in the eastern and south-eastern territories of occupation in particular, this argument would apply to all territories under German occupation, since in all of these cases the same institutions, in particular the Wehrmacht, were present. On the face of it, this argument is convincing, as documents, statements, measures, and institutional structures of the German occupation regimes across Europe clearly reflect an 'abhorrence of irregular warfare'. Yet, it is questionable whether this was a typically German trait. In spite of the unprecedented ruthlessness with which Nazi Germany tried to suppress armed resistance, the German notions of irregular warfare and the appropriate measures to counter it had their roots in a punitive approach to irregular warfare that had emerged before the Second World War. The following order by Lieutenant General Dr Walter Hinghofer, commander of the 342 Infantry Division in Serbia, is illustrative of this continuity:

> Anyone who raises a weapon *or* supports corresponding resistance is an *irregular*. Accordingly, *every member* of a rebellious band against whom the division fights is to be treated as an *irregular*. ... Our opponents should be regarded by us without exception as irregulars ... Only (those) who do not oppose us in *any way* are to be excepted.[108]

[106] Philip W. Blood, *Hitler's Bandit Hunters: The SS and the Nazi Occupation of Europe* (Dulles, VA: Potomac Books, 2006). Cf. Shepherd, *War in the Wild East*, 41 ff.; Shepherd, *Terror in the Balkans*, 25; Mazower, *Inside Hitler's Greece*, 157. This argument can be partly traced back to Isabel Hull, *Absolute Destruction: Military Culture and the Practices of War in Imperial Germany* (Ithaca, NY: Cornell University Press, 2005). Edward Westermann made a similar argument with respect to 'organizational culture' as an explanatory factor for the role played by German police forces in the occupation regimes in the east: cf. Westermann, *Hitler's Police Battalions*, 235.

[107] Shepherd, *Terror in the Balkans*, 246.

[108] Order of 25 September 1941, quoted in Shepherd, *Terror in the Balkans*, 126.

The requirement that civilians had to show absolute loyalty to the occupation regime in order to be exempt from collective punishment actions can be traced back to Lieber's three-tiered classification of 'civilians' in civil war situations.[109] It was not a genuine German invention. The pejorative notion of the irregular and its linkages to a punitive approach towards the civilian population was a discourse that had been typical for western military culture, especially since the French Revolution. It was not simply an issue that had arisen from German militarism.[110]

However, what was unprecedented in previous occupation policies was the extent to which the leadership and the security forces fused the irregular discourse with racist notions of the enemy. If racism was not per se a causal factor for the German 'militarization of the partisan issue'[111] and the escalation of violence, it did play a part in the selection of the German forces' victims. Most importantly, it was the equation of irregular fighters with Jews that made the racist infusion of the irregular discourse particularly dangerous. 'Where there's a Jew there's a partisan, and where there's a partisan there's a Jew' was a concept that the German occupation forces applied not only in the Soviet Union, but also across Europe:[112] in all occupied territories Jews were the main targets for reprisals and other measures intended to repress armed resistance. This discursive fusion of the concept of the irregular with anti-Semitic motives tagged racial stereotypes onto existing anti-irregular prejudices. Anti-irregular prejudices existed more or less in all western armed forces in the mid-twentieth century. However, only when merged with racism and anti-Semitism did the anti-irregular discourse become particularly virulent, turning Wehrmacht soldiers into 'willing executioners' of Nazi Germany's genocide.[113] Hence, although the 'racialization' of the irregular was by no means singular or new,[114] the extent to which Nazi Germany merged anti-partisan warfare with genocide was unique in history.

[109] Cf. Chapter 3.

[110] This argument is undermined further by Shepherd's account of anti-partisan operations in Yugoslavia, which showed that those officers who conducted particularly ruthless anti-partisan operations tended to be Austrian-born rather than German-born: Shepherd, *Terror in the Balkans*, 252.

[111] Lutz Klinkhammer, 'Der Partisanenkrieg der Wehrmacht 1941–1944', in Rolf-Dieter Müller and Hans-Erich Volkmann (eds), *Die Wehrmacht: Mythos und Realität* (Munich: Oldenbourg, 1999), 829.

[112] Quoted in Shepherd, *War in the Wild East*, 89.

[113] See Blood, *Hitler's Bandit Hunters*, for a similar argument. However, Blood argues that the close connection between Nazi anti-irregular warfare and the Holocaust was the result of a specific German tradition of anti-irregular warfare, which he calls 'Bandenbekämpfung' and which he considers as prone to exterminatory designs.

[114] Cf. Chapters 2 and 5.

Armed Resistance and the Revival of the Irregular as the Auxiliary

The concept of the irregular fighter as an auxiliary to the regular armed forces had been the central aspect of the invention of *la guerre irregulière* in the mid-eighteenth century.[115] This interpretation lost a great deal of its relevance during the French Revolution with the emergence of the idea of the *levée en masse* in its double meaning as national conscription, on the one hand, and popular uprising, on the other. However, the irregular arguably never ceased to carry forward some features of his initial auxiliary character. For instance, during the Peninsular War Spanish guerrillas fought partly in conjunction with the British forces under Wellington, and during the American Civil War irregular fighters sometimes acted as scouts for regular formations.[116] Yet these instances of regular–irregular partnership were mostly spontaneous and not underpinned by concepts or arguments at the level of strategic or operational theory.

In the Second World War, this was different. In particular British strategic thinking in the interwar years and during the first years of the war was still strongly influenced by the experience of the First World War; it was characterized by the search for an indirect approach to defeat Nazi Germany: 'Britain's strength, it was believed, lay in her ability to wage economic warfare and to cause economic collapse in Germany.'[117] Alongside air force and navy, engagement with continental European resistance movements was perceived as a means to achieve this. In line with this idea, the creation of the SOE on 22 July 1940 was intended to enable liaison between Britain and European resistance movements. Hugh Dalton, who had lobbied for the establishment of the SOE in his role as Minister of Economic Warfare, described the mission of the SOE as follows:

> We must organise movements in every occupied territory comparable to the Sinn Fein movement in Ireland, to the Chinese guerrillas now operating against Japan, to the Spanish Irregulars who played a notable part in Wellington's campaign or—one might as well admit it—to the organisations which the Nazis themselves have developed so remarkably in almost every country of the world. We must use many different methods, including industrial and military sabotage, labour agitations and strikes, continuous propaganda, terrorist acts against traitors and German leaders, boycotts and riots.[118]

[115] Shepherd, *War in the Wild East*, 89.
[116] Shepherd, *War in the Wild East*, 89; see also Chapter 3.
[117] Stafford, *Britain and European Resistance*, 11.
[118] Quoted in Stafford, *Britain and European Resistance*, 25.

However, the idea behind the creation of the SOE was not quite as radical as Winston Churchill's famous of dictum of 'now to set Europe ablaze' might have indicated.[119] Churchill's enthusiasm for supporting irregular warfare was partly inspired by a romanticized idea of T. E. Lawrence, the 1916–18 Arab revolt, and some limited personal experience of irregular fighting on the Northwestern Frontier.[120] Dalton initially envisaged a mission for the SOE in which it would foment continuous subversion in the occupied territories of continental Europe, mainly by supporting political forces on the left and trade unions.[121] However, there was institutional competition between SOE and the Secret Intelligence Service (SIS) that had been established in 1909. SIS officials viewed the launch of SOE with suspicion, not only because they feared competition over resources and manpower, but more importantly because they worried that SOE's direct action mission could compromise SIS's intelligence-gathering efforts.[122] In addition, despite Churchill's radical rhetoric and Dalton's drastic political plans, the SOE's revolutionary outlook soon gave way to a more conventionally minded mission. With the German invasion of the Soviet Union and the entry of the United States into the war, the strategic picture in Europe changed. Now it was possible to defeat Germany with conventional forces.[123] This had an impact on the SOE's mission: under the header of the 'detonator concept', the SOE was to support the formation of 'secret armies' across Europe. Civil unrest and guerrilla warfare were to be discouraged. Instead the secret armies were to lie low and eventually strike in conjunction with the Allied invasion forces.[124] In short, armed resistance in occupied Europe was envisaged as an auxiliary tool in the framework of Allied strategy.[125]

Most governments in exile, many of which were located in London, welcomed this view of armed resistance as an auxiliary force.[126] On the one hand, they were concerned that a direct action approach could precipitate reprisals

[119] Quoted in Stafford, *Britain and European Resistance*, 26.

[120] Stafford, *Britain and European Resistance*, 206.

[121] Stafford, *Britain and European Resistance*, 29.

[122] Keith Jeffery, *MI6: The History of the Secret Intelligence Service, 1909–1949* (London: Bloomsbury, 2010), 357.

[123] Evan Mawdsley, 'Anti-German Insurgency and Allied Grand Strategy', *Journal of Strategic Studies*, 31/5 (2008), 699.

[124] Mawdsley, 'Anti-German Insurgency', 30.

[125] Britain also successfully used native irregular forces in an auxiliary role in their 1940–1 campaign against Italian East Africa. Abyssinian tribes were closely integrated with the British operations. They fought partly as guerrillas and partly as raiding forces. This regular–irregular partnership had also been inspired by T. E. Lawrence and the Arab revolt; cf. Dawn M. Miller, '"Raising the Tribes": British Policy in Italian East Africa, 1938–41', *Journal of Strategic Studies*, 22/1 (1999), 96–123.

[126] However, the relationship between SOE, governments in exile, and resistance movements on the ground was often fraught with difficulties that cannot be covered in detail here, but see M. R. D. Foot, *SOE: The Special Operations Executive, 1940–1946* (London: Pimlico, 1999), 246 ff.

against civilians in their countries. On the other hand, too much activism among the local resistance movements could also jeopardize the political legitimacy of the governments in exile and hinder their return after the liberation.[127] For armed resistance organizations in continental Europe this meant that they were under pressure to conform to this auxiliary role or else risk losing Allied military support. To some groups, the auxiliary role came naturally, in particular to those that were composed of former regular officers and troops, whose conventional mindset meant that they could easily accept a subordinate role.[128] However, those resistance organizations that had the ambition of self-liberation were often disillusioned by the fact that the Allies supported them only as long as they kept to their auxiliary function or as long as the strategic situation continued to render their contributions useful to the Allies.[129]

While Britain had thus assigned an auxiliary role to irregular fighters, the military leadership in the Soviet Union continued to debate about whether partisans ought to serve as auxiliaries to Red Army operations or whether they signified a true popular uprising. This debate had its roots in Red Army thinking and doctrine on irregular warfare going back to the 1920s and 1930s. The founding father of this theoretical and doctrinal tradition was Mikhael Frunze. He had argued that in the event of war the Red Army had to compensate for its technological inferiority towards its main potential opponents by using partisan warfare.[130] This tradition partly harked back to the Civil War period.[131] Frunze had regarded partisan warfare as 'the second means of struggle', supplementing and supporting the regular Red Army operations.[132]

As indicated above, the Soviet leadership had broken with this tradition of irregular warfare theory and doctrine in the second half of the 1930s. However, soon after the German invasion, it began to make efforts to control the partisan war that had started in the German-occupied territories. The changing institutional responsibilities regarding partisan warfare reflected the Soviet leadership's ambivalence towards this issue. First, the NKVD–NKGB (Narodnyy Komissariat Vnutrennikh Del–Narodnij Komissariat Gossudarstwennoi Besopasnosti; People's Commissariat for Internal Affairs/People's

[127] Stafford, *Britain and European Resistance*, 33.

[128] The AS *maquis* in Haute-Savoie who conducted the operation at Glières is a good example of a group with a conventional mindset.

[129] This was painfully experienced by the AK during the Warsaw uprising. It also threatened to happen to the Italian partisans in the autumn of 1944.

[130] Maslov, 'Concerning the Role of Partisan Warfare', 886.

[131] Eberhard Müller, 'Autonome Bewegungen des Volkskriegs in Sowjetrußland nach der Revolution von 1917', in Gerhard Schulze (ed.), *Partisanen und Volkskrieg: Zur Revolutionierung des Krieges im 20. Jahrhundert* (Göttingen: Vandehoeck & Ruprecht, 1985), 36–56.

[132] Maslov, 'Concerning the Role of Partisan Warfare', 886.

Commissariat for State Security) was in charge of the partisans, then the Party and the army shared responsibility, and finally in 1942 Stalin established the partisan Central Staff, under the leadership of General Panteleimon Ponomarenko. At the same time, Stalin appointed Marshal Kliment Voroshilov as supreme commander of the partisan forces.

Subsequently, a clash of ideas about the nature of partisan warfare ensued between Ponomarenko and Voroshilov. Whereas the latter saw partisans as auxiliary forces to the Red Army and hence demanded that they ought to be under Red Army command, the former regarded partisans as an expression of the people's war:

> The partisan movement is a people's movement; it is constructed on a different basis than the Red Army.... The people's masses are included in the active partisan struggle not according to orders, but according to the call of the Party, and as a result of much political work, conducted in the rear by underground party organizations and the partisan detachments. Enrollment in the partisan detachments is conducted only among citizens who have voluntarily decided to become partisans.[133]

The clash between Voroshilov and Ponomarenko reflected the Soviet leadership's dilemma with respect to partisan warfare: on the one hand, it was anxious to safeguard the political reliability of the partisan movement; on the other, it was also aware that too rigid control might decrease the partisans' effectiveness and cut the movement off from the mass basis and mass support that it had gained by 1943.[134] In the case of the Soviet Union, the tension between the concepts of the partisan as the auxiliary and of the partisan as the expression of a popular uprising were never fully resolved.

Nazi Germany also used a multitude of auxiliaries and native *Hilfswillige* (*Hiwis*) in the occupied territories. Beginning in 1940, the Waffen-SS started a recruitment programme for *Volksdeutsche* (ethnic German minorities) in the occupied territories.[135] Because of their native cultural knowledge and linguistic skills, these units were particularly well suited for anti-partisan warfare. A prime example of such an ethnic German SS division was the 'Prince Eugen' division, which was created at the beginning of 1942 in Yugoslavia. Its main task was anti-partisan warfare, and it conducted itself in a particularly brutal way.[136] But the Waffen-SS not only recruited ethnic Germans. It also raised native units in a number of allied or occupied countries, such as Hungary,

[133] Ponomarenko in a note to Stalin in January 1943, quoted in Slepyan, *Stalin's Guerrillas*, 163.

[134] Slepyan, *Stalin's Guerrillas*, 105, 186 ff.

[135] Valdis O. Lumans, *Himmler's Auxiliaries: The Volksdeutsche Mittelstelle and the German National Minorities of Europe, 1933–1945* (Chapel Hill, NC: University of North Carolina Press, 1993), 213.

[136] Cf. Shepherd, *Terror in the Balkans*, 218.

Romania, France, Belgium, the Baltic states, and the Ukraine. Ethnic minorities in the occupied territories were often the preferred targets of recruitment into the Waffen-SS, as the German occupation forces hoped to exploit ethno-political rifts in this way. In 1943, the Waffen-SS formed the 'Handschar' division composed of Bosnian Muslims and the 'Galicia' division in western Ukraine, whose members were mainly Ukrainian nationalists.[137]

Nazi Germany also recruited volunteers in the Soviet Union, in particular from their POW camps. However, recruits from the pre-1939 Soviet territory were not integrated into SS divisions, but formed the so-called *Osttruppen* under Wehrmacht command. As was the case with the native SS divisions, the recruits for the *Osttruppen* stemmed mainly from disaffected minorities in the Soviet Union such as the Cossacks and Tatars.[138] Besides the *Osttruppen*, there was the so-called Vlasov Army, a quasi-regular formation raised by Major General Andrey Vlasov as a Russian nationalist army of Russian POWs that was supposed to fight against the Red Army as an independent German ally.[139]

But there were also a number of less formal collaboration arrangements in the occupied territories. In the Baltic states local paramilitary organizations and nationalist partisans participated in German pogroms against the Jewish population.[140] In Belarus, as elsewhere, the German occupation authorities also recruited native Belarusians into the local auxiliary police forces (*Schutzmannschaften*).[141]

Despite the Germans' wide-ranging and varied use of ethnic Germans, ethnic minorities, or local natives in auxiliary roles, these arrangements were never tied to a proper strategic idea about how to combine regular and irregular warfare. On the face of it, it seemed as though this question was never pertinent, as the German authorities had a clear preference for integrating their local recruits into existing 'regular' formations such as the Waffen-SS and the Wehrmacht. Yet the lines between war-fighting, anti-partisan operations, policing, and genocide were often deliberately blurred in German thinking and practice. Frequently SS, Wehrmacht, police, and native auxiliaries formed joint mobile anti-partisan detachments that resembled the 'irregularized' responses common in anti-irregular conflicts.[142] However, there was never unity among the political and military leadership over the precise role and mission of local auxiliaries: Hitler himself was deeply suspicious about

[137] Mazower, *Hitler's Empire*, 457 ff.
[138] Rieber, 'Civil Wars in the Soviet Union', 148.
[139] Rieber, 'Civil Wars in the Soviet Union', 155.
[140] Rieber, 'Civil Wars in the Soviet Union', 143.
[141] Chiari, *Alltag hinter der Front*, 162.
[142] Cf., e.g., the Union's anti-guerrilla units in the American Civil War; see Chapter 3.

their reliability.[143] For Heinrich Himmler, head of the SS, the increase of SS manpower through local recruits was a means to augment his power and influence within the labyrinthine Nazi state bureaucracy. For the Wehrmacht and local occupation authorities, local recruits were mostly a pragmatic solution to the endemic problem of manpower shortages.

Inversions: Discipline, Legitimacy, and the *levée en masse*

Irregular warfare inverted established practices of military organization, social structures, and patterns of political legitimacy. The topic of discipline is a recurrent feature of the irregular discourse. It also played a role with respect to irregular warfare during the Second World War. Ill discipline was an endemic problem on both the occupying forces' and the armed resistance side. Regarding the occupying forces, Mazower, for instance, described how Wehrmacht officers and troops in Greece engaged in military actions with the sole purpose of private gain:

> The efforts of the military authorities to preserve a distinction between two forms of violence—between an acceptable, disciplined use of force for the sake of the common good and public order on the one hand, and on the other, a reprehensible, wild violence, which aimed at private gain—had broken down.[144]

Native auxiliaries often contributed to this situation of lawlessness and plunder. In Belarus, for instance, local police recruits abused their role in order to settle private scores or to rob the civilian population—a behaviour that was tolerated and even supported by the German authorities.[145] However, undisciplined behaviour led to the paradoxical situation that the occupying forces saw their mission as upholding law and order, while at the same time their actions made a mockery of this aim and were indeed proof of the opposite.

Discipline was also a problem on the side of the partisans. A special challenge lay in upholding the ambiguous difference between requisitioning goods from the local population, as sanctioned by a partisan commander, and looting. Partisan commanders were keenly aware of this challenge. Often there were severe punishments for looting, but 'for partisans provided with even fewer comforts than their compatriots at the front looting was difficult to resist'.[146] In the Soviet discourse, this kind of undisciplined, autonomous, and politically and ideologically deviant behaviour was labelled

[143] Mazower, *Hitler's Empire*, 460.
[144] Mazower, *Inside Hitler's Greece*, 218.
[145] Chiari, *Alltag hinter der Front*, 184.
[146] Hill, *The War behind the Eastern Front*, 145.

'partizanchina'.[147] It was the epitome of the Soviet leadership's fears about the partisan movement.

Partisan tactics were often just as ruthless as those of the occupiers. Partisans were notorious for conducting murderous campaigns against villages and collaborators who opposed them.[148] Infighting among partisan groups was also a major problem. In many cases political polarization underpinned such infighting. In the western borderlands of the Soviet Union various nationalist partisan forces fought against both the German occupation forces and the Soviet partisans.[149] In both Greece and Yugoslavia communist-oriented partisans also fought against their republican or monarchist counterparts.

Yet, despite this pull towards ill discipline, infighting, and civil war, the armed resistance movements also featured strong contrasting trends towards discipline, authority structures, and quasi-legality. Many partisan formations implemented military training for their recruits. In Major Stepan Stebelski's Ukrainian UPA unit, for instance, 'discipline was rigorously followed' and training included both political and historical lessons, marching exercises, and instructions about communications techniques.[150] Aris Velouchiotis, the founder of the Greek ELAS, was renowned as 'a strict disciplinarian' and meted out severe punishments for offences such as stealing and looting among his men.[151] Marshal Tito insisted on a strictly hierarchical command structure within his Partisan forces.[152]

Discipline was often fostered and facilitated by a charismatic style of leadership within the irregular formations. Tuvia Bielski, head of a group of Jewish partisans in the Soviet Union, for instance, was described by his followers as having a magnetic attraction and as conveying a sense of security and determination that instilled unconditional trust and respect in the members of his group.[153] However, beside charismatic leadership as a source of discipline, accounts of partisan warfare also emphasize the value of self-discipline as a specific virtue of the irregular fighter. As partisans had to act independently of fixed command structures, they often had to operate autonomously, and this required self-discipline.[154] Both charismatic leadership and self-discipline were depicted as conceptual opposites of the type of discipline that prevailed in regular military formations, where the model of authority is more akin to Weber's legal–rational form of authority and discipline is synonymous with

[147] Slepyan, *Stalin's Guerrillas*, 136.

[148] Slepyan, *Stalin's Guerrillas*, 82; Hill, *The War behind the Eastern Front*, 145 ff.

[149] Rieber, 'Civil Wars in the Soviet Union', 150; Statiev, 'Was *Smuglianka* a Lunatic', 762.

[150] Larysa Zariczniak, 'Major Stepan Stebelski of the Ukrainian Insurgent Army: Examples of Insurgency Leadership and Tactics', *Small Wars & Insurgencies*, 22/3 (2011), 441.

[151] Mazower, *Inside Hitler's Greece*, 125 ff.

[152] Michel, *Shadow War*, p. 277.

[153] Slepyan, *Stalin's Guerrillas*, 231.

[154] Slepyan, *Stalin's Guerrillas*, 167.

obedience. Hence, irregular formations inverted the regular military forces' concept of ideal authority structures and discipline.

Moreover, the partisan movement featured the inversion of social roles and social structures. In the Soviet partisan movement, for instance, a reversal of social status often occurred within the groups: those who had held a superior position in the pre-war Soviet society, such as 'urban dwellers, white collar workers, and members of the intelligentsia', often faced a hard life in the partisan gangs, as they lacked the practical knowledge and skills necessary for survival in the forests and swamps. Their more apt fellow combatants from a rural background regarded the urban members of the movement as '"useless" people'.[155]

Some scholars claimed that this social inversion also applied to gender roles, where it occurred in the form of partial emancipation rather than complete inversion. Henri Michel, for instance, outlined that women participated at all levels and in various roles in the resistance movements across Europe and argued that 'the part women played in the resistance was undoubtedly an important factor in their post-war social advance and emancipation'.[156] However, Slepyan contended that women who participated in the Soviet partisan movement remained socially marginalized within the partisan groups, although by the end of the war they accounted for almost 10 per cent of the overall number of partisans.[157] However, even if women in fighting roles never gained full emancipation within their own groups, the German occupying forces saw in them the epitome of the partisan's perfidious character, attacking them in the disguise of female innocence and weakness. The German occupation forces denounced female partisans as 'Flintenweiber' (women with rifles) and had orders to 'deal with them ruthlessly'.[158]

Inversion also occurred in the fields of legality and legitimacy. Across the occupied territories, armed resistance movements saw themselves as 'the judge, jury and executioners of popular justice'.[159] In Greece, EAM/ELAS 'was like a mirror image of the official state administration'.[160] It provided local jurisdictional structures by running village courts and exerted political control in the villages and areas under its control.[161] In rural France, local *maquis* groups often took on quasi-police functions in order to protect the local population from 'false' *maquisards* who were trying to steal from the population under the guise of resistance. The *maquis* also issued '*maquis*

[155] Slepyan, *Stalin's Guerrillas*, 256. [156] Michel, *Shadow War*, 190.
[157] Slepyan, *Stalin's Guerrillas*, 195.
[158] Klinkhammer, 'Der Partisanenkrieg der Wehrmacht', 834 ff.
[159] Klinkhammer, 'Der Partisanenkrieg der Wehrmacht', 154.
[160] Mazower, *Inside Hitler's Greece*, 130. [161] Mazower, *Inside Hitler's Greece*, 271 ff.

vouchers' to the local population in return for requisitioned goods, thereby staking their claims for both legitimate authority and ultimate success (as reimbursement would be possible only after the liberation of France).[162] In the discourse of armed resistance, occupiers and resisters had thus switched roles: the occupiers and their local collaborators were no longer seen as the custodians of law and order. Instead, irregular fighters claimed that role for themselves: 'It was the *Milice* [a paramilitary auxiliary force created by the Vichy government] which was increasingly accused of being the real terrorists, a sign that the *maquis* assertions that real justice lay outside the law was [*sic*] growing in acceptability.'[163]

Also the understanding of military sacrifice and the conditions for political representation were inverted. Especially towards the end of Nazi occupation, many armed resistance movements across Europe hoped and sometimes demanded to have 'earned' a say in the political reconstruction of their countries. This, in effect, amounted to a reinterpretation of the concept of the *levée en masse*. The historical situation that had initially led to the marginalization of the irregular fighter had been the French Revolution and the subsequent *levée en masse* of 1792. The political and discursive agenda in this situation had been to differentiate the 'good', government-inspired *levée* from the 'bad' *levée* in the sense of a popular uprising. In this historical precedent, the revolution had given effect to the *levée*.[164] In the case of many armed resistance movements in the Second World War, the sequence of revolution and *levée* was inverted: the act of defending the fatherland preceded the revolution that the resisters, in particular the communist resistance groups in western Europe, intended. In other words, when 'the people' took up arms, it was not because they formally already had a stake in the political fate of their country; rather, they claimed to have earned that stake by taking up arms and by risking their lives. Thus the meaning of the *levée* as a popular uprising gained renewed relevance and legitimacy as opposed to the notion of the *levée* as a universal national draft.

This inversion of the concept of the *levée en masse* was nowhere more visible than in the attempted national uprisings, which were both attempts at self-liberation and expressions of a group's or movement's claim to political representation. The most tragic case without doubt was the Warsaw uprising, which lasted from July to October 1944 and was ultimately

[162] Kedward, *In Search of the Maquis*, 158. [163] Kedward, *In Search of the Maquis*, 101.

[164] In Daniel Moran's words, the *levée* of 1792–3 was depicted 'as a spontaneous, voluntary expression of the French people's ideals and enthusiasm, to which a revolutionary regime had merely given effect' (Daniel Moran, 'Introduction: The Legend of the *Levée en masse*', in Daniel Moran and Arthur Waldron (eds), *The People in Arms: Military Myth and National Mobilization since the French Revolution* (Cambridge: Cambridge University Press, 2003), 2.

crushed by German forces. In spite of the huge sacrifices made by the Polish resistance and Polish civilians—it is estimated that 60,000 people were killed in the uprising, and Warsaw was completely destroyed—it could neither liberate the city nor prevent Poland from becoming part of the Soviet sphere of influence.[165] Thus, the *levée*-as-popular-uprising had definite limits when it did not tie in with the view of armed resistance as an auxiliary to the Allied forces.[166]

The communist emphasis on direct action as the preferred method of resistance across Europe followed a similar rationale. With respect to communist resistance in France, M. R. D. Foot stressed the more sinister side of this approach:

> Useful as this policy of terrorism was in making the Germans jumpy, it undoubtedly—indeed deliberately—attracted severe reprisals, usually wreaked on the neighbourhood where the killing had taken place and not on the men who had done the job. This did not worry the communists, who believed that they were thus 'precipitating a revolutionary situation', a jargon term carrying conviction to them alone.[167]

Yet, it must not be forgotten that the communist activists in the resistance movements also risked their own life and not only that of innocent bystanders. It was precisely this risk and potential sacrifice that they regarded as an act of staking their claim in the political future of their countries.[168]

Even in the Soviet Union, where the partisans had fought on behalf of the existing regime under Stalin that ultimately survived the war, there was a conflict over the appreciation of the contribution and the sacrifice made by the partisans. Many partisans felt betrayed by the official Soviet discourse after the war that emphasized the role of Stalin and the Party in achieving the Soviet victory. The autonomy and emancipation from political control that many partisans had experienced during the war were reversed after the partisans were no longer needed, and 'the leadership worked hard to reestablish and maintain control over its population and its discourse'.[169] Hence the reinterpretation of the *levée en masse* in the sense of a popular uprising that had animated many resistance movements during the war hardly prevailed anywhere after the war.

[165] Norman Davies, *Rising '44: The Battle for Warsaw* (London: Penguin, 2005).

[166] Yugoslavia was the reverse case: despite their dislike for communist resistance movements, the Allies found it strategically useful to support Tito's Partisans.

[167] M. R. D. Foot, *SOE in France* (London: Frank Cass, 2006), 385

[168] Cf. Kedward, *In Search of the Maquis*, 203.

[169] Slepyan, Stalin's Guerrillas, 271.

Conclusion

The preceding discussion of irregular fighters during the Second World War left many issues untouched. Given the comprehensiveness of the Second World War and the multitude of actors and countries involved, it was necessary to be selective. The aim was to focus on those aspects of armed resistance and irregular fighting that would prove to be relevant for the genealogy of the concept of the irregular and for its further development in the post-war era.

The first objective of this chapter was to locate armed resistance and partisan warfare during the Second World War within the trajectory of the concept of the irregular fighter. This contextualization is by no means self-evident. The historiography of the Second World War often placed armed resistance to Nazi occupation and, in particular, Nazi Germany's attempts to repress it in the context of the Holocaust. Given the close factual links between German anti-partisan warfare and the extermination of Europe's Jewish population, this is both understandable and appropriate. Yet, owing to the singularity of the Holocaust, this perspective had the side effect of obscuring the continuities between the pre-existing marginalization of the irregular in the western military discourse and the German response to irregular armed resistance. The emphasis on the illegality of this response, which was reinforced by the post-war Nuremberg trials, concealed the extent to which German orders on conduct towards partisan warfare and the 'civilian population' in the occupied territories at least had their roots in the contemporary legal framework. As shown in the second section of this chapter, some orders even echoed Lieber's stipulations on the punishment of 'civilians' disloyal to the occupying forces in civil wars. Such a contextualization does not relativize the Holocaust. However, it is a stark illustration of the genocidal potential of the marginalization of the irregular: if a very generalized notion of the irregular is sufficiently fused with racist stereotypes and a bellicose narrative that links it to an imagined all-out war between different races, the consequences are likely to be catastrophical.

This contextualization is also valuable because it enables us to trace the trajectory of the concept of the irregular in the twentieth century. As we have seen, during the Second World War the irregular was, on the one hand, interpreted as the auxiliary of the conventional armed forces. This was an interpretation especially put forward by Britain and the SOE and many governments in exile, but also by parts of the Soviet leadership, and even by some resistance groups themselves. On the other hand, the self-perception of many groups within the armed resistance was that of a popular uprising, a proper people's war. These two interpretations coexisted uneasily and sometimes clashed, with disastrous consequences for the irregular resistance movements, as was the case in the Warsaw uprising.

The clash between these two interpretations was palpable at the post-war negotiations over the Geneva Conventions. The most prominent indication of this was that the issue of irregular fighters was not tackled head-on in the Geneva Conventions. The requirements for legal belligerency as set out in the Third Geneva Convention on Prisoners of War remained largely identical to those of the 1907 Hague Convention.[170] The main innovation in the Geneva Conventions, then, was the extensive protection of civilians in occupied territory, which dominates the Fourth Geneva Convention, the civilian convention: civilians were not to be deported en masse from occupied territory, the destruction of civilian property was forbidden unless absolutely necessary for military operations, occupying forces were not allowed to draft civilians into forced labour schemes, and, crucially, civilian reprisals were explicitly prohibited.[171] Yet, while the Fourth Convention did include far-reaching protections for civilians in occupied territory and was intended to curb the excesses of the suppression of armed resistance and anti-partisan warfare, it did not spell out conditions for lawful acts of resistance against the occupier. The French delegation to the Geneva Conference, which included many former members of the resistance, had lobbied for such provisions, but the US and the UK delegations had no interest. In the end, France had to realize that it had by now switched sides: it was no longer occupied, but part of the Allied occupation forces in Germany.[172]

However, just as the Hague Convention had introduced some vague and minimal standards for military conduct, regardless of the opponent that a state was facing, in common article 3 the Geneva Conventions introduced minimum standards of behaviour applicable in 'armed conflict not of an international character'.[173] Common article 3 prohibited inhumane treatment, mutilation, murder and torture, the taking of hostages, and extra-legal executions and punishments. Common article 3 was a compromise solution, as Europe's main colonial powers, Britain and France, were wary of any development that could curtail their use of military force in the colonies.[174] As the next chapter will show, even though common article 3 contained only

[170] Cf. Annex to the Hague Convention IV, articles 1–3; and Geneva Convention III, article 4, printed in Adam Roberts and Richard Guelff (eds), *Documents on the Laws of War* (Oxford: Oxford University Press, 2000), 73, 245 ff.

[171] Geneva Convention IV, article 33, printed in Roberts and Guelff (eds), *Documents*, 49–78; Annex to the Hague Convention IV, articles 1–3; Geneva Convention III, article 4, printed in Roberts and Guelff (eds), *Documents*, 312 ff.

[172] Cf. Karma Nabulsi, *Traditions of War: Occupation, Resistance, and the Law* (Oxford: Oxford University Press, 1999), 14; Best, *War and Law since 1945*, 118 ff.

[173] Common article 3 to the Geneva Conventions, printed in Roberts and Guelff (eds), *Documents*, 198 ff.

[174] Best, *War and Law since 1945*, 174.

minimum protections for irregular fighters in colonial wars, it did cause worries in London and Paris during the wars in Malaya, Kenya, and Algeria. The introduction of common article 3 as well as the inclusion of the Martens clause in the 1907 Hague Convention indicates that, beneath the trend towards the exclusion of irregular fighters in the codified law of armed conflict, there was a weak but visible counter-current that was inclusive towards all sorts of fighters. However, this more inclusive strand in legal thinking would become more central only with the codification of the Additional Protocols to the Geneva Conventions in 1977.

5

Wars in the Colonies

Orientalism and the Social Production of Colonial Subjects

This chapter breaks with the timeline followed so far and addresses wars in the colonies of European states from the end of the eighteenth century to the second half of the twentieth century. Thus it establishes a parallel narrative to the one presented in the previous three chapters. At the same time, it tries to trace links between warfare in the centre and warfare in the peripheries of the international system. Irregular warfare is often equated with war in the peripheries and with historic instances of colonial war in particular. The previous three chapters have shown that this view is not accurate—irregular warfare was a common occurrence in Europe and North America. What remains to be shown is how the interaction between the peripheries and the centre affected both the practice of irregular and anti-irregular warfare and the political and legal conceptualization of the irregular.

The chapter focuses on two themes in particular. First it discusses if and to what extent colonial warfare was less restrained and more violent than wars in Europe and North America. There is no clear-cut, yes-or-no answer to this question. On the one hand, colonial warfare often took up specific, anti-irregular 'fashions' originating from the centre—that is, the metropolitan power in question. Specific practices and measures travelled back and forth between the centre and the peripheries, which should not be surprising given that militaries did so as well, hence providing an institutional memory in which these practices could be stored. On the other hand, some measures taken within the framework of colonial wars did not travel back to the centre, presumably because most policymakers and military officers deemed them too extreme. First among these were concepts and measures that aimed at exterminating the opponent. German occupation policies in the Second World

War are the only exception in this respect, in that they did aim at extermination. But, even in this case, it is difficult to draw a clear line from the camps in German South West Africa to Auschwitz. The question is what led to the implementation of extreme measures in the colonies. Was it racism? Or was it a specific legal perspective on the colonies, often referred to as the 'standard of civilization', that justified the denial of all legal protections for colonial populations? The argument presented in this chapter is that, while such general factors did play a certain role, it is more useful to analyse the extreme violence in these conflicts in the context of other political, economic, and situative factors.

The second theme of this chapter is the impact of the Second World War on irregular warfare in the colonies. As discussed in the previous chapter, the Second World War saw a revival of the concept of irregular fighters as auxiliary forces. Of course, in the colonies auxiliary fighters had never ceased to play a central role. They were one of the building blocks of colonialism itself.[1] But resistance fighters in the territories occupied by the Axis powers saw themselves mainly as liberation movements. In Asia, this led partly to a direct political and personal continuity between anti-Japanese resistance and anti-colonial liberation movements. But the moral appraisal of irregular groups fighting against oppression had changed more generally. The 1955 Bandung Conference denounced both Nazism and colonialism as equally linked to a system of oppression. While the legal conditions for lawful belligerency had not changed with the drafting of the Geneva Conventions in 1949, it was now far more difficult for colonial governments to denounce irregular fighters as illegal rebels. In many wars of decolonization, irregular fighters were hence decried as 'communist terrorists'. Alternatively, their motivations were pathologized as a mix of mental illness and religious primitivism, as happened with the Mau Mau fighters in Kenya. Irregular movements such as the Front Libération Nationale (FLN) in Algeria fought back by condemning colonial repression measures as 'Gestapo methods'. The terms of the moral discourse on irregular fighters had changed, and this would affect the legal development after the wars of decolonization, in particular the drafting of the Additional Protocols to the Geneva Conventions in 1977.

The wars on which the analysis in this chapter is focusing are the Haitian Revolution (1791–1804), the Anglo-Zulu War of 1879, the Boer War (1899–1902), the war in German South West Africa (1904–7), the Maji Maji uprising in German East Africa (1905–7), the Malaya Emergency (1948–60), the Mau Mau uprising in Kenya (1952–60), and the war in Algeria (1954–62).

[1] Karl Hack with Tobias Rettig, 'Imperial Systems of Power, Colonial Forces and the Making of Modern Southeast Asia', in Tobias Rettig and Karl Hack (eds), *Colonial Armies in Southeast Asia* (London: Routledge, 2009), 3–36.

The decision on which instances of colonial warfare to analyse was a difficult one. The wars selected here were chosen, first, because they provide a—somewhat uneven—historical parallel to the wars analysed in the previous three chapters. This allows for a reconstruction of the links between irregular warfare in the centre and in the peripheries. Also, the examples chosen include colonial wars fought by Europe's two major colonial powers—Britain and France. Germany plays an important role, because much has been written on German practices in the colonies and their potential impact on Nazi military policies. Finally, the selection covers a broad geographical scope, from Asia through Africa to the Americas. The Anglo-Zulu War stands out, as it was tactically speaking not an instance of 'irregular' warfare. Nonetheless, the British used some of the repressive measures that were usually applied in the context of anti-irregular campaigns. Thus the question is why they deemed the Zulus a 'quasi-irregular' opponent. A final word on the Boer War: the Boer War is not usually counted among the 'classical' type of colonial warfare, inasmuch as it was fought between two 'colonizers', and hence between whites of European origin. Nonetheless, the Boer War was for the most part fought as an irregular war in the periphery of the international system. Its inclusion in the list is particularly helpful as it enables us to analyse the impact of factors such as racism and the so-called standard of civilization on colonial warfare, as those factors should have been absent from the Boer War.

The chapter proceeds in five steps. The next section is devoted to the Haitian Revolution. The second part looks at colonial warfare around 1900, and the third part is focused on the wars of decolonization. The fourth and fifth parts take stock of the analysis with a particular focus on the two central themes outlined above.

1791–1804: The Haitian Revolution

The series of events referred to as the 'Haitian Revolution' were a sequence of uprisings and violent conflicts, the political outcome of which was the declaration of Haitian independence on 1 January 1804. The political and military dynamic of the Haitian Revolution was extremely complex. Independence was neither the predetermined result nor was it the explicit aim of the leaders of the slave rebellion that started in 1791. The French Revolution had no doubt a large impact on the events that unfolded in the French colony of Saint Domingue. However, this impact was by no means a straightforward translation of republican ideals into anti-slavery rebellion. Many of the leaders of the 1791 slave rebellion were royalists, not republicans. They had heard a rumour according to which the French king had granted all slaves in Saint

Domingue (later Haiti) three free days per week to work for themselves and assumed that the local settlers were withholding that privilege from the slaves.[2]

The slave rebellion of 1791 sparked in turn the mobilization of Saint Domingue's *gens de couleur* (people of colour who had either been freed or had never been enslaved) population. Some of the *gens de couleur* initially fought alongside the slaves, though not for the same political aims: they wanted the same political rights as white French settlers. On 4 April 1792 the French National Assembly finally granted equal rights to free citizens of colour. With this decision, the tactical alliance between the slaves and the *gens de couleur* came to an end: free citizens of colour in Saint Domingue had no interest in the abolition of slavery. On the contrary, many of them were slave-holding property-owners, especially in the south of the colony.[3]

The relationship between slaves and *gens de couleur* demonstrates that the Haitian Revolution entailed a complex intersection of race conflicts, class conflicts, and conflicts of political ideologies. In addition, the violent conflicts in Saint Domingue overlapped and intersected with the French Revolutionary Wars and their resulting conflicts among European forces in the West Indies. By 1793, France was at war with Britain and Spain in Europe. Both countries also had political and strategic interests in the colonial territories in the Caribbean, including Saint Domingue. Spain recruited auxiliaries from among the slave rebels living in exile on the Spanish side of the island, Santo Domingo, to fight against French rule in Saint Domingue. Among these rebels was Toussaint L'Ouverture, who would become the leading figure of the Haitian Revolution. He was himself a freed slave at the time of the slave rebellion. Also the British were involved in Saint-Domingue, supporting royalist planters in the fight against the new influx of republican representatives from France. In order to counter the growing influence and impending take-over of Saint Domingue by Britain and Spain, in February 1793, the newly arrived French Civil Commissioner for Saint Domingue, Léger-Félicité Sonthonax, girondist and abolitionist, offered freedom to those slaves who joined the fight for the republic against England and Spain.

Sonthonax then pushed his abolitionist programme further. On 29 August 1793, he declared the emancipation of slaves in the northern Province of Saint Domingue. On 4 February 1794, the French National Convention declared the abolition of slavery in all French colonies. This convinced Toussaint L'Ouverture that his political agenda could be realized in collaboration with the

[2] David Patrick Geggus, *Haitian Revolutionary Studies* (Bloomington, IN: Indiana University Press, 2002), 12, 104.

[3] Carolyn E. Fick, *The Making of Haiti: The Saint Domingue Revolution from Below* (Knoxville, TN: University of Tennessee Press, 1990), 153.

French republican forces. He switched sides. His army had by then become the strongest force in the colony.[4] Only André Rigaud's army of *gens de couleur*, which dominated in the south and the west of the colony, came close to L'Ouverture's military weight and manpower.

For the three subsequent years, Saint Domingue was relatively stable owing to the well-balanced triangle between Sonthonax, L'Ouverture, and Rigaud. In the beginning of 1797, with Sonthonax's backing, L'Ouverture was appointed commander-in-chief and governor of the colony. Curiously, in August 1797, L'Ouverture forced Sonthonax to leave Saint Domingue. The reasons for this expulsion are unclear, but L'Ouverture might have feared that Sonthonax's radical politics were about to cause a conflict with the Directory in Paris.[5]

On 28 March 1798, General Gabriel Hédouville arrived in Saint Domingue as the representative of the French government. His mission was to re-establish French control over the colony. Hédouville attempted to achieve this by playing off L'Ouverture against Rigaud, thereby using the underlying political conflicts between the *gens de couleur* and the ex-slaves. These conflicts pertained to issues of both race and class. Hédouville's intrigues proved successful, although possibly not in the way that he had anticipated: in 1799 civil war broke out between the ex-slaves and the *gens de couleur*, ending in July 1800 when Rigaud fled the colony.

At the same time, there were new developments in France's approach to its colonies. Napoleon Bonaparte's new constitution called into question the emancipation of slaves that had been granted six years before. 'Special laws' were to govern the colonies. In December 1801, a French expedition led by Napoleon's brother-in-law, Charles Victor Emmanuel Leclerc, landed in Saint Domingue. Its aim was, again, to re-establish metropolitan control over the colony. Although Leclerc claimed that the aim of the expedition was not to re-establish slavery, L'Ouverture did not believe him and started preparing for war. On 7 June 1801, L'Ouverture was arrested and deported with his family. However, this did not defuse the situation. To make matters worse, in May 1802 Napoleon publicly declared the restoration of slavery in France's Caribbean colonies. This only hardened the resistance of the ex-slaves. On 18 November 1803, the French forces were defeated in the Battle of Vertière in the north of the island near Cap Haïtien, and, on 1 January 1804, L'Ouverture's successor, Jean-Jacques Dessalines, declared independence. Saint Domingue was rechristened as Haiti.

[4] Fick, *Making of Haiti*, 191.
[5] C. L. R. James, *The Black Jacobins* (London: Penguin, 1980), 152 ff.

1791–1794: Irregular Fighting and Repression

The slave uprising of 1791 was entirely self-constituted. Its leaders, Jean-François, Georges Biassou, and Boukman Dutty, were slaves, although many of them had held privileged positions within the slave system before the revolt. The early European historiography of the Haitian Revolution tended to emphasize that vodou was a driving force behind the uprising and that the ceremony at Bois Caïman on the night of 14 August 1791 was important for the start of the rebellion. However, this interpretation was subsequently criticized as an instance of 'barbarisme imaginaire'.[6] What was more important than the religious impetus for the early success of the revolt in the north of the colony and its subsequent spread to other parts of Saint Domingue was a sound political organization.[7]

Initial violence was unsurprisingly directed against white planters and their property. Slave rebels killed the inhabitants of plantations, among them also loyal slaves, burned down their property, and looted the plantations.[8] Against French forces, they used guerrilla tactics and retreated into the hills when attacked by a superior force. Some slave rebels, such as Jeannot, were notorious for the atrocities they committed, which were a 'terrifying mirror image of the tortures long inflicted by white masters upon slaves'.[9] Jeannot's superior in the rebel group, Jean-François, eventually arrested and killed him for his brutality. 'Most of the insurgents, however, were committed to the "complete destruction of the whites".'[10] While revenge may be part of the explanation for the extreme levels of violence that the slaves used, brutality was also endemic among the *gens de couleur* insurgents under Rigaud.

Many accounts of the Haitian Revolution attributed the use of guerrilla tactics to the African, in particular Kongo, origin of many of the insurgent slaves.[11] It is plausible that the experience of fighting in the civil wars of the kingdom of the Kongo was useful for the slave rebels, and contemporaries also acknowledged this.[12] However, to label these tactics as 'African' and to argue that the French were not used to them, is not accurate. Not only did the French have experience with irregular fighting both at home and abroad; their initial allies, local militias composed of *gens de couleur*, also had experience in

[6] Geggus, *Haitian Revolutionary Studies*, 82.

[7] Laurent Dubois, *Avengers of the New World: The Story of the Haitian Revolution* (Cambridge, MA: Harvard University Press, 2004), 101.

[8] Fick, *Making of Haiti*, 109 ff.

[9] Dubois, *Avengers of the New World*, 123.

[10] Dubois, *Avengers of the New World*, 124.

[11] Dubois, *Avengers of the New World*, 109; Robin Blackburn, 'Haiti, Slavery, and the Age of Democratic Revolution', *William and Mary Quarterly*, 63/4 (2006), 664.

[12] Dubois, *Avengers of the New World*, 109.

fighting in the *maréchaussée*, which had been used to hunt down fugitive slaves, so-called maroons, before the revolt.[13]

In areas fully occupied by the rebel forces, the slaves created self-governed communities, often referred to as 'kingdoms'. The most well known of these was the 'Kingdom of Platons', formed in early 1792. The labels 'king' and 'queen' were often used for leaders or persons of high social standing. They did not necessarily indicate that the rebel slaves were royalists. Rather, these labels went back to African traditions of kingship. Those traditions were not necessarily conservative, but could be combined with revolutionary aims.[14]

French counter-measures displayed equal levels of brutality as those used by the slaves and the *gens de couleur*. French troops were initially thinly spread and poorly trained, struggling with the climate and with tropical diseases.[15] Six thousand additional troops arrived in 1792, but even this reinforcement did not bring a decisive edge. The white planters, sometimes aided by loyal slaves, increasingly organized themselves into militia forces. The French forces and the planter militia declared the slave insurgency into a 'war of brigandage' in which no rules applied. 'No-quarter' orders were given, even towards black women, children, and the elderly. Atrocities such as cutting off ears and heads were common.[16] As Laurent Dubois stated: 'It had become an exterminating war.' Extermination, however, was not a new mindset among the French colonists. Rather the whole system of slavery before the revolt had allowed for, if not actually intended, the extermination of slaves, insofar as their working conditions and the provision of food and shelter had often been so inadequate that widespread death was the result.[17] It had simply been cheaper to purchase new slaves than to enhance the living conditions of the existing ones.

Auxiliaries in the Haitian Revolution

As noted above, the Haitian Revolution was by no means a simple confrontation between white colonizers and black slave rebels. Rather, the revolution's cleavage lines had multiple fissures and fragmentations. This was mainly reflected in the use of auxiliaries.

Most notorious in this respect was the recruitment of black slaves into the ranks of the 'confederates', meaning the *gens de couleur* uprising in 1791, led by Rigaud and allied with white royalists. The 'confederates' promised

[13] Cf. Chapter 2; see also Dubois, *Avengers of the New World*, 110.

[14] John K. Thornton, '"I am the Subject of the King of Congo": African Political Ideology and the Haitian Revolution', *Journal of World History*, 4/2 (1993), 181–214.

[15] Dubois, *Avengers of the New World*, 40 ff.

[16] Dubois, *Avengers of the New World*, 116, 140; Fick, *Making of Haiti*, 132.

[17] Dubois, *Avengers of the New World*, 116, 40 ff.

freedom after the war to slaves who would fight as auxiliaries. With this promise, they were able to recruit approximately 300 slaves, who were called *Les Suisses*, in reference to the Swiss mercenaries who served the French king. When the *gens de couleur* reached an agreement with the French republicans over citizenship rights, the *Suisses* became a sticking point. For the French republicans, freedom for the slaves was out of the question, as it would undermine the institution of slavery. Hence the *gens de couleur* had to renege on their promise. The compromise agreement was that the *Suisses* were to be deported, but eventually—and it is unclear whether this was the initial plan or simply the effect of the coincidence of adverse circumstances—they were killed on the ship that was supposed to deport them or left to die on board, 'with the exception of about twenty, who were spared and sent back by the whites to the West to convince the blacks that the mulattoes had betrayed them'.[18]

The *Suisses* incident is a perfect illustration of the way in which the recruitment of auxiliaries served the dual purpose of augmenting the colonists' manpower, which was almost always insufficient for keeping colonial rebellions in check, while at the same time engaging in divide-and-rule politics that were supposed to prevent rebellions from happening in the first place. Not surprisingly, the massacre of the *Suisses* left a deep rift in the relations between the slaves and the *gens de couleur*, which was one of the main reasons for the outbreak of the civil war in 1799. As we shall see, there are many parallels to this situation in other instances of colonial war.

However, auxiliaries could also enter into more autonomous arrangements, which usually lowered the risk of exploitation and manipulation. The rise to power of Toussaint L'Ouverture and his army and its services to the Spanish as *tropas auxiliares* eventually forced the French to emancipate all slaves for fear of losing the colony to Spain.[19] This demonstrates that auxiliaries could have an important impact on a political situation that was hard to control for the colonial governments. But, more importantly, it also shows how closely related the two manifestations of irregular fighters—the insurgent and the auxiliary—were in colonial warfare.

The Haitian War of Independence

As noted above, in 1794 L'Ouverture changed sides from being a Spanish auxiliary to joining the French republican forces. He was subsequently made

[18] Fick, *Making of Haiti*, 125; Fick uses the term 'mulattoes' to refer to the *gens de couleur*, which is not entirely accurate, as there were many blacks among them—which underlines the point that this was not mainly a race conflict.

[19] Geggus, *Haitian Revolutionary Studies*, 175.

Brigadier General of the French army. Technically speaking, he and his army ceased to be irregulars and became part of the regular French armed forces in Saint Domingue. However, the Leclerc expedition challenged their status: Leclerc demanded that L'Ouverture and his officers cede control of the colony to Leclerc and leave Saint Domingue. L'Ouverture and his men refused, which in Leclerc's eyes made them once again 'partisans' and 'brigands'.[20]

In terms of its tactical approach, faced with the Leclerc expedition, L'Ouverture's army practised once again guerrilla warfare. When faced with imminent defeat, several of its units, among them the one led by Jean-Jacques Dessalines, changed sides and fought alongside Leclerc. This probably had little to do with undisciplined behaviour or with a lack of confidence in the political cause of the war. More important was the prospect of being executed by the French. Leclerc and his successor Rochambeau also staged mass drownings of individuals suspected of aiding L'Ouverture's army.[21] The drownings specifically seem to echo French repressive practices in the war in the Vendée, although it is difficult to trace how these practices had travelled from mainland France to Saint Domingue, since neither Leclerc nor Rochambeau had been deployed in the Vendée.

Extermination was also a potential aim of the Leclerc expedition. On 7 October 1802, Leclerc wrote to Bonaparte: 'Here's my opinion on this country: we must destroy all the blacks of the mountains—men and women—and spare only children under twelve years of age. We must destroy half of those in the plains and must not leave a single colored person in the colony who has worn an epaulette.'[22] The only option to safeguard the colony, according to Leclerc, was to start afresh with slaves newly imported from Africa. For the ex-slaves, the potential confrontation with exterminatory policies, epitomized in the *Suisses* episode, increased the stakes in the conflict and hardened their resistance. Eventually, the threat of extermination was also the decisive factor for independence: if France threatened the ex-slaves' bare existence, then no negotiated settlement could be reached.

Africa around 1900: The Anglo-Zulu War, the Boer War, German South West Africa, and the Maji Maji Uprising

The following section addresses war in colonial Africa around 1900. It starts by offering brief overviews of the four selected conflicts and then turns to a summary analysis of military operations and repressive measures. It identifies

[20] Dubois, *Avengers of the New World*, 254, 283.
[21] Dubois, *Avengers of the New World*, 284, 289.
[22] Dubois, *Avengers of the New World*, 291 ff.

a catalogue of repressive measures that characterized colonial powers' reactions to armed conflict in the colonies.

1879: The Anglo-Zulu War

The source of the conflict that eventually led to the Anglo-Zulu War was a dispute over a strip of territory between Transvaal and Zululand. In July 1878 British High Commissioner Sir Henry Bartle Frere charged a commission with deciding the dispute. The commission subsequently confirmed the award of the land to the Zulu king Cetshwayo. However, Frere ignored the commission's decision, as it contravened his plan of creating a confederation of all South African territories as a British dominion. He communicated an ultimatum to Cetshwayo to cede the claim to the territory in question by January 1879. When Cetshwayo refused, Britain declared war.

Since its formation under Shaka Zulu in the mid-1810s, the Zulu kingdom had become an advanced military power.[23] Shaka had introduced several innovations. He had replaced the traditional *assegai*, the long hurling spear, with a shorter stabbing spear, the *iklwa*. He had also introduced a system of military organization and training that placed a premium on loyalty and discipline. In addition, he had increased the mobility and the tactical prowess of his army, which was capable of conducting complicated manœuvres such as envelopment operations.[24] So the British forces were up against an opponent who did not resemble the stereotype of 'backward' African fighters.

The British commander-in-chief, Baron Chelmsford, invaded Zululand on 11 January with three columns. In spite of the fact that the Zulus had a formidable reputation as a warrior nation, British forces felt confident in their superiority towards an enemy that was armed mainly with *iklwas* and cowhide shields and possessed only few rifles.[25] This sense of superiority was an important factor for what was to develop into 'the most humiliating British defeat since the start of the century'.[26] On 20 January, the main British column had pitched camp at Isandhlwana, but had made no efforts to secure the camp against a potential Zulu attack. Part of the column then went on a scouting mission guided by Chelmsford. On 22 January, the main Zulu force attacked the insufficiently guarded camp and annihilated the British forces and their native auxiliaries. Chelmsford returned to camp to find his men killed and their bodies mutilated with the ritual slash across the abdomen.

[23] Cf. Richard J. Reid, *Warfare in African History* (Cambridge: Cambridge University Press, 2012), 120 ff.

[24] Donald R. Morris, *The Washing of the Spears: The Rise and Fall of the Zulu Nation* (London: Pimlico, 1994), 50 ff.

[25] Thomas Pakenham, *The Scramble for Africa* (London: Weidenfeld & Nicolson, 1991), 67.

[26] Pakenham, *Scramble for Africa*, 70.

Fighting continued after the battle of Isandhlwana, with the British forces unable to gain a decisive victory. On 23 January, Zulu forces attacked the mission station at Rorke's Drift. The British narrowly escaped another defeat and managed to defend the station. Subsequently Zulu forces cut off the right flank column under Colonel Charles Pearson from their supply lines at Eshowe. The left flank was involved in smaller battles. Given their precarious position, the British forces staged a complete retreat from Zululand in early April.[27] In June 1879, after the arrival of reinforcements, they started the second invasion. On 4 July 1879, they finally defeated the Zulu forces at the battle of Ulundi.

1899–1902: The Boer War

Enduring British ambitions to achieve a position of political dominance in South Africa formed the political background to the Boer War. The independent Boer Republics of the Transvaal and the Orange Free State were obstacles to British grand designs in the region. British ambitions became even more pressing with the discovery of gold in 1886 in the Witwatersrand area. Moreover disputes over so-called *Uitlander* rights, the rights of British citizens living in the Transvaal, led to increasing tensions between the British and the Boers.[28] The first attack took place in 1895 with the 'Jameson Raid', a small-scale British invasion of the Transvaal that was meant to spark off an *Uitlander* rebellion. The raid was planned and financed by Cecil Rhodes, who evidently had his own economic motives for interfering with the growing conflict between the British and the Boers. In the years between 1895 and 1899 the political conflict escalated further. In October 1899 the Boer Republics declared war on Britain.

The first phase of the Boer War was fought as a 'regular' war by the conscripted militia of the Boer Republics, on the one hand, and the British army, several of its 'regular', 'irregular',[29] and colonial regiments, and its native auxiliaries, on the other hand. At first the Boer forces achieved tactical victories at the battles of Stormberg, Magersfontein, and Colenso and in the sieges of Ladysmith, Kimberley, and Mafeking. However, by the end of February 1900, British forces had taken over Ladysmith and Kimberley and had prevailed at the battle of Paardeberg. Finally, in May they overcame Mafeking. Subsequently, they invaded the Transvaal, and in June 1900 they captured Pretoria.

[27] Morris, *Washing of the Spears*, 421 ff.

[28] Dennis Judd and Keith Surridge, *The Boer War: A History* (London: I. B. Tauris, 2002), 17 ff.

[29] On the category of 'irregular' units in colonial warfare, see the section 'Repression in Colonial Wars around 1900'.

In March 1900, increasingly forced into the defensive, the Boer leadership decided to adopt irregular tactics instead of continuing to fight in regular formations. The tactics chosen are often described as 'guerrilla warfare', but a more accurate term would be 'raiding warfare', designed to cut off the British from their communications.[30] The British forces under Lord Kitchener reacted to this change in tactics by trying to strengthen their control over both territory and populations by cutting off the Boer commandoes from food, shelter, and mobility. In addition, they conducted offensive sweeps to flush out the remaining Boer commandoes. This approach eventually undermined Boer resistance. A peace treaty was signed on 31 May 1902.

1904–1907: The War with Herero and Nama in German South West Africa

The conflict in German South West Africa started with the revolt of the Herero tribe in January 1904. The main reasons for the revolt were economic hardship and political repression resulting from the increasing colonial domination of the Herero.[31] The Herero conducted mainly raids on white settlers and killed about 100 men, although women and children were spared on the explicit orders of the Herero chief Samuel Maherero. In spite of the arrival of German reinforcements in early 1904, the Herero could not be defeated. Responsibility for this German failure to end the revolt was quickly pinned on Colonel Theodor Leutwein, governor of German South West Africa. As a result, he had to cede control of the German troops to Lieutenant General Lothar von Trotha, who arrived in the colony on 11 June 1904.

As the Herero fighters, together with their families, had retreated to the Waterberg, the German operational plan developed under Leutwein was pinned on a decisive battle in that area. Von Trotha adopted this plan, and on 11 August the battle of Waterberg took place. The plan had envisaged crushing the Herero force with an envelopment manœuvre. The envelopment failed, however, and the Herero broke through the weakest point in the German lines into the Omaheke desert.[32] Von Trotha ordered pursuit, which failed, as German logistics were overstretched. In September 1904, von Trotha gave orders to cordon off the Omaheke desert, thereby denying the Herero and their families any access to food or water. On 2 October 1904, von Trotha

[30] Cf. Judd and Surridge, *Boer War*, 187; Thomas Pakenham, *The Boer War* (New York: Random House, 1979), 387, uses the term 'raiding strategy'.

[31] Susanne Kuß, *Deutsches Militär auf kolonialen Kriegsschauplätzen: Eskalation der Gewalt zu Beginn des 20. Jahrhunderts* (Berlin: Ch. Links, 2010), 81.

[32] For a detailed discussion of the historiography of the battle of Waterberg, see Kuß, *Deutsches Militär*, 88 ff., and Isabel V. Hull, *Absolute Destruction: Military Culture and the Practices of War in Imperial Germany* (Ithaca, NY: Cornell University Press, 2005), 33 ff.

publicized an order that was to become notorious as the epitome of German policies in German South West Africa: in his *Vernichtungsbefehl* (extermination order), von Trotha made explicit his aim to exterminate the Herero nation. After a political outcry in Berlin, Kaiser Wilhelm II rescinded the *Vernichtungsbefehl* in December 1904. As of early 1905 the remaining Herero were detained in camps.

But the situation in German South West Africa was far from being under control. The Herero revolt indirectly sparked a revolt of the second major tribe in the colony, the Nama, located in the southern part of German South West Africa. Starting in September 1904, the Nama conducted raids on German farms and their inhabitants. Although their leader, Hendrik Witbooi, was killed in 1905, they continued a small-scale guerrilla war until 1907. As with the Herero, German forces interned captured Nama in large camps. The camps were characterized by inadequate living conditions, forced labour, and correspondingly high death rates. In 1908, when internment was ended, both the Herero and the Nama population were decimated to a fraction of their pre-1904 size, and their social structures were entirely destroyed.

1905–1908: The Maji Maji Uprising in German East Africa

The Maji Maji revolt was rooted in the increasing assertion of German authority over the colony of German East Africa since the 1890s. German authority smothered both cultural and economic pursuits of the indigenous populations.[33] In early 1905 Maji Maji started as a transethnic religious cult centred around a spiritual belief in an admixture of water, castor oil, and millet seeds called 'Maji' ('water' in Swahili).[34] In July 1905, followers of the cult started attacking German farms, colonial representatives, and garrisons. In the beginning, Maji fighters often attacked in formations, with some success, although they failed to capture the fortified garrison of Mahenge in an attack on 30 August 1905. However, the initial intertribal coordination and the fight in larger formations ended towards the end of 1905; the subsequent fighting mostly took place in the form of ambushes or small raids.[35] In October 1905, the German governor, Gustav Adolf von Götzen, received reinforcements in the form of two Marine cruisers and 1,000 regular infantry plus hired local auxiliaries. Götzen organized his forces into columns, which he ordered to march south to subdue the last remnants of the revolt. On 18 June 1908, the war eventually ended with the capture of the last Maji Maji leader, Mpangire.

[33] Kuß, *Deutsches Militär*, 102 ff. [34] Pakenham, *Scramble for Africa*, 616.
[35] Pakenham, *Scramble for Africa*, 621; Kuß, *Deutsches Militär*, 109.

'African way of war' and Western 'military orientalism'

The four conflicts discussed here have in common that they did not start as an outright campaign of irregular fighting on the part of the indigenous or the settler forces (in the case of the Boer War) against superior colonial forces. This is important to note, as colonial warfare is often equated with irregular warfare. This is an instance of 'military orientalism'[36] that beset not only many European commanders fighting in the colonies, but also parts of the historiography of colonial warfare. It obscures the importance of 'regular' warfare in colonial encounters. The most obvious case in that respect is the Boer War, which in J. F. C. Fuller's retrospective assessment was a 'gentlemen's war'.[37] In the case of the Boer War, the tendency is to overemphasize its 'regular' aspects, as the Boers were white and of European origin; the reverse is true for colonial warfare in which one side constituted native inhabitants of the colonies. But equating African warfare with irregular warfare is inaccurate. Even if the Zulu forces did not wage war in the same way as European armies, they did engage in elaborate campaigns and used a sophisticated military organization.[38] And, while the Herero did attack isolated settler farms, killing and mutilating their male inhabitants, they never adopted full-blown irregular tactics. In fact, according to Isabel Hull, the Herero failed to achieve strategic superiority in the first phase of the conflict precisely because they did not carry out more systematic attacks on the German lines of communication. Instead, they restricted their attacks to small military outposts.[39] In the first stages of the conflict, the Maji Maji took a similar approach, although they increasingly switched to guerrilla tactics towards the end of the war. The Nama rebellion is thus the only case considered here in which indigenous fighters relied solely on irregular tactics.

However, even if African forces often did not engage in irregular tactics, ritualized practices, such as the mutilation of the bodies of killed opponents,[40] led colonial forces to believe that they were dealing with opponents whose violence knew no boundaries. In addition, colonial forces and settlers often amplified the perception of savagery with fabricated 'horror stories', such as

[36] On the notion of 'military orientalism', see Patrick Porter, *Military Orientalism: Eastern War through Western Eyes* (London: Hurst, 2009).

[37] Quoted in Judd and Surridge, *Boer War*, 228.

[38] Reid, *Warfare in African History*, 120 ff.; for a general critique of the equation of African warfare with 'primitive warfare', see also John K. Thornton, *Warfare in Atlantic Africa, 1500–1800* (London: Routledge, 1999).

[39] Hull, *Absolute Destruction*, 11.

[40] Morris, *Washing of the Spears*, 35, described the religious meaning of mutilations for the Bantu tribes, from which the Zulu nation originated: 'A warrior who killed an opponent stood to develop a serious ailment, culminating in madness, unless he took immediate preventive action. The first step in the cleansing process was to rip open the fallen foe's abdomen with a single slash of the assegai, in order to allow his spirit to escape.'

the rumour spread by Germans claiming that the Herero had tortured and killed women and children.[41] What was true, however, was that many of the Europeans' opponents did not take prisoners: the Zulu fighters killed all British and auxiliary forces they encountered at Isandhlwana.[42] Herero, Nama, and Maji Maji fighters killed missionaries and settlers. And, even though the Boers did not kill British prisoners, they executed civilians (mostly *Uitlanders*) whom they suspected of having collaborated with the British.[43] In addition, they routinely killed black British auxiliaries.[44]

Against the background of these partly real and partly imagined deviations from what contemporaries perceived as 'civilized warfare', many European sources justified the ruthless repression of violence in the colonies as a natural response to an alleged 'African way of war'—in other words, as a necessary adaption to local ways of fighting.[45] This notion was underpinned by the fact that colonial authorities relied to a greater or lesser extent on native or colonial auxiliaries. During the Anglo-Zulu War, a third of the British forces consisted of Natal natives. During the Boer War, Britain relied on a wide variety of forces: some were colonial auxiliaries from Canada and Australia, but the British also raised *Uitlander* units among British citizens living in the Boer Republics. Numerically the largest group of auxiliaries were black or mixed-race natives, who numbered 30,000 by the end of the Boer War. However, at the start of the war, the British were hesitant to recruit non-white natives. In line with the perceptions of the 'uncivilized' nature of African warfare outlined above, 'there was concern that the war should be a "white man's war" and that black participation would turn a civilized war into a barbaric struggle'.[46] Nonetheless, the need for more manpower soon weighed more heavily than such reservations.

German colonial authorities also relied on native auxiliaries, although this varied between the different colonies. In German South West Africa the Germans had recruited a small number of Nama and Baster into the *Schutztruppe*. Native auxiliaries were far more important in German East Africa, which was perceived as a less prestigious, non-settler colony and hence did not receive the same level of security forces.[47] However, the auxiliaries deployed in East Africa were mainly recruited from outside the colony, as

[41] Hull, *Absolute Destruction*, 10 ff. [42] Morris, *Washing of the Spears*, 387.
[43] Pakenham, *Boer War*, 480. [44] Judd and Surridge, *Boer War*, 87.
[45] Tanja Bührer, *Die Kaiserliche Schutztruppe für Deutsch-Ostafrika: Koloniale Sicherheitspolitik und transkulturelle Kriegführung, 1885–1918* (Munich: Oldenbourg, 2011), 261 ff. Bührer quotes many contemporary German sources, but unfortunately there is almost no critical distance in her account. Rather, she confirms contemporary perceptions and explains escalating levels of repression as an 'Africanization of violence'.
[46] Judd and Surridge, *Boer War*, 85. [47] Kuß, *Deutsches Militär*, 160 ff.

the authorities mistrusted the loyalty of the local tribes, in particular after the outbreak of the intertribal Maji Maji rebellion.[48]

Some auxiliaries were called 'irregular forces', hence reaffirming the eighteenth-century 'auxiliary roots' of the concept of the irregular. For instance, in the Boer War, 'irregular' units were those that had been 'raised by private individuals or British officers', such as the Imperial Light Horse or Rimington's Guides, which were *Uitlander* units raised by Aubrey Woolls-Sampson and Major M. F. Rimington respectively.[49] German colonial authorities distinguished between Askaris, who were mainly recruited in Sudan and Ethiopia and received regular pay and training, and Ruga-Ruga, who were often recruited on the spot and were hence untrained. The latter were referred to as 'irregulars', meaning that in this case the 'irregular' label was based on the fact that they did not receive regular pay.

Contemporary European sources as well as some historians blamed escalating levels of violence in wars in the colonies on the participation of auxiliaries, even though native fighters were more likely than any other group to become the victim of 'barbaric' methods. Attributing the escalation of violence in warfare in the colonies to the native fighters had the added advantage that Europeans could charge them with the most 'uncivilized' tasks or simply blame them for atrocities.[50] Moreover, colonial authorities also deliberately manipulated local conflicts in accordance with their divide-and rule approach, thereby often fuelling an escalation of violence.[51]

Repression in Colonial Wars around 1900

Repressive measures in wars in the colonies varied according to the local situation. They ranged from operational designs for decisive battles in the Anglo-Zulu War and in German South West Africa to scorched-earth policies in the later stages of the Boer War and in German East Africa during the Maji Maji rebellion. However, it is possible to identify a more or less established toolkit of repressive measures on which European forces relied: executions, massacres, internment, torture, deportation, hostage-taking, fines, and the destruction of property and livelihoods.

Executions of enemy fighters were legally sanctioned during the suppression of the Herero and the Nama revolts in German South West Africa, as well as in German East Africa during the Maji Maji uprising.[52] The legal requirements applying to executions in these conflicts were neither unequivocal nor

[48] Bührer, *Kaiserliche Schutztruppe*, 286. [49] Judd and Surridge, *Boer War*, 83.

[50] See, e.g., Bührer, *Kaiserliche Schutztruppe*, 256.

[51] Kuß, *Deutsches Militär*, 163; cf. Michael Lieven, '"Butchering the Brutes All Over the Place": Total War and Massacre in Zululand, 1879', *History*, 84/276 (1999), 625.

[52] Hull, *Absolute Destruction*, 26, 146; Pakenham, *Scramble for Africa*, 620.

particularly strict, which shows that the boundaries between executions and massacres were often unclear. Executions could be ordered by unit commanders and did not have to be preceded by a court martial, although courts martial did take place occasionally. It is unlikely that executions were restricted to combatants, since the recorded number of executions was very high—for instance, the summary executions of up to 100 persons in German East Africa in one incident.[53] During the Boer War, the question of whether 'rebels', meaning inhabitants of the Cape Colony fighting with the Boers, could be court-martialled and executed was contested.[54] Some executions of Cape Colony rebels did take place—for example, in 1901 Commander Johannes Lötter was executed. Some Boer officers, such as Gideon Scheepers, were executed after courts martial as well, although not for their combatant status, but rather for atrocities committed during the war.[55] Apart from these legally sanctioned executions, there were numerous allegations of Boer prisoners being shot, so that Kitchener eventually felt compelled to deny publicly that he had ordered such shootings.[56]

When it comes to massacres, the case that stands out among the conflicts analysed here is without doubt the extermination of the Herero subsequent to the battle of Waterberg. The physical destruction of the Herero fighters and their families became the explicit aim of German actions.[57] But large-scale massacres also took place in other conflicts. While the Anglo-Zulu War is frequently 'presented as the epitome of chivalric combat' between two distinguished military opponents, massacres did take place during the war.[58] As noted above, Zulus fighters did not take prisoners during the battle of Isandhlwana. What is less frequently discussed in the historiography is that British forces also carried out massacres or, rather, used their native auxiliaries for this work. As Michael Lieven pointed out, there is evidence that British officers ordered the killing of wounded and fleeing Zulu fighters, ostensibly justified by the fact that 'the Zulu gave not quarter and expected none'.[59] Massacres were also common in German East Africa. The fact that German forces sometimes executed large numbers of 'enemy fighters' without trial demonstrates that the lines between executions and massacres were blurred. The shooting of prisoners after they had been tortured for intelligence was also common.[60] During the Boer War, British forces and their auxiliaries did not carry out massacres, but Boer commandoes did routinely kill African auxiliaries and even civilians if they had aided the British. The most notorious incident was

[53] Hull, *Absolute Destruction*, 146.
[54] Pakenham, *Boer War*, 490.
[55] Judd and Surridge, *Boer War*, 235.
[56] Keith Surridge, 'An Example to Be Followed or a Warning to Be Avoided? The British, Boers, and Guerrilla Warfare, 1900–1902', *Small Wars & Insurgencies*, 23/4–5 (2012), 619.
[57] Hull, *Absolute Destruction*, 55.
[58] Lieven, '"Butchering the Brutes"', 615.
[59] Lieven, '"Butchering the Brutes"', 624.
[60] Kuß, *Deutsches Militär*, 116.

the destruction of Liliefontein and the massacre of its inhabitants on 31 January 1902.

Internment and deportations played a role in German South West Africa and in the Boer War, but not in East Africa or the Anglo-Zulu War. While internment in German South West Africa is often regarded as a direct continuation of von Trotha's extermination policy towards the Herero, the role of internment during the Boer War has been subject to historical debates.[61] Internment was a technique in colonial warfare that was not entirely unprecedented: European settlers had used the concentration of the Native American populations in reservations as a means to facilitate their domination of North America. However, the massive use of camps in colonial warfare started around 1900 and was practiced by all European powers.[62] The internment camps for Herero and Nama in German South West Africa were located in Windhuk, Karibib, Okahandja, and on Shark Island. They were characterized by forced labour and minimal food rations. The death rate in these camps was 45 per cent over the duration of the conflict.[63] During the Boer War, the first camps were erected in 1900. The initial intention was to protect the 'hensoppers' from the 'bittereinders'—that is, to protect those Boers who wanted to give up fighting after the British had formally annexed the two Boer Republics from those who wanted to continue the war.[64] However, the internment camps soon became a central part of the anti-irregular strategy: the British plan was to intern 'all men, women and children and natives from the Districts which the enemy's bands persistently occupy'.[65] This comprehensive plan also comprised punitive aspects, insofar as Kitchener stipulated that the camps' inmates should be classified into two groups:

> The women and children brought in . . . should be divided in two categories, viz. 1st: Refugees, and the families of neutrals, non-combatants and surrendered Burghers. 2nd: Those whose husbands, fathers and sons are on commando. The preference in accommodation etc., should[,] of course, be given to the first class. It should be clearly explained to the Burghers in the field that, if they voluntarily surrender, they will be allowed to live with their families in the camps until it is safe for them to return to their homes.[66]

[61] See Kuß, *Deutsches Militär*, 97, for a discussion of German internment policy in German South West Africa.

[62] Ian F. Beckett, *Modern Insurgencies and Counterinsurgencies: Guerrillas and their Opponents since 1750* (London: Routledge, 2001), 36.

[63] Kuß, *Deutsches Militär*, 98 ff.; Hull, *Absolute Destruction*, 70 ff.

[64] Judd and Surridge, *Boer War*, 194.

[65] Kitchener, quoted in Adam Roberts, 'Detainees: Misfits in Peace and War', in Sibylle Scheipers (ed.), *Prisoners in War* (Oxford: Oxford University Press, 2010), 264.

[66] Kitchener, quoted in Roberts, 'Detainees', 264.

This approach bore a strong resemblance to Francis Lieber's three-tier scheme regarding the treatment of what we would today call civilians in the American Civil War. Whether Kitchener was directly influenced by Lieber and the practices of the American Civil War or whether this resemblance was accidental, it shows once again that there were strong parallels between anti-irregular warfare in the centre and in the peripheries of the international system.

Hence, despite the insistence of Iain Smith and Andreas Stucki that 'these were not "punishment" camps for a defeated enemy as was the case for the Herero and Nama in German South West Africa',[67] the camps in South Africa did have punitive aspects. Yet, they were not designed as extermination camps. There was no forced labour in the Boer camps. As Isabel Hull pointed out, while the low rations in the camps in German South West Africa were deliberately aimed at starving the inmates, this was not the case in the Boer camps.[68] The abominable conditions in the Boer camps were usually—and plausibly—attributed to British negligence rather than exterminatory aims. The death rate of the camps for the duration of the war was estimated at 22 per cent and was thus considerably lower than in the camps in German South West Africa, where the death rates stood at 45 per cent.[69] The conditions in the camps were improved after public criticism was sparked in June 1901 by the philanthropist Emily Hobhouse and by the subsequent report of the Fawcett commission in December 1901.[70]

The first deportation in German South West Africa took place in 1904, when 119 Witbooi fighters were deported first to Togo and later to Cameroon. Von Trotha's successor, Friedrich von Lindquist, toyed with the idea of a complete swap of Herero and Nama populations and their respective territories, which soon proved unfeasible. Later, he planned to deport another 300 Witbooi fighters to New Guinea, but this too proved impracticable, owing to the prohibitively high cost.[71] During the Boer War, no deportations occurred other than the detention of Boer POWs in oversees camps. However, Kitchener considered the possibility of threatening commando leaders with banishment and even phantasized about the expulsion of all Boers to the 'Dutch East Indies, Fijis or Madagascar'.[72] But these radical solutions were never realized, and the detained POWs returned to the Boer Republics after the war.

[67] Iain R. Smith and Andreas Stucki, 'The Colonial Development of Concentration Camps (1868–1902)', *Journal of Imperial and Commonwealth History*, 39/3 (2011), 427.

[68] Hull, *Absolute Destruction*, 153. However, others pointed to the fact that German South West Africa was resource-starved at the time of the Herero rebellion and argued that the low food rations were predominantly due to this resource scarcity: Kuß, *Deutsches Militär*, 426.

[69] Cf. Keith and Surridge, *Boer War*, 196; Kuß, *Deutsches Militär*, 98 ff.; Hull, *Absolute Destruction*, 70 ff.

[70] Pakenham, *Boer War*, 515; Judd and Surridge, *Boer War*, 194; but see Smith and Stucki, 'The Colonial Development', 429, who argue that 'incremental reforms in the camps were already underway at the time of the Fawcett Commission's report'.

[71] Kuß, *Deutsches Militär*, 100. [72] Pakenham, *Boer War*, 500.

With the exception of German South West Africa, the destruction of property and livelihoods, fines, and the taking of hostages were a staple of repression in the wars discussed here. During the Anglo-Zulu War, the British engaged in the systematic burning of Zulu kraals and the seizure of Zulu cattle. Initially the seizure of cattle, which was of central economic and cultural importance to the Zulu, aimed at putting pressure on King Cetshwayo to cede the Zulu claim to the disputed territory. However, after the humiliating defeat of the British at Isandhlwana, a negotiated settlement was out of the question, and destruction became the sole aim of British actions.[73] In the Boer War, farm-burning and the destruction of property were a central part of Kitchener's anti-irregular strategy, matched by the erection of the camps.[74] What sets the Boer War apart from the other conflicts in this respect, however, is the fact that the peace agreement that ended the war awarded £3,000,000 to the Boers for reconstruction.[75] German repressive measures in German South West Africa did not include scorched-earth policies. The reason for this was that German South West Africa was a colony in which resources were scarce anyway, and trying to quell native rebellions with such a policy of denial would have had adverse consequences for the settlers and the colonial representatives as well.[76] However, in German East Africa the systematic destruction of the opponents' livelihoods and the confiscation of all moveable property and artefacts was the foundation of the German approach to the suppression of the Maji Maji uprising. Women and children were often taken hostage and were released only once their relatives had submitted themselves to the German authorities. Hostages were frequently forced to work for the German authorities. Low-level rebels were also compelled to do forced labour when captured. Sometimes fines were levied on them as a punishment.[77] Repressive measures in German East Africa were hence more indirect, and the proclaimed aim of these measures was not the extermination of the rebelling tribes, but rather the integration of the native inhabitants into the economic system of the colony. Nonetheless, the death toll in terms of absolute numbers was higher in German East Africa than in German South West Africa.

Wars of Decolonization after 1945: Malaya, Kenya, and Algeria

1948–1960: The Malaya Emergency

The insurgency in Malaya was rooted in the broader dynamic in South East Asia in the late 1940s. The central aspect of this dynamic was the repercussions of

[73] Lieven, '"Butchering the Brutes"', 619.
[74] Pakenham, *Boer War*, 493.
[75] Judd and Surridge, *Boer War*, 292 ff.
[76] Kuß, *Deutsches Militär*, 425 ff.
[77] Kuß, *Deutsches Militär*, 117 ff.

the Second World War in Asia. After the withdrawal of the Japanese forces, Britain re-established control over its South East Asian colonies. The colonial administration disbanded the anti-Japanese militias, which had formed during the Second World War. In Malaya this was the Malayan People's Anti-Japanese Army (MPAJA), which had its political roots in the Malayan Communist Party (MCP), even if not all of the MPAJA's members were communist. Both the MPAJA and the MCP were dominated by the ethnic Chinese part of the Malayan population. Although the MCP had been banned before the Second World War, the British army had trained and equipped the MCP-connected MPAJA during the war.[78] As in the European theatre of war, in particular in Yugoslavia, Britain supported communist irregular forces if this seemed expedient.

The British colonial administration disbanded the MPAJA in December 1945. It honoured Chin Peng, the most senior MPAJA member, in a victory parade held on 6 January 1946 in Singapore. However, in 1948 armed resistance arose anew. The MCP under Chin Peng launched the Malayan National Liberation Army (MNLA). The unrest grew out of increasing economic and political frustration. The Malayans' initial hopes for political reform in early 1946 were disillusioned by the British colonial administration under Mountbatten, which increasingly relied on repressive measures to quell labour protest in Malaya.[79] Violence first erupted in June 1948 in the Kinta valley in the Perak district, as large proportions of Chinese squatters lived in this area. Chinese squatters felt discriminated against by both the British administration and the Malay majority, as they did not possess equal economic and political rights.[80] However, the MNLA itself did not only have Chinese members, but also included some Malays, Indians, and Indonesians. The MNLA carried out attacks on security forces, economic infrastructure, and indigenous loyalists.

Initially the British colonial administration relied mainly on punitive measures to quell the insurgency, but this approach had little success. It then introduced emergency measures that included detention without trial for up to two years, the suspension of due process regulations, the reintroduction of the death penalty, and sweeping police powers.[81] The MNLA's most notorious action was the killing of High Commissioner Sir Henry Gurney on 6 October 1951. However, in 1951 the MNLA had already passed its apex. In 1950 the British colonial administration had started to implement a new approach that

[78] Abu Talib Ahmad, 'The Impact of the Japanese Occupation on Colonial and Anti-Colonial Armies in Southeast Asia', in Tobias Rettig and Karl Hack (eds), *Colonial Armies in Southeast Asia* (London: Routledge, 2009), 225.

[79] Christopher Bayly and Tim Harper, *Forgotten Wars: The End of Britain's Asian Empire* (London: Penguin, 2008), 197 ff.

[80] Bayly and Harper, *Forgotten Wars*, 421.

[81] Bayly and Harper, *Forgotten Wars*, 441.

concentrated on spatial and population control as first envisaged in the Briggs Plan and later refined under High Commissioner Gerald Templer. Squatters were resettled from jungle frontier locations to 'new villages'. This approach proved effective, since they effectively separated the MNLA from the population and thus also from access to resources.[82] In 1957 Malaya declared itself independent under Prime Minister Tunku Abdul Rahman. In 1960 the emergency was declared over.

1952–1960: The Kenya Emergency

Kenya had developed into a settler colony from the late nineteenth century to 1920, when Britain formally declared it a colony. It saw repeated revolts against colonial domination from the late nineteenth century onwards. The Kikuyu tribe, which would later become central in the Mau Mau uprising, undertook several rebellions between 1920 and 1940. The growing unrest that would eventually lead to the Mau Mau rebellion was a result of long-term trends of land expropriation and cultural clashes, mainly over the practice of clitoridectomy.[83] But it would be wrong to interpret the Mau Mau rebellion as a mere resistance movement against colonial oppression. As with most wars in the colonies, anti-colonial violence and civil war were deeply interwoven.[84]

In the initial stages of the Mau Mau rebellion oathing played an important role. Oaths were taken in secret ritual ceremonies.[85] They were mainly instruments for building networks and for securing intelligence and resources for the Mau Mau fighters. However, British officials frequently misunderstood the practice of oathing and reduced it to its ritualistic content, thereby 'presenting Mau Mau as the product of witchcraft and as a psychological condition'.[86] From 1946 to 1948 the Rift Valley had already seen some localized unrest. In 1952, violence against European settlers and African loyalists became widespread. On 20 October 1952, Governor Evelyn Baring declared a state of emergency and introduced emergency powers, including the power to detain suspects without trial. Britain's military response to the rebellion drew upon a range of different forces: the British army and the Royal Air Force (RAF), the King's African Rifles (KAR), consisting of locally recruited colonial forces, and the Kenya Regiment, mostly comprising European settlers, the Kenya Police,

[82] Karl Hack, 'Everyone Lived in Fear: Malaya and the British Way of Counter-Insurgency', *Small Wars & Insurgencies*, 23/4–5 (2012), 683 ff.; Karl Hack, '"Iron Claws of Malaya": The Historiography of the Malayan Emergency', *Journal of Southeast Asian Studies*, 30/1 (1999), 102 ff.

[83] David Anderson, *Histories of the Hanged: Britain's Dirty War in Kenya and the End of Empire* (London: Weidenfeld & Nicolson, 2005), 13 ff.

[84] 'The Mau Mau war is most usefully conceptualised as a helix, with the strands of anti-colonial and civil war violence intertwined,' writes Daniel Branch (*Defeating Mau Mau, Creating Kenya: Counterinsurgency, Civil War, and Decolonization* (Cambridge: Cambridge University Press, 2009), 8).

[85] Anderson, *Histories of the Hanged*, 43 ff.

[86] Branch, *Defeating Mau Mau*, 38.

and the Home Guard, a loyalist militia. The KAR were officered by British officers on secondment; and both KAR and Kenya Regiment were within the British army chain of command. These forces embarked on a campaign of fierce repression against Mau Mau based on mass detention, abuses, and compulsory resettlements. These tactics proved effective, and by 1956 Mau Mau was a spent force. In 1960 emergency measures were lifted, and in December 1963 Kenya declared its independence.

1954–1962: The War in Algeria

Algeria was France's foremost settler colony. In 1834 France had declared it a colony, and in 1848 Algeria was formally assimilated into French territory, although the indigenous population of Algeria did not enjoy the same rights as metropolitan citizens. Colonial law allowed for the dispossession of land of the Muslim population in favour of French settlers. Violent resistance against colonial oppression started during the First World War. Despite French attempts at reforms in the interwar years, resistance grew and was increasingly underpinned by Algerian nationalism. The Second World War altered the political situation of Algeria. After the Allied landing in Operation 'Torch' on 8 November 1942, British and American Forces were present in Algeria, and it became the main base for the Free French forces under Charles de Gaulle. In 1944, de Gaulle extended citizenship rights to Muslims. However, that did not assuage Algerian nationalists.[87]

The historiography of the Algerian war usually indicates 1954 as the start of the war. However, the first outbreak of violence had happened already in May 1945 in the Sétif and in Guelma. Algerian nationalists killed and mutilated 102 Europeans. In response, the French army and French settlers killed between 6,000 and 10,000 Algerian Muslims.[88] More coordinated efforts at resistance began on 1 November 1954 when the FLN and its armed wing, the Armée Libération Nationale (ALN), organized a series of attacks on economic infrastructure, security forces, and European settlers, but also against Muslim loyalists, particularly in the Aurès and in Kabylia.[89] The most notorious attack happened in August 1955 and became known as the Philippeville massacre, in which ALN fighters killed 123 settlers and loyalists, including children and the elderly. Security forces involved in the repression of FLN/ALN violence included the French army, the Foreign Legion, and locally recruited auxiliaries, the *harkis* and the *mokhasnis*. The auxiliary forces grew into an important part of the security forces. In 1958 they numbered 79,000. The French

[87] Martin Evans, *Algeria: France's Undeclared War* (Oxford: Oxford University Press, 2012), 79.
[88] Evans, *Algeria*, 91. [89] Evans, *Algeria*, 127.

approach focused on punitive measures to suppress the insurgency and on population control and resettlement.

In 1958 the French Fourth Republic fell owing to the deteriorating security situation in Algeria. De Gaulle returned to power as prime minister. While French settlers hoped that de Gaulle's return signified enhanced military measures, de Gaulle also embarked on reforms in Algeria and offered an amnesty to ALN fighters who laid down their arms.[90] De Gaulle's agenda rapidly moved away from keeping Algeria within France at all costs and towards containing the security problems posed by the war, which de Gaulle saw predominantly from the perspective of the global strategic situation of the cold war.[91] This triggered increasing hostility towards de Gaulle among the settlers and parts of the army. 1960 and 1961 saw the creation of the Organisation Armée Secrète (OAS), an underground terrorist organization launched by parts of the army leadership. It carried out bombings in Algeria as well as metropolitan France with the aim of obstructing the peace process. In April 1961, a number of army generals attempted a coup against de Gaulle, but this failed. In May 1961, a ceasefire was negotiated in Evian. Despite the Evian accords, however, violence actually increased in 1962 owing to a wave of OAS bombings and an increasing number of attacks on settlers. As a result, many settlers as well as *harkis* left Algeria in the summer of 1962. Algerian independence was declared in July 1962.

Violence and Repression in Wars of Decolonization

A common trait of the three wars analysed here is that anti-colonial warfare and civil war were intertwined: insurgent violence was aimed at representatives of the colonial system, meaning at the colonial authorities and settlers alike, as well as at loyalist populations within the country. Violence against loyalists often became the most notorious aspect of the war. On the one hand, it was politically exploited in propaganda efforts by the colonial authorities; on the other, it often had an indisputably polarizing impact on local communities. One example of this is the Lari massacre on 26 March 1953, in which Mau Mau fighters attacked and killed dozens of loyalists and their families. The Lari massacre led to a public outcry and severely damaged the reputation of Mau Mau within Kikuyu society.[92] The same pattern could be observed in Malaya: 'MNLA exactions of "justice" by summary shooting, executions in front of families, or tossing grenades into the shops of alleged collaborators'

[90] Alistair Horne, *A Savage War of Peace: Algeria 1954–1962* (New York: New York Review of Books, 2006), 303 ff.

[91] Evans, *Algeria*, 267.

[92] Branch, *Defeating Mau Mau*, 59; Anderson, *Histories of the Hanged*, 77 ff.

were widespread and undermined the MNLA's standing among the population.[93] Once the British colonial authorities had recognized that the dynamic of inter-communal violence was damaging to the MNLA, they became more successful in exploiting it for their own purposes.[94] In Algeria, in April 1955 alone the ALN killed eighty-eight native loyalists. The most notorious ALN attack on loyalists was the Mélouza massacre: on 29 May 1957 the ALN killed and mutilated 300 inhabitants of the village of Mélouza as a 'reprisal' for the village's anti-ALN stance.[95] Inter-communal violence reached such dimensions during the Algerian war that, according to Matthew Evans, it was Algerians who 'suffered the most from FLN attacks'.[96]

However, as already noted, it is important to put the dynamics of inter-communal violence and civil war into the perspective of the colonial system. The civil war aspects of the wars of decolonization were fuelled by the various ways in which colonial administrations had coopted local elites into their system of power before the war and continued to do so during the war. In fact, the civil war dimension received a continued impetus from the binary dynamic of colonial repression and selective cooptation during the war. According to Daniel Branch, this could be observed during the Kenya Emergency, where the selectively repressive measures transformed local cleavage lines and alliances, reinforcing the civil war dynamic.[97] In Algeria, the recruitment of *harkis* auxiliaries dramatically increased the number of potential loyalist targets and led to a pattern of inter-communal violence that persisted even after the end of the war.[98]

Regarding the reaction of colonial powers to anti-colonial violence and civil war, four observations are central. First, the established 'catalogue' of repressive measures that had emerged around 1900 persisted. Standard practices included executions, massacres, deportation, internment, torture, fines, and the destruction of livelihood and property. Second, colonial repression did not reach the same levels as in colonial warfare around 1900. That is, in none of the three cases did it reach genocidal dimensions. Third, there are important differences between the three wars analysed here: in Kenya and Algeria the levels of indiscriminate violence were much higher than in Malaya.

Executions were a prominent tool in all three wars. French security forces carried out summary executions of alleged ALN fighters in Algeria. In Kenya, British security forces and their native auxiliaries shot numerous Mau Mau suspects under the pretext that they had been 'attempting to escape'. These shootings became so endemic that General Erskine issued explicit orders to stop the practice, which, however, seemed to have little impact on

93 Hack, '"Iron Claws of Malaya"', 103. 94 Hack, 'Everyone Lived in Fear', 679.
95 Evans, *Algeria*, 127, 216 ff. 96 Evans, *Algeria*, 337.
97 Branch, *Defeating Mau Mau*, 9, 15. 98 Evans, *Algeria*, 325 ff.

the ground.[99] Executions were also carried out after trials, such as the trials of the perpetrators of the Lari massacre. Technically, these executions were legally sanctioned. However, the investigation and trial following the Lari massacre were 'deeply flawed'.[100] Under emergency regulations in Malaya, even minor offences such as possession of arms carried the death penalty.[101]

Even though executions continued to represent a central aspect of the catalogue of repressive measures, the legal framework for their justification had changed profoundly. The respective colonial powers treated the wars in Malaya, Kenya, and Algeria as domestic affairs, to be dealt with by imposing emergency regulations. They argued that the Geneva Conventions did not apply in domestic affairs nor to the states' right to develop regulations for dealing with domestic emergencies. This had implications for both the terminology they used to describe and label their opponents and for the way in which they used repressive measures. In the first years of the war in Malaya, the British authorities called the MNLA fighters 'bandits'. In 1952 the official label changed to 'CTs', meaning 'communist terrorists'. The reason for avoiding the terms 'rebel' or 'insurgent' was that the British estates and mines had no insurance protection against 'rebellion or insurrection'; they were protected only against 'riot and civil commotion'.[102] The British authorities also used the term 'terrorists' in Kenya. Again the background was their legal strategy: the colonial administration was determined not to apply common article 3 of the Geneva Conventions, which actually applied to 'armed conflict not of an international character' and expressly prohibited 'the carrying out of executions without previous judgment pronounced by a regularly constituted court'.[103] The French authorities also used the term 'terrorist' to refer to ALN fighters and insisted that the war was a 'domestic affair'.[104] The French, as the British, were wary that the war could come within the remit of common article 3, not only because this would mean that the scope of available repressive measures would be limited, but also because they feared that this could confer a degree of political legitimacy on their opponents.[105]

[99] Huw Bennett, *Fighting the Mau Mau: The British Army and Counter-Insurgency in the Kenya Emergency* (Cambridge: Cambridge University Press, 2012), 217.

[100] Anderson, *Histories of the Hanged*, 176.

[101] Bayly and Harper, *Forgotten Wars*, 441.

[102] Bayly and Harper, *Forgotten Wars*, 436.

[103] Geneva Conventions I–IV, common article 3, printed in Adam Roberts and Richard Guelff (eds), *Documents on the Laws of War* (Oxford: Oxford University Press, 2000); on the British legal strategy, see Bennett, *Fighting the Mau Mau*, 66 ff.

[104] Raphaëlle Branche, 'The French in Algeria: Can there be Prisoners of War in a "Domestic" Operation?', in Sibylle Scheipers (ed.), *Prisoners in War* (Oxford: Oxford University Press, 2010), 175.

[105] Branche, 'The French in Algeria', 175, although common article 3 explicitly states: 'The application of the preceding provisions shall not affect the legal status of the Parties to the conflict' (printed in Roberts and Guelff (eds), *Documents*, 198).

Emergency regulations hence gave a certain veneer of legality to the repressive measures carried out in the wars of decolonization. It is plausible to assume that colonial authorities' stronger consideration of legal issues during the wars of decolonization in comparison to the peak of their colonial power around 1900 had a mitigating effect on the levels of violence.[106] Still, emergency regulations were indeed only a thin legal façade. Even if executions were prohibited, there were ways to justify them, either by covering them up as instances of shooting persons 'attempting to escape' or by holding sham trials. Finally, in Algeria in spite of the French insistence that this was not a war, rules of engagement harked back to the established practice in anti-irregular warfare of executing 'rebels' without trial. According to a decree of 1 July 1955: 'Every rebel using weapons or seen carrying weapons or in the middle of carrying out an exaction must be immediately shot . . . every suspect who tries to escape must the opened fire upon.'[107]

Similar paradoxes are visible with respect to massacres. Although colonial authorities insisted that the wars in Malaya, Kenya, and Algeria were domestic 'emergencies', they used deadly violence in a massive, punitive, and exemplary way. The war in Algeria had essentially started in 1945 with the massacres in Sétif and Guelma. Throughout the war, the French security forces continued to kill alleged FLN suspects and their supporters during punitive raids.[108] Although levels of indiscriminate violence in the war in Malaya were low in comparison to Kenya and Algeria, massacres did happen in the first phase of the conflict.[109] The most notorious instance happened on 12 December 1948, when a company of 2nd Scots Guards killed twenty-four male workers on a rubber plantation at Batang Kali.[110] One notorious example in Kenya was the Chuka massacre: in June 1953, members of the KAR B company killed twenty Mau Mau suspects during a search and capture mission. The commanding officer, Major Gerald Selby Griffiths, was later court-martialled for his role in the massacre, but the trial failed to address the deeper and more systematic roots of the massacre: the existence of a 'keen competitive spirit regarding terrorist kills' among the security forces.[111] Hence security forces, in particular in Kenya and Algeria, were subject to contradictory communication

[106] Bennett (*Fighting the Mau Mau*, 159) argues that setting out a clear legal framework in Kenya was one of the factors that 'pushed the army away from genocide or even attrition, when many settlers in Kenya called for harsher repression'.

[107] Quoted in Evans, *Algeria*, 138.

[108] Horne, *Savage War of Peace*, 120 ff.

[109] Huw Bennett, '"A Very Salutary Effect": The Counter-Terror Strategy in the Early Malayan Emergency, June 1948 to December 1949', *Journal of Strategic Studies*, 32/3 (2009), 433 ff.

[110] Bayly and Harper, *Forgotten Wars*, 455.

[111] Bennett, *Fighting the Mau Mau*, 190. This was probably also a factor in Malaya; see Bayly and Harper, *Forgotten Wars*, 456.

on the part of the authorities, who insisted on carrying out repression within certain legal limits, but usually did not do much to rein in excesses.[112]

Deportations were particularly widely used in Malaya. This was partly due to the nature of the conflict: the colonial administration regarded Chinese squatters as the main instigators of the conflict, and many of them did not have citizenship. In November 1948, emergency regulation 17C sanctioned the banishment of non-Federal citizens along with their dependents.[113] In the early years of the conflict the colonial authorities deported a minimum of 20,000 ethnic Chinese inhabitants to China. According to Karl Hack, mass deportations were one factor in explaining the lower levels of indiscriminate violence in the repression of the insurgency in Malaya, as they worked like a 'safety valve'.[114]

While internment and detention had been prominent measures during the wars in the colonies around 1900, they became the hallmark of colonial repression in the wars of decolonization. However, their features and functions partly differed from the camps used around 1900. In addition, they varied across the three wars analysed in this section. In Malaya, the implementation of the 'Briggs Plan' starting in May 1950 ended the 'counter-terror' phase of the early years of the conflict and introduced population and spatial control as the main focus of the campaign.[115] A central measure of the Briggs Plan was the resettlement of 600,000 squatters from the margins of the jungle to 400 resettlement areas between June 1950 and the end of 1951. The situation in these so-called new villages was at first grim and characterized by a lack of resources, but it improved over time. For the colonial authorities, the villages were a tactical success: they made it more difficult for the MNLA fighters to acquire supplies. Moreover, they enabled the authorities to use both coercion and inducement towards the population in a more targeted way.[116] According to Hack, the 'new villages' provided a 'context of rising security and "confidence"' for their inhabitants, who no longer had to fear insurgent reprisals as long as they collaborated with the government.[117] Apart from the resettlement programmes that essentially interned large parts of the population, the colonial authorities also detained 30,000 persons without trial during the conflict.[118]

Internment and detention were also a central part of colonial repression in Kenya. In the early stages of the conflict, colonial authorities evicted the

[112] See also David Anderson's discussion of this double bind in Kenya, 'British Abuse and Torture in Kenya's Counter-Insurgency, 1952–1960', *Small Wars & Insurgencies*, 23/4–5 (2012), 700–19. In Malaya, the situation was different in that the implementation of the 'Briggs Plan' starting in 1950 signified a more radical break with indiscriminate violence, even if punitive elements persisted until the end of the conflict.

[113] Bayly and Harper, *Forgotten Wars*, 449.

[114] Hack, 'Everyone Lived in Fear', 693.

[115] Hack, 'Everyone Lived in Fear', 683.

[116] Hack, 'Everyone Lived in Fear', 688.

[117] Hack, 'Everyone Lived in Fear', 688.

[118] Hack, 'Everyone Lived in Fear', 692.

Kikuyu population from the Rift Valley to reserves in the Central Province. This eviction process first started as a quasi-official measure by the settlers in the Rift Valley with the aim to protect lives and property and was later adopted as official policy.[119] Following Operation 'Anvil' in 1954 that had targeted Mau Mau in Nairobi, the colonial authorities started a villagization programme that was inspired by experiences in Malaya. Its scope was even more extensive than that of the resettlement programme in Malaya: over 1,000,000 people were resettled into 845 villages. However, in contrast to the 'new villages' in Malaya, reward and punishment were not handled in a flexible way. Rather, villages were distinguished into 'punitive' and 'reward' villages, and security forces placed individuals into either category on the basis of an initial assessment. Moreover, the living conditions in the villages did not improve over time, but stayed at a comparatively low level.[120] Apart from the villagization programme, the detention of Mau Mau suspects in the narrower sense became the epitome of the war in Kenya. The detention camps attracted the label of 'gulags'.[121] The detention programme also grew exponentially subsequent to Operation 'Anvil'. At its peak in 1954, the detention camps had more than 70,000 inmates.[122] The rationale for detention was that the camp inmates were supposed to be 'rehabilitated' step by step. The rehabilitation process was supposed to begin with a confession. The rehabilitation agenda was based on 'research' by the anthropologists and psychologists Louis Leaky and J. C. Carothers, whose interpretation of the Mau Mau movement as a ritualistically framed mental disorder had gained disproportionate influence at the start of the conflict.[123] Conditions in the resettlement villages were often grim, and health and nutrition shortfalls were endemic; low food rations and inadequate medical provisions were part of a deliberate system of punishment in the detention camps.[124] Forced labour was also a feature of both the resettlement villages and the detention camps, though in the former it consisted only of limited communal labour, whereas in the detention camps heavy labour was designed as part of the rehabilitation process.

Population control in the form of resettlement was also practised in Algeria, with up to two million Algerians being relocated into so-called *camps de regroupement* during the war. These camps became particularly prominent as part of the Plan de Constantine reforms launched after de Gaulle's return to power in 1958. As in Malaya and Kenya, the main rationale behind the camps was to make the insurgents' lives more difficult by cutting them off from their sources of supply. The French colonial authorities publicly depicted the camps

[119] Bennett, *Fighting the Mau Mau*, 172 ff. [120] Bennett, *Fighting the Mau Mau*, 224.
[121] Caroline Elkins, *Britain's Gulags: The Brutal End of Empire in Kenya* (London: Pimlico, 2005).
[122] Anderson, *Histories of the Hanged*, 5, 344.
[123] Anderson, *Histories of the Hanged*, 314.
[124] Anderson, *Histories of the Hanged*, 317; Bennett, *Fighting the Mau Mau*, 224.

as protected areas for the population and as a flagship of Gaullist reformism in Algeria.[125] Yet the actual conditions in the camps were bleak, and inhabitants lacked basic provisions in terms of food and medical care.[126] Alongside the *camps de regroupement*, there were also detention camps of various sorts. The most important among them were the so-called *centres de tri et de transit* (*CTTs*), which were meant as screening camps, but often developed into sites of long-term detention. They were notoriously overcrowded and under-resourced. Over time, women and children were also detained in *CTTs*. As in Kenya, the detention system foresaw re-education for some of its inmates. These measures were referred to as *action psychologique*. They included hard labour and the successful completion of a three-step process. Towards the end of the war, in March 1958, the French authorities also erected a special kind of camp for ALN fighters who could prove that they had not engaged in 'terrorist' activities. This category of inmates was referred to as *pris les armes à la main* (*PAMs*) and enjoyed a status that came close to that of a POW, although France insisted that the Third Geneva Convention did not apply to them.[127]

Both Kenya and Algeria became notorious for the use of torture by security forces. Neither in Kenya nor in Algeria was torture ever explicitly sanctioned, but in both cases officials in France and Britain knew that torture occurred on a systematic scale and did little to stop it. In Algeria, senior officers encouraged the use of torture informally.[128] Reports about torture became public in 1957 and caused an outcry in mainland France. After these revelations the French government made some hesitant efforts at enhancing legal protections for detainees, but these measures were not sufficiently thorough to be effective. In fact, there is some evidence that they provided additional incentives for the unlawful killing of detainees in order to 'destroy' all evidence of torture.[129] In Kenya, according to David Anderson:

> The key decision in the institutionalization of torture came in the authorization of Regulation 17 of the Emergency (Detained Persons) Regulations... By offering legalized definitions of 'compelling force' and 'punitive force', and by allowing officers in detention camps to use violence so defined in the normal and routine treatment of detainees, the British took a step that could not be reversed.[130]

Both in Kenya and in Algeria the line between torture and unlawful executions was often blurred, not only because many members of the security forces thought it opportune for torture victims to disappear, but also because torture often ended in the death of the victim. Although the quasi-official

[125] Evans, *Algeria*, 250; cf. Merian Helen Davis, 'Restaging *mise en valeur*: "Postwar Imperialism" and the Plan de Constantine', *Review of Middle East Studies*, 44/2 (2010), 176–86.

[126] Evans, *Algeria*, 254.

[127] Branche, 'The French in Algeria', 175 ff.

[128] Evans, *Algeria*, 206.

[129] Branche, 'The French in Algeria', 176.

[130] Anderson, 'British Abuse', 707.

justification or, rather, excuse for torture was that violent techniques were indispensable in order to gain intelligence and to extract confessions, the line between this supposedly 'rational' application of torture and the destruction of the opponent by all means became blurred. As Raphaëlle Branche argued, torture was both a means of repression and a symbol of colonial power:

> It was necessary for the French to keep the Algerian people off-balance, while confronting the nationalists operating in their midst. Torture was not, therefore, simply one means of obtaining information. It was an essential weapon, allowing the French to fulfill their fundamental war aims. Torture was not merely a weapon against terrorists, but also a political weapon. Torture was used not only to force people to talk, but to make them understand, and to remind them who wielded power.[131]

Levying fines and destroying livelihoods and property were common practices in Malaya, Kenya, and Algeria. In the early phase of the war of Malaya, the British approach was focused on punitive sweeps and the destruction of property and habitations of alleged insurgents and their supporters.[132] The legal basis for the destruction of property was Emergency Regulation 18 B of 13 November 1948. The aim of these measures was to instil fear in the population by means of collective punishment.[133] While in 1950 the counter-terror approach was abandoned, and colonial authorities adopted the Briggs Plan that focused on spatial and population control, targeted collective punishment was still practised in the later stages of the conflict. Villages that were suspected of helping insurgents or of withholding intelligence could be punished by the reduction of their food rations or the imposition of curfews. However, in the later stages of the conflict, punitive measures were accompanied by the promise of rewards for cooperation.[134] In Kenya, fines and the destruction of property were also used in the context of collective punishment. Fines were not a new measure; they had been an established means of colonial policing even before the war.[135] During the uprising, the British first imposed fines in April 1952. The confiscation and destruction of property were new measures, introduced only during the uprising. According to Huw Bennett, 'by May 1956 the number of confiscations approached 200 bicycles, 200 donkeys, 12,000 cattle, and 44,000 sheep and goats'.[136] In addition to the confiscations, the security forces in Kenya had orders to destroy all property and housing in a 10-mile radius of Mau Mau attacks. In Algeria, collective reprisals against villages suspected of harbouring or aiding ALN fighters were legally sanctioned.[137] The security forces in Algeria operated on the basis of an

[131] Branche, 'The French in Algeria', 178.
[132] Hack, 'Everyone Lived in Fear', 680.
[133] Bennett, '"A Very Salutary Effect"', 431.
[134] Hack, 'Everyone Lived in Fear', 688.
[135] Bennett, *Fighting the Mau Mau*, 220.
[136] Quoting Sorrenson, in Bennett, *Fighting the Mau Mau*, 222.
[137] Evans, *Algeria*, 138.

explicit doctrine of 'collective responsibility'. The official version of this doctrine envisaged, for instance, that the inhabitants of villages close to attacks on French lines of communication had to rebuild destroyed telegraph poles and other installations. Governor General Jacques Soustelle tried to spell out limits to the doctrine. For instance, he insisted that 'collective responsibility' did not warrant indiscriminate killings. However, in practice the doctrine soon proved to be a slippery slope, with villages in the vicinity of FLN attacks not only being forced to repair the damage; the villagers were also evicted, and their homes were subsequently destroyed by aerial bombing. In the later stages of the war, aerial bombings were also carried out without previous evacuations.[138]

Was Colonial Warfare War without Restraint?

It is often assumed that war in the colonies was by definition more savage and less restrained than war in the centre of the international system—that is, in Europe and North America. Racism and the so-called standard of civilization, including the idea that the law of armed conflict applied only among 'civilized' peoples, seem obvious explanations for the escalation of violence in the colonies.[139] However, accounts focusing on military culture, such as Isabel Hull's *Absolute Destruction*, provide a counter-argument to this line of thinking. While Hull conceded that the colonial context provided the Imperial German military with greater degrees of freedom from legal constraints, she insisted that there were continuities between German military practices in Europe and in the colonies: 'Germans approached colonial warfare from inside the frames of their military culture as it had developed in Europe. The colonial situation merely provided the opportunity to practice on Africans or Chinese what the military experts took to be the immutable precepts of war.'[140]

The account of colonial warfare presented in the preceding sections suggests a reading of war in the colonies between these two positions; in some ways it is similar to war in the centre, and in some ways it is unique. On the one hand, there were remarkable continuities between the practices of anti-irregular warfare in the centre and in the peripheries. The French forces used collective

138 Horne, *Savage War of Peace*, 113 ff.

139 Cf. Paul Kennedy and George J. Andreopoulos, 'The Laws of War: Some Concluding Reflections', in Michael Howard, George J. Andreopoulos, and Mark R. Shulman (eds), *The Laws of War: Constraints on Warfare in the Western World* (New Haven: Yale University Press, 1994), 215; Gerrit W. Gong, *The Standard of Civilization in International Society* (Oxford: Oxford University Press, 1984).

140 Hull, *Absolute Destruction*, 3.

drowning during the Haitian Revolution, which had already been used in the Vendée. Kitchener classified the Boer civilians into two groups depending on their political position towards the war, thereby echoing the approach that had been used during the American Civil War. Although it is difficult to trace a direct 'transmission vessel' for these practices, either in the form of a commanding officer or in the form of quasi-doctrinal texts, these instances are strong indications that colonial powers did not approach irregular warfare in the colonies in a substantially different way from irregular warfare in the centre. The tactical toolkit of irregular warfare in the colonies was similar to that used in Europe and North America: collective punishment by means of summary executions, massacres, arrests, deportations, fines, and the destruction of property. The institutionalization of camps around 1900 was indeed an innovation that first evolved in the colonies, but it quickly travelled to Europe, where internment camps for 'enemy aliens' as well as mass camps for prisoners of war became a hallmark of the First World War.[141] However, while means and techniques such as camps and interment first cropped up in the colonial context, it is difficult to draw a direct line from the camps in German South West Africa to the concentration camps in the Second World War, as Benjamin Madley and Jürgen Zimmerer tried to do.[142] Both the political contexts and the functions of the various types of camps were too different to construct such a historical lineage.[143]

At the same time, warfare in the colonies did in some cases reach levels of violence that were unique to colonial warfare. While the war in the Vendée was characterized by atrocities, it had not been a war of extermination, as the French campaign conducted in Haiti had been. The wholesale slaughter of wounded and defeated fighters in the Anglo-Zulu War was not a standard practice of the British army. The genocidal extent of the war in German South West Africa had not been reached in modern Europe before 1900. Equally, the extent of destruction during the suppression of the Maji Maji revolt had no precedent in irregular warfare in modern Europe.

[141] Alan Kramer, 'Editorial', *Mittelweg 36*, 20/4 (2011), 36.

[142] Benjamin Madley, 'From Africa to Auschwitz: How German South West Africa Incubated Ideas and Methods Adopted and Developed by the Nazis in Eastern Europe', *European History Quarterly*, 35/3 (2005), 429–64; Jürgen Zimmerer, *Von Windhuk nach Auschwitz? Beiträge zum Verhältnis von Kolonialismus und Holocaust* (Münster: LIT Verlag, 2011).

[143] Cf. Andreas Stucki, 'Aufbruch ins Zeitalter der Lager? Zwangsumsiedlung und Deportation in der spanischen Antiguerilla auf Kuba, 1868–98', *Mittelweg 36*, 20/4 (2011), 31. Moreover, the assertion of such a lineage overlooks the fact that the structure and function of the Nazi concentration camps themselves were not uniform. Rather, they evolved from the pre-war camps, which were intended as an instrument of 'domestic political terror', to the wartime camps, which were set up to facilitate the mass persecution and extermination of Europe's Jewish population. Cf. Christian Goeschel and Nikolaus Wachsmann, 'Before Auschwitz: The Formation of the Nazi Concentration Camps, 1933–9', *Journal of Contemporary History*, 45/3 (2010), 521.

Racism and the 'standard of civilization' go some way in explaining these levels of violence, but on the whole such explanations are too unspecific. In the Haitian Revolution racism did play a role, but it was only one of several cleavage lines that dominated the conflict. Economic cleavages between slave-holders (who could be black or mixed race) and slaves played a more important role. The notorious massacre of the *Suisses* was carried out not primarily because the victims were black, but because they were slaves, and freeing them would have set a dangerous precedent. Racism did play a role in colonial warfare, but not in a simple or primordial fashion. The category of 'race' is a social construction that can be manipulated and tagged on to different groups of people. It can be charged with negative as well as positive attributes. Racism in colonial warfare was not a simple determinant factor of a confrontation between white Europeans and black natives (or Asian natives).

The Boer War is an instructive case in this respect. When compared to the war in German South West Africa, it is clear that the British war aims in the Boer War were far less extreme than von Trotha's objectives. The camps in South Africa were not designed as means of extermination, and the death rates—22 per cent compared to 45 per cent in German South West Africa—were due to neglect rather than design. No doubt, the issue of race played a role: it is hard to conceive of a lobbying campaign on the same scale and with the same repercussions of that conducted by Emily Hobhouse and the Fawcett Commission had the inmates of the South African camps been black. And yet racist stereotypes also cropped up in Britain's 'Gentlemen's war' in South Africa, where Kitchener famously denounced the Boers as 'uncivilized Afrikander savages with a thin white veneer'.[144] British newspapers often harked back to the stereotype of the 'dirty' Boer, which seemed to indicate that, even if the Boers were white, their European racial origin was concealed under layers of dirt accumulated through their 'uncivilized' habits and customs. Finally, the issue of race was given a particular twist with respect to the Boers, as popular culture often depicted the Boers as being particularly cruel towards black native inhabitants. In other words, the Boers were seen as 'uncivilized' colonizers, a deceptive hybrid who could not claim the same civilizational achievements as the British.[145]

On the other hand, British contemporaries during and after the Anglo-Zulu War did not stereotypically describe the Zulu fighters as 'primitive savages'. To be sure, the slaughter of Isandhlwana was a major justification for waging war without restraint at Ulundi. At the same time, the British could explain their

[144] Quoted in Pakenham, *Boer War*, 500.

[145] Simon Popple, 'From "Brother Boer" to "Dirty Boer": Colonizing the Colonizers through the Popular Representations of the Boer in the British Illustrated Journal 1899–1902', *Journal of War & Culture Studies*, 5/2 (2012), 137–56.

defeat at Isandhlwana in a face-saving way only by acknowledging that the Zulus were in fact an advanced warrior caste who had fought the British in a 'splendid spirit'.[146] The trope of the 'noble savage' was as important as the concept of the 'uncivilized savage' in colonial warfare. In fact, the former was the foundation of the idea of 'martial races' and hence the basis for the recruitment of colonial auxiliary forces, in particular in the context of the British Empire.[147] Hence orientalism is probably a more useful concept for describing the Europeans' perceptions of the native inhabitants of the colonies. According to Patrick Porter, orientalism attributes both inferior and superior traits to the 'Other'. With respect to the 'martial races', this meant that they were described as 'Superior, because the British believed them to be untamed primitives who had not been polluted by commerce, and that their savagery endowed them with honour and simplicity'. At the same time, they were seen as 'Inferior, as colonial rhetoric infantilized them, seeing their stupidity as the key to their obedience'.[148]

Neither does the 'standard of civilization' provide us with a sufficiently specific explanation for the escalation of violence in colonial warfare. At the heart of the concept of the 'standard of civilization' was the idea that the law of armed conflict applied only among 'civilized' peoples. As the native inhabitants of the colonies allegedly had not yet reached the requisite levels of civilization, the law could not be applied to them and protect them.[149] Yet it is not entirely true that the colonies were a lawless zone as far as constraints on the use of military force were concerned. Haiti was under metropolitan French laws from 1794 to 1800. Indeed, when Napoleon dispatched the Leclerc expedition to Saint Domingue, his government publicly pandered to the discourse of European civilization being re-established in the colony that had been taken over by 'black barbarism'.[150] Yet his reason for the introduction of different laws for the colonies was linked not so much to considerations of the degree of 'civilization' of the inhabitants of the colonies, but to questions of economic viability. Napoleon had realized that planter colonies such as Haiti would thrive economically only if slavery could be reintroduced.[151] When it comes to colonial warfare around 1900, a time that is usually described as the heyday of the 'standard of civilization', the colonies were not a zone of lawlessness either. It is true that legal texts written at the time as well as military handbooks such as C. E. Callwell's *Small Wars*, which

[146] Colonel Henry Harford, quoted in Lieven, '"Butchering the Brutes"', 616.

[147] Hack and Rettig, 'Imperial Systems of Power', 11 ff.

[148] Porter, *Military Orientalism*, 42.

[149] Gong, *The Standard of Civilization*; Frédéric Mégret, 'From "Savages" to "Unlawful Combatants": A Post-Colonial Look at International Humanitarian Law's "Other"', in Anne Orford (ed.), *International Law and its 'Others'* (Cambridge: Cambridge University Press, 2006), 265–317.

[150] Dubois, *Avengers of the New World*, 256.

[151] Dubois, *Avengers of the New World*, 241.

was influential among officers across Europe, repeatedly referred to an alleged lack of 'civilization' as the compelling reason for a specific approach to ruling and policing the colonies; however, the impact of the 'standard of civilization' should not be overestimated.[152] Contemporaries did indeed assume that the Hague Conventions did not apply in the colonies, but military penal codes did.[153] It is true that transgressions were not as strictly enforced as they may have been in Europe,[154] but this did not necessarily make the colonies a lawless zone. The case in which the 'standard of civilization' does explain Europeans' escalation of violence is the Anglo-Zulu War: the practice of giving 'no quarter' in the battle of Ulundi was justified by the fact that the Zulu fighters had done the same in Isandhlwana, and had thus demonstrated that they were not sufficiently 'civilized' to be treated with restraint in war. The alleged lack of 'civilization' could thus be substituted for irregular characteristics in the narrower sense of the term: the Zulu fighters may have been well organized militarily and may have engaged in 'campaigning war', but, from the point of view of the British, that did not mean that they were 'regular' fighters.

Even more interesting with respect to the 'standard of civilization' were the wars of decolonization after the Second World War. From the perspective of proponents of the 'standard of civilization' thesis, the Second World War was the watershed event after which the 'standard' became untenable in international law and international relations.[155] In fact, as observed in the previous section, the European legal approach to warfare in the colonies did change: European colonial authorities were now eager to declare all kinds of armed conflict in the colonies to be 'domestic emergencies', since these do not fall within the jurisdiction of the law of armed conflict. This was quite clearly due to the law's expansion envisaged in the Geneva Conventions, most importantly common article 3. Yet the introduction of extensive emergency regulations allowed colonial authorities to operate within a sophisticated construct of twisted legality: on the one hand, it gave the situation a legal façade by making legal standards explicit; on the other, emergency regulations were often so extensive and so vague that they provided next to no legal protections for the inhabitants of the colonies. Hence, even after the supposed

[152] C. E. Callwell, *Small Wars: Their Principle and their Practice* (Lincoln, NB: University of Nebraska Press, 1996 [1896]).

[153] Kuß, *Deutsches Militär*, 198 ff.

[154] It is difficult to obtain a systematic overview of the empirical evidence on this point. Kuß observes that the German military leadership enforced strict legal standards among the troops in the colonies with respect to offences affecting internal discipline, but took a lenient view with respect to offences against civilians, and in some cases even encouraged or sanctioned the latter. Kuß, *Deutsches Militär*, 205.

[155] Edward Keene, *Beyond the Anarchical Society: Grotius, Colonialism and Order in World Politics* (Cambridge: Cambridge University Press, 2002).

downfall of the 'standard of civilization', the colonies' special legal status persisted.[156] However, the complexity of the legal structure and the ambivalence in its application also indicated that the era of decolonization was a period of transformation in the concept of the irregular. While the moral assessment of irregulars as well as of the civilians associated with them had definitely changed after the Second World War, the legal standing of irregular fighters was still strongly influenced by pre-Second World War standards, and this still had an impact on the treatment of civilians.

Possibly the most important factor for the explanation of the escalation of violence in colonial warfare is the extent to which Europeans saw the colonies as sites of social engineering. Colonial wars were waged with the aim of making the colonies governable in the long run, and colonial authorities did not shrink back from even the most extreme of measures. They justified exterminatory designs, such as Leclerc's plan to kill all black ex-slaves in Haiti or von Trotha's *Vernichtungsbefehl*, by pointing out that only extreme measures could keep particular colonial territories under colonial domination in the long run. The large-scale destruction of villages and agricultural structures in German East Africa during the Maji Maji revolt was not only aimed at subduing the insurgents; otherwise the civilians would not have been affected at such a high scale. More importantly, it destroyed the traditional social structure of the native societies and their economies. As a result, the native population had no other choice but to become integrated into the new colonial economy, where their labour force was desperately needed. After the end of the Second World War, genocidal violence did not occur any more in the colonies. Social engineering in the form of extermination of whole populations or radical destruction of their social and economic structures had given way to 'softer' forms of social engineering: the camps were sites no longer of mass death, but of re-education. This approach was most explicit in Kenya, where the British colonial authorities interpreted Mau Mau predominantly as a mental condition that could be 'treated' by rehabilitation. But the rehabilitation concept was also prominent in the camps in Malaya and in Algeria. The idea was that good colonial subjects could be 'created' by use of force. Hence, in sum it was not so much dynamics of exclusion such as the 'standard of civilization' that caused the escalation of violence, but rather attempts at including the colonies, at making them governable and amenable to the political, economic, and social designs of the colonial powers.[157]

[156] Cf. Oren Gross and Fionnuala Ní Aoláin, *Law in Times of Crisis: Emergency Powers in Theory and Practice* (Cambridge: Cambridge University Press, 2006), 181.

[157] Of course, inclusion in the form of a 'civilizing mission' was the flip-side of the 'standard of civilization'. However, in the academic discussion the exclusive dimensions of the 'standard' have dominated.

Regarding the question of whether specific national approaches among the colonial powers caused greater levels of violence, the evidence is mixed. While Hull is certainly correct that the level of violence in German colonial warfare at the beginning of the twentieth century was characterized by a particularly ruthless and partly exterminatory approach, the German approach to anti-irregular warfare was definitely not unique. Rather, it drew upon an established toolkit that the British and the French used as well.[158]

Finally, it is important to note that global factors such as racism, the 'standard of civilization', the impact of social engineering plans, and military culture cannot sufficiently explain the differences between different colonial wars. Why, for instance, did the Germans resort to internment in German South West Africa, but not in East Africa?[159] Why did the British abandon the counter-terror approach two years into the conflict in Malaya, but never revised their punitive policies in Kenya?[160] These questions have become increasingly prominent in recent debates over warfare in the colonies, and they point to the importance of situative factors. However, more research needs to be done in order to provide a clearer understanding of these factors.

The Impact of the Second World War on the Concept of the Irregular Fighter

The experience of the Second World War, in particular that of resistance against the occupation by Axis powers, left a lasting impact on the concept of the irregular fighter. After this experience, no one could deny that irregular fighters could be morally vindicated, even if their way of fighting did not meet the requirements of lawful belligerency under the law of armed conflict. In cases such as Malaya, the impact of the Second World War was evident, as the protagonists of the anti-colonial insurgency were essentially the same armed groups who had already fought against Japanese 'imperialism'. In Kenya and in Algeria, the Mau Mau and the FLN/ALN leadership included many former members of British and French colonial troops who had fought alongside the metropolitan armies during the Second World War. On the one hand, this obviously provided them with military training and experience.[161] On the other hand, even if they may not have consciously fought against 'Axis

[158] Cf. Peter Lieb, 'Suppressing Insurgencies in Comparison: The Germans in the Ukraine, 1918, and the British in Mesopotamia, 1920', *Small Wars & Insurgencies*, 23/4–5 (2012), 627–47.

[159] Cf. Kuß, *Deutsches Militär*, 424 ff.

[160] Cf. Hack, 'Everyone Lived in Fear', 692 ff.

[161] J. 'Bayo Adekson, 'The Algerian and the Mau Mau Revolts: A Comparative Study in Revolutionary Warfare', *Comparative Strategy*, 3/1 (1981), 81.

imperialism' during the Second World War, with hindsight the lesson of the moral vindication of irregular resistance fighters was not lost on them.[162]

As noted above, the legal framework regarding irregular fighters had not changed much as a consequence of the Second World War. The 1949 Geneva Conventions had focused most of the attention on the plight of civilians in occupied territory (even if colonial powers disputed that even the minimum protections entailed in common article 3 applied to armed conflicts in the colonies). However, the criteria for lawful belligerency were essentially unchanged. If anything, irregular fighters had gained minimum protections as listed in common article 3. However, what had changed was the moral discourse on irregular fighters. For the Malayan communists, anti-Japanese nationalism could easily translate into anti-British nationalism. The 1955 Bandung conference 'made an explicit link between Nazism and colonialism' and condemned both.[163]

The experience of repression and resistance during the Second World War had the most visible impact in the war in Algeria. This was most likely due to the fact that France embodied both experiences and roles: that of the resister and that of the oppressor.[164] The discursive battle over legitimacy of both the insurgent forces and the French forces harked back almost exclusively to the moral universe of the experience of the Second World War. The FLN repeatedly denounced torture and other French repressive measures as 'Gestapo methods'. Instances of collective punishment were decried as a French version of the 1944 SS massacre in Oradour-sur-Glane. Even the perpetrators used the analogy. A corporal of the Foreign Legion who was shocked by the extent of systematic brutality used by the French security forces in Algeria wrote in a letter: 'If one day there was a new Nuremberg trial, we will all be guilty: Oradours, we are carrying them out every day.'[165] Conversely, the French colonial authorities condemned the FLN/ALN for using Nazi methods. They decried the Mélouza massacre as the FLN's Oradour.[166] In Kenya, the use of Second World War analogies was much less prominent than in Algeria. However, there were isolated references—for instance, a notorious loyalist was denounced as the 'Hitler of Kinyona'.[167]

In sum, the experience of legitimate irregular resistance during the Second World War made it more difficult for colonial powers to vilify insurgents in the colonies out of hand. The declaration that Malaya, Kenya, and Algeria

[162] However, the Malay Regiment, which was a colonial unit that had existed since 1933 and had distinguished itself during the Second World War in the fight against the Japanese, remained loyal to the British colonial authorities during the insurgency in Malaya.

[163] Evans, *Algeria*, 139.

[164] Cf. Martin Evans, 'French Resistance and the Algerian War', *History Today*, 41/7 (1991), 38–45.

[165] Quoted in Evans, *Algeria*, 169.

[166] Evans, *Algeria*, 217.

[167] Branch, *Defeating Mau Mau*, 61.

were 'domestic emergencies' and the introduction of emergency legislation helped the colonial powers in this context: after all, they were democratic regimes bound by the rule of law and not totalitarian states such as Nazi Germany. However, in this context, as we have seen, insurgents could no longer simply be denounced as 'rebels'. Instead, they became 'communist terrorists' in official terms. Alternatively, the colonial authorities depicted them as victims of a 'mind destroying disease', as the British did with the Mau Mau in Kenya.[168] Against this background, it comes as no surprise that a psychiatrist, Frantz Fanon, became one of the leading voices of the anti-colonial movements. According to Fanon, resistance was not a sign of mental disorder; rather, colonial oppression produced mental illness in its victims.[169]

Of course, the wars in Malaya, Kenya, and Algeria led to the end of colonialism and to independence for these countries. This had two consequences. First, as with the resistance fighters against the Axis occupation forces, with decolonization the cause of national liberation movements (as they were called after their respective victories) appeared to be morally vindicated.[170] Second, their success bestowed them with agency in the international realm. This agency was important in the framework of the Conférence Diplomatique sur le Droit Humanitaire (CDDH) held from 1974 to 1977 and for the negotiation of the Additional Protocols I and II to the Geneva Conventions. The CDDH and its outcomes were paradoxical. On the one hand, they represented a deliberate attempt at de-westernizing the law of armed conflict.[171] One of the most pressing issues on the agenda of the decolonized states was protections for irregular fighters. Additional Protocol I effectively abandoned the notion of lawful belligerency and made the protections of POW status dependent on procedural criteria. Article 43, 1, Additional Protocol I stated that the 'armed forces of a Party to a conflict consist of all organized armed forces, groups and units which are under a command responsible to that Party for the conduct of its subordinates'.[172] Article 44 made the protections of POW status dependent on the condition that fighters carry their arms openly '(a) during each military engagement, and (b) during such time as he is visible to the adversary while he is engaged in a military deployment preceding the launching of an attack in which he is to participate'.[173] This legal innovation reflected the fact that insisting on more comprehensive criteria systematically skews the military balance in favour of more powerful opponents, which are

[168] Michael Blundell, quoted in Anderson, *Histories of the Hanged*, 279.

[169] Frantz Fanon, *The Wretched of the Earth* (London: Penguin, 2001 [1961]).

[170] The exceptions were, of course, communist insurgencies in Asia, at least from a Western point of view.

[171] Geoffrey Best, *War and Law since 1945* (Oxford: Oxford University Press, 1994), 416.

[172] Additional Protocol I, article 43 (1), printed in Roberts and Guelff (eds), *Documents*, 445.

[173] Additional Protocol I, article 44 (3), printed in Roberts and Guelff (eds), *Documents*, 445.

essentially organized as states. National liberation movements often could not satisfy the criteria of wearing a uniform, for instance, even if they had wanted to.

On the other hand, many western states were apparently shocked by the extent of the de-westernization of the law of armed conflict envisaged in the Additional Protocols. The result was that 'acceptance of the texts [of the Additional Protocols] was made conditional, by many States, on the attachment of specific "reservations", "declarations", and "interpretive statements", over and above such qualifications and expressions of doubt as were voiced in the debates [at the CDDH]'.[174] The USA did not ratify Additional Protocol I, precisely because of the protections it gave to irregular fighters. The upshot of this is that the legal status of irregular fighters continued to be unclear and contested.

Conclusion

This chapter has shown that the concept of the irregular in colonial warfare was not qualitatively different from war in the centre of the international system. In wars in the colonies before the Second World War, irregular fighters were denounced as rebels and were liable to be executed. The 'catalogue' of repressive measures did not differ in the colonies from that in Europe or North America. One exception was the introduction of internment camps, which were first institutionalized around 1900 in the colonies. However, the implementation of repressive measures was in many cases more extreme in colonial warfare than it was in wars in the centre. With the exception of Nazi German policies during the Second World War, repressive measures against irregular fighters in Europe were usually not carried to the extreme of extermination. The primary reason for the escalation of violence in colonial warfare was probably not so much racism or the application of the 'standard of civilization'. Rather it was the attempt by colonial powers to 'produce' suitable colonial subjects by force, even if this cynically implied the extermination of the resisting population and its substitution with European settlers or newly imported slaves, as General Leclerc envisaged it for Saint Domingue.

The Second World War proved to be a watershed moment in the concept of the irregular. Irregular resistance could no longer be assumed to be morally wrong by definition. The national liberation movements consciously built on this insight and denounced colonial repression as 'Nazi methods'. The liberators of the Second World War, Britain and France (even if this central

[174] Best, *War and Law since 1945*, 418.

identification claim for the Fourth Republic was mainly based in myth), found themselves accused of playing the role of the oppressor. Their answer was to try to transform the discourse on their own terms: either colonial insurgents were demonized as 'communist terrorists', or they were pathologized as suffering from a mental condition. However, the important lesson that the West did not learn from the relation between the Second World War and the wars of decolonization is that even powerful actors cannot control auxiliary forces. During the Second World War the Allies themselves had rehabilitated irregular fighters and reconceptualized them as auxiliary armies, and supported them accordingly. In some cases, such as Malaya, this backfired directly after the end of the Second World War, when the MPAJA transformed into the MNLA and fought against British colonial authority. But, even if the effects of the rehabilitation of the irregular qua auxiliary did not have a direct impact, such as in Algeria and Kenya, it proved impossible to turn back the clock on the indirect impact related to the increase of legitimacy and international recognition of irregular resistance.

6

Irregular Fighters in the Twenty-First Century

Between 'Unlawful Combatants' and 'Rebel' Auxiliaries

At the beginning of the twenty-first century the 'unlawful combatant' is without doubt the most prominent as well as the most contested incarnation of the irregular fighter. An outgrowth of the so-called war on terror, the label 'unlawful combatant' was attached by the United States to a variety of persons ranging from suspected members of the al-Qaeda leadership to low-level, local insurgents. This label was a stepping stone on the route to the most disastrous aspects of the 'war on terror', such as *incommunicado* detention, rendition, torture, and inhumane treatment—not only of the more spontaneous and politically unsanctioned type such as in Abu Ghraib, but also of the politically more directly sanctioned type that was (and that some claim still is[1]) practised in Guantánamo Bay.

The overwhelming assessment of the critics of these practices is that the detention, rendition, and torture debacle came about as a result of the legally dubious attempts of the political elites of the world's most powerful state, the USA, and its allies to bend the legal rules to suit their political interests. The narrative on which this assessment is based is that the shock of the 9/11 attacks led to a security hysteria in the USA and to a lesser extent in Europe. This, in turn, precipitated the demonization and dehumanization of 'terrorists' and their potential supporters and a concomitant loss of restraint in their

[1] Judge Gladys Kessler of the Federal District court in Washington DC, for instance, stated that the force-feeding procedures performed on Guantánamo detainees who were on hunger strike in the summer of 2013 was 'painful, humiliating and degrading' (Reprieve, press release, 9 July 2013 <http://www.reprieve.org.uk/press/2013_07_09_Guantánamo_judge_force_feeding_obama/> (accessed 23 July 2013).

treatment. In such situations, the narrative goes on, international law proves to be weak, as it is underpinned only by few and imperfect institutions to enforce it. Optimistic versions of the narrative tend to end on the positive note that, even if the law and its institutions are weak, the underlying norms, such as the norm against torture, are strong: in the face of international criticism and public activism, the abuse of political power will be unsustainable in the long run. In fact, some of the developments since 2004 point towards such a slow but steady reassertion of legal norms.

At first glance, this narrative is a convincing description of what happened in the context of the emergence of the category of the 'unlawful combatant'. Public criticism and debate over the moral, political, and strategic costs of the misguided detention policy after 9/11 are laudable. However, this chapter aims to go beyond this narrative and to dig deeper into the configuration of the irregular fighter at the beginning of the twenty-first century. It focuses on the inconsistencies and paradoxes involved in this configuration. It aims to show that there are inconsistencies and contradictions both within the concept of the 'unlawful combatant' and between this concept and other incarnations of the irregular fighter that emerged at the same time: the irregular fighter as the target of drone strikes and other forms of targeted killings as well as the irregular as the 'rebel' auxiliary. The existence of such inconsistencies ultimately undermines the impression that the detention regime was a rational, if morally dubious, policy on the part of the USA and its Western allies in the face of a challenging security situation. In reality, the emergence of the detention regime was underpinned by various and partly contradictory ideological assumptions and rationales.

As already mentioned, the concept of the 'unlawful combatant' is the most prominent embodiment of the irregular fighter in the first decades of the twenty-first century. It has been widely discussed from the perspective of international law as well as of ethics and political philosophy. The bulk of this chapter is devoted to a reconstruction of the 'unlawful combatant'. However, the chapter does not proceed by treating the 'unlawful combatant' as an aberration. Rather, it explores various paradoxes that are at the heart of the concept. First, the 'unlawful combatants' narrative operated as a series of legal exclusions of irregular fighters from the law. The justifications for these exclusions were not only argued from a temporal perspective, such that transnational terror networks of irregular fighters were labelled a 'novel' phenomenon. They were also argued from a spatial perspective, such that it was said that the US Constitution did not apply in places such as Guantánamo Bay. And yet it is not entirely convincing to characterize the detention regime in the 'war on terror' as creating a 'legal vacuum' or as placing detainees in a 'legal black hole', as many commentators have claimed. Rather, the detention regime was based on a plethora of legal memos and consultations and justified in legal terms. Moreover, some of the arguments for legal exclusion harked back to long-established traditions in the treatment of irregular fighters. This brings us to the second paradox: the main argument for the introduction of the category of the 'unlawful combatant' was that the 'war on terror' was a novel conflict, the features of which were so different from previous wars that the law of armed conflict was obsolete. Yet the 'unlawful combatants' narrative drew upon century-old legal arguments on irregular fighters, hence showing more signs of continuity than of a categorical break with the past. Third, despite the fact that both its proponents and its critics presented the introduction of the 'unlawful combatants' category as a direct reaction to 9/11, the problem of the treatment of irregular fighters and terrorists had already been 'predicted' in late-twentieth-century accounts of the future of warfare such as 'fourth-generation warfare' and the 'new wars' thesis. Thus, strictly speaking, the invention of the concept of 'unlawful combatants' did not come about as a spontaneous suspension of the law in a situation of an acute emergency. Instead, it bore some qualities of a self-fulfilling prophecy.

But the concept of the 'unlawful combatant' not only suffers from internal inconsistencies and paradoxes; it also sits uneasily with the two other manifestations of the irregular fighter at the beginning of the twenty-first century: the irregular fighter as the target of drone strikes and of other forms of targeted killings as well as the irregular fighter as the 'rebel' auxiliary. These are the subject of the second and third parts of this chapter.

Targeted killing has come under increased legal, political, moral, and strategic scrutiny since 9/11. In contrast to the legal/extralegal regime underpinning detention and interrogation in the 'war on terror', the legal opinions guiding the US use of this counter-terrorism instrument have not been widely released as of 2014. In fact, the lack of transparency is one of the most frequently voiced criticisms regarding targeted killings. However, from the arguments that are currently in the public domain, we can infer that the practice of targeted killing draws on legal arguments that are different in structure and content from those belonging to the 'unlawful combatants' narrative. Rather than attempting to exclude irregular fighters from the jurisdiction of various legal instruments and protections, the justification for targeted killings operates on the basis of widening the category of combatants so as to include, on a prolonged basis, civilians who are directly participating in the hostilities as well as 'hostile' individuals who do not operate in the direct context of a territorially limited armed conflict such as the war in Afghanistan, for instance.

Finally, the legal and political focus on the issue of 'unlawful combatants' obscured the extent to which the West continues to rely on irregular auxiliaries in the 'war on terror' and beyond. In today's parlance, such irregular auxiliaries are usually referred to as 'rebels', which is an ironic twist in political rhetoric given that the term 'rebel' was once a pejorative term implying something akin to an outlaw status. 'Rebel' auxiliaries have played a vital role in Western strategy in Afghanistan, Iraq, Libya, and Syria. Even though the political rhetoric regarding the use of auxiliaries has changed, the problems related to the use of auxiliaries have not. The term 'rebel' may now designate a positive image, but the reality is that many of those fighting since 2001 in conjunction with the West in the context of the Northern Alliance in Afghanistan, for instance, have had a record of atrocities that was just as comprehensive as that of the Taliban. The Western press hailed the success of the Western-supported 'rebels' in Libya as a victory of national self-determination against the tyrannical regime of Muammar Gaddafi, but its coverage of the political turmoil and the deteriorating human-rights situation in Libya as a result of the 'rebel' takeover is rather muted. Only in the context of the civil war in Syria did the West begin to debate the question of whether the anti-government 'rebels' are really 'good' irregulars: are they worthy of Western support, or are they in fact 'unlawful combatants' in disguise—in other words, Islamist extremists with connections to al-Qaeda. In reality, of course, irregular auxiliaries can be useful, but they are also extremely hard to control and subject to rapid changes in the political context. The fact that some of the most notorious 'unlawful combatants' after 9/11 had started their military careers as auxiliary 'rebels'—such as the anti-Soviet mujahidin in Afghanistan in the 1980s—should give the West pause for thought.

The Emergence of the Term 'Unlawful Combatants'

The 'War on Terror'

On 11 September 2001 the terror network al-Qaeda attacked various targets in the USA. The death toll of the attack, almost 3,000, was by far the highest for any recorded terrorist incident. As a reaction to the attacks, on 12 September NATO considered the invocation of article 5 of the Washington Treaty, which would trigger measures of collective self-defence by NATO. The invocation of article 5 was later confirmed. On 20 September the US administration under George W. Bush issued an ultimatum to the Taliban leadership in Afghanistan demanding the extradition of Osama bin Laden, who was identified as the leading figure behind the attacks. The Taliban refused. On 7 October US armed forces in conjunction with the British military began the invasion of Afghanistan within the framework of Operation 'Enduring Freedom'. A second military mission, the International Security Assistance Force (ISAF), under the auspices of NATO and based on UN Security Council resolution 1386, was set up in December 2001. Initially the two missions were designed to fulfil separate but complementary tasks: Operation 'Enduring Freedom' was charged with counter-terrorist operations, while ISAF was designed to carry out nation-building measures. However, soon these tasks overlapped. ISAF's responsibility was gradually extended to the entire territory of Afghanistan. ISAF's mission in Afghanistan will come to an end in 2014–15 with the withdrawal of British forces and the bulk of the US combat forces, while other contributing nations started to withdraw their forces as of 2011.

The second large theatre of military operations in the framework of the 'war on terror' was the war in Iraq. The Bush administration and the British government under Tony Blair suspected the Iraqi leadership of developing weapon-of-mass-destruction (WMD) capabilities with the potential intention of using these against the West or handing them over to terrorist groups. Although the USA was unsuccessful in its attempts to gain UN Security Council support for immediate military action against Iraq, on 20 March 2003 a US-led coalition attacked and invaded Iraq. Saddam Hussein was ousted and captured and later executed. However, the Iraq Study Group, tasked with investigating the alleged WMD programmes that had led to the decision to invade, found no evidence of an imminent danger to the West or its local allies. On 30 June 2004 the occupation forces formally handed over sovereignty to the Iraqi interim government. Coalition partners started to withdraw their forces as of 2004. US forces were the last to leave Iraq in 2011.

Beyond the two large theatres of operations in Afghanistan and Iraq, the USA conducted a number of smaller counter-terrorist operations. Under the umbrella of Operation 'Enduring Freedom', the US dispatched forces to the Philippines, the Horn of Africa, and the Maghreb, and even sent a small contingent of Special Forces to the Caribbean.[2] The CIA also carried out anti-terrorist operations on a global scale.

However, it soon became clear that the 'war on terror', which US policymakers and military planners had initially envisaged largely as a counter-terrorism mission, required a more substantial commitment on the part of the Western coalitions. The wars in both Afghanistan and Iraq had in common that what initially seemed like a quick and overwhelming victory of the Western forces was followed by the emergence of lengthy insurgencies. In Afghanistan Taliban fighters had initially withdrawn to the south-western regions of Afghanistan and across the border into Pakistan, where they regrouped and mounted a steadily increasing insurgency starting in 2002.[3] In Iraq, the fall of Saddam Hussein and the Baath regime had left a power vacuum that the coalition forces were slow to fill. The respective causes for the

[2] A sixth component of Operation 'Enduring Freedom' was a small US training force in Georgia tasked with training the Georgian army.

[3] Antonio Giustozzi, *Koran, Kalashnikov and Laptop: The Neo-Taliban Insurgency in Afghanistan* (New York: Columbia University Press, 2008), 2.

insurgencies that developed in Afghanistan and Iraq varied: the insurgency in Afghanistan was more a case of the Taliban reorganizing to gain back political control over the territory of Afghanistan, whereas in Iraq the sectarian aspect played a larger role, with both Sunni and Shia irregular groups fighting against each other and against the coalition occupation forces. Moreover, a number of jihadists associated with al-Qaeda had fled Afghanistan in the wake of the Western invasion, among them Abu Musab al-Zarqawi, who would gain notoriety as the leader of al-Qaeda in Mesopotamia (AQM), a self-professed 'branch' of al-Qaeda in Iraq. The Iraqi insurgency was partly fuelled by an influx of such transnational Salafi jihadists, who fought under the banner of AQM, siding with other Sunni insurgent groups.[4]

With respect to both Afghanistan and Iraq, the upsurge in violence was blamed by observers and commentators on a failure of post-invasion planning and a lack of resources and manpower on the part of the respective US-led coalitions.[5] Slowly, the USA and, to a greater or lesser extent, its coalition partners reacted to the growing insurgencies in Afghanistan and Iraq by complementing the counter-terrorism approach with counter-insurgency measures. The immediate focus of these measures was the Iraqi insurgency, which by 2006 had developed into a full-blown civil war. In January 2007, President Bush appointed General David Petraeus as the Commander of Multi-National Forces in Iraq. In addition, Bush pledged an additional 30,000 troops for operations in Iraq.

In 2007, security in Iraq started to improve. Part of the reason for this improvement was the troop 'surge' and the concomitant reorientation of operational aims on a 'population-centric' counter-insurgency approach. However, the 'surge' alone would have achieved less had it not coincided with the so-called Anbar Awakening, a movement led by Sunni insurgents in the violence-ridden Anbar province with the aim to evict AQM fighters from the region.[6] AQM violence had become increasingly extreme over the years since the start of the insurgency. AQM had also started to terrorize Sunni civilians. While the Awakening itself was a local, bottom-up development, the US command soon designed the 'Sons of Iraq' scheme, a top-down programme intended to enlist former Sunni insurgents in the effort to expel AQM from other parts of Iraq and to enhance overall security. As a result of

[4] Ahmed Hashim, *Iraq's Sunni Insurgency* (Oxford: Oxford University Press for the International Institute for Strategic Studies, 2008) [Adelphi Paper 48/402], 23 ff.

[5] On Afghanistan, see Cyrus Hodes and Mark Sedra, *The Search for Security in Post-Taliban Afghanistan* (Oxford: Oxford University Press for the International Institute for Strategic Studies, 2007) [Adelphi Paper 47/391].

[6] For an extended discussion of the reasons for the reversal in Iraq starting in 2007, see Stephen Biddle, Jeffrey A. Friedman, and Jacob N. Shapiro, 'Testing the Surge: Why did Violence Decline in Iraq in 2007', *International Security*, 37/1 (2012), 7–40.

this improvement in security, the last coalition forces were able to withdraw in a situation of relatively low levels of violence.

Also in Afghanistan, policymakers and military planners advocated concentrating more strongly on counter-insurgency and state-building. However, manpower and resources were even scarcer in this theatre than in Iraq. In contrast to Iraq, pre-invasion Afghanistan had no solid infrastructure to speak of and no functioning state institutions. In addition, the Afghan insurgents faced a more benign strategic situation in that they were able to use Pakistan's Federally Administered Tribal Areas (FATA) as a refuge. In 2010, the Obama administration dispatched an additional 33,000 US troops to Afghanistan as a short-term surge with the aim of turning around the security situation in the country. General Stanley McChrystal took over command of ISAF and the American forces in Afghanistan. However, this 'Afghan surge' failed to achieve results similar to those in Iraq in 2007. While some areas in Afghanistan have been largely pacified over the last decade and while the training of the Afghan National Army (ANA) has seen rapid progress over the past few years, the overall military and political situation in the country remains unstable; it is unclear how it will develop after the withdrawal of the remaining Western forces.

Detention and Interrogation in the 'War on Terror'

Soon after the start of military operations in Afghanistan, a debate emerged within the Bush administration over the detention and treatment of terrorist suspects and their supporters. The main players in this debate were the Department of Justice (DOJ), the Department of Defense (DOD), and the White House. On 13 November 2001, the president issued the administration's first executive order regarding detainees and their treatment. It stated that terrorist suspects were to be 'detained, and, when tried, to be tried for violations of the laws of war and other applicable laws by military tribunals'. It also stipulated 'that it is not practicable to apply in military commissions under this order the principles of law and the rules of evidence generally recognized in the trial of criminal cases in the United States district courts'.[7] On 28 December 2001, the DOJ's Office of Legal Counsel (OLC) produced a memorandum arguing that detainees held at the military base in Guantánamo Bay, Cuba, would not be able to apply for a writ of habeas corpus in a US court.[8]

[7] 'Military Order of November 13, 2001: Detention, Treatment, and Trial of Certain Non-Citizens in the War against Terrorism', printed in Karen J. Greenberg and Joshua L. Dratel (eds), *The Torture Papers: The Road to Abu Ghraib* (Cambridge: Cambridge University Press, 2005), 25 ff.

[8] Memorandum for William J. Haynes, II, General Counsel, Department of Defense: Possible Habeas Jurisdiction over Aliens Held in Guantánamo Bay, Cuba, 28 December 2001, printed in Greenberg and Dratel (eds), *The Torture Papers*, 29 ff.

In January 2002, a number of memos were exchanged between the DOJ, DOD, State Department, and the White House on the subject of the application of international law to al-Qaeda and Taliban detainees. Based on these evolving legal opinions, on 7 February 2002 the White House issued a memorandum stating 'that the provisions of Geneva will apply to our present conflict with the Taliban' more generally; however, 'none of the provisions of Geneva apply to our conflict with al Qaeda in Afghanistan or elsewhere throughout the world' and 'common article 3 of Geneva does not apply to either al Qaeda or Taliban detainees'. Finally, the memo stated 'that the Taliban are unlawful combatants and, therefore, do not qualify as prisoners of war under Article 4 of Geneva'. Nonetheless, the memo indicated that, 'as a matter of policy, the United States Armed Forces shall continue to treat detainees humanely, and, to the extent appropriate and consistent with military necessity, in a manner consistent with the principles of Geneva'.[9] The essence of these memoranda was that suspected al-Qaeda and Taliban detainees had no legal protections whatsoever: no rights or protections evolving from POW status, no minimum protections as stipulated in common article 3 of the Geneva Conventions, no habeas corpus rights, and no rights to due process in case of criminal prosecution. The extent to which their treatment was constrained by international human rights law was not explicitly discussed.[10]

However, the first set of memoranda had remained silent on the question of the treatment of detainees during interrogation. The US Field Manual guiding interrogation and intelligence gathering was FM 34-52 of 1992. The manual distinguished between two different categories of detainees: enemy prisoners of war (EPW) and 'protected persons', which it largely equated with civilians. It referenced and included the relevant sections of the Geneva Conventions on the treatment of either of these categories.[11] FM 34-52 did not include any reference to 'unlawful combatants', but it explicitly applied in 'low-intensity conflicts' and to the handling of insurgents and terrorists.[12] It stated that the

[9] Memorandum for the Vice President, the Secretary of State, the Secretary of Defense, the Attorney General, Chief of Staff to the President, Director of Central Intelligence, Assistant to the President for National Security Affairs, Chairman of the Joint Chiefs of Staff, 7 February 2002, printed in Greenberg and Dratel (eds), *The Torture Papers*, 134 ff.

[10] The OLC's 'Memorandum for Alberto R. Gonzales and William J. Haynes II' of 22 January 2002 briefly touches upon the question of whether customary international law applies to al-Qaeda and Taliban detainees. Customary international law would cover aspects of the law of armed conflict such as common article 3 and article 75 of the Additional Protocol I (1977), but also elements of human rights law, such as the absolute prohibition of torture. However, the memorandum concludes that 'Customary international law... cannot bind the executive branch under the Constitution because it is not federal law' (printed in Greenberg and Dratel (eds), *The Torture Papers*, 111).

[11] Field Manual FM 34–52: Intelligence Interrogation, Headquarters, Department of the Army, Washington DC, 28 September 1992, sections 1–11, D-1 <https://www.fas.org/irp/doddir/army/fm34-52.pdf> (accessed 25 July 2013).

[12] Field Manual FM 34–52, sections 1–17, 2–17.

Geneva Conventions and US policy 'expressly prohibit acts of violence, including physical or mental torture, threats, insults or exposure to inhumane treatment as a means of or aid to interrogation'.[13]

In a press briefing given on 22 June 2004 at which three high-level administration legal counsels were present, Dan Dell'Orto, DOD Deputy General Counsel, stated: 'The first [Guantánamo] detainees were interrogated under FM 34-52, but by the summer of 2002 it was clear that some had been trained to resist these techniques.'[14] The briefing was held after the treatment of detainees at the Abu Ghraib prison on Iraq had become public. The subject of the briefing was the release of several legal memoranda written by legal counsels within the Bush administration in the summer of 2002. The upshot of these memos was actually that the Convention against Torture was deemed to apply to 'unlawful combatants'. However, the memo of 1 August 2002, written by Assistant Attorney General Jay S. Bybee, 'redefined torture in such a way as to allow a huge range of cruel and inhuman treatment to escape the definitional net of United States and international legislation against torture'.[15] The trail of legal memos also included a list of 'counter-resistance techniques', the details of which had been developed by lawyers at Guantánamo Bay. These techniques were grouped in three categories. Category I techniques comprised yelling at the detainee and using deception. Category II techniques included the 'use of stress positions (like standing), for a maximum of four hours' and the 'removal of clothing'. Category III techniques were intended to be used on 'a very small percentage of the most uncooperative detainees' and for their application, prior authorization by the commander at USSOUTHCOM was required. These techniques included 'exposure to cold weather or water' and the 'use of a wet towel and dripping water to induce the misperception of suffocation', which later became known as 'water-boarding'.[16] Secretary of Defense Donald Rumsfeld subsequently approved the suggested techniques in Category I and II, plus the 'use of mild, non-injurious physical contact' as listed in Category III. William J. Haynes II, General Counsel to the Secretary of Defence, wrote in the memo that the other techniques listed in Category III, while deemed to be 'legally available', were not recommended for approval as a matter of policy: 'Our Armed Forces are trained to a standard of interrogation that reflects a

[13] Field Manual FM 34–52, sections 1–8.

[14] Quoted in Philippe Sands, *Torture Team: Deception, Cruelty and the Compromise of Law* (London: Allen Lane, 2008), 24.

[15] Adam Roberts, 'Detainees: Misfits in Peace and War', in Sibylle Scheipers (ed.), *Prisoners in War* (Oxford: Oxford University Press, 2010), 273. The Bybee memo, 'Memorandum for Alberto R. Gonzales Counsel to the President', 1 August 2002, is printed in Greenberg and Dratel (eds), *The Torture Papers*, 172 ff.

[16] Memorandum for Commander, Joint Task Force 170, 'Request for Approval of Counter-Resistance Strategies', printed in Greenberg and Dratel (eds), *The Torture Papers*, 227 ff.

tradition of restraint.' Rumsfeld's handwritten note on the memo, saying 'However, I stand for 8–10 hours a day. Why is standing limited to 4 hours?', indicated that he had apparently not partaken in that tradition.[17]

The final piece in this mosaic of legal exclusions pertained to the legal situation of Central Intelligence Agency (CIA) personnel. One of the first memos on the legal status of detainees in the 'war on terror' made clear that the CIA claimed exemptions, if not from the law, which the memos argued did not apply anyway, but from the detention policy: 'CIA lawyers believe that, to the extent that GPW's [Geneva Convention on Prisoners of War] protections do not apply as a matter of law but those protections are applied as a matter of policy, it is desirable to circumscribe that policy so as to limit its application to the CIA.'[18] After the Bush administration had been forced to withdraw the notorious Bybee memo on torture in 2004 and Congress had passed the 2005 Detainee Treatment Act banning the use of cruel, inhumane, and degrading treatment, Vice-President Dick Cheney argued that CIA personnel were exempt from the new restrictions and cited the 2002 memo. At the signing of the 2006 Military Commissions Act, which was intended to provide a firmer legal basis for the US detention regime in the 'war on terror', Bush insisted that it did not affect the CIA's interrogations of high-level terrorist suspects.[19]

Apart from the obvious moral and legal issues, these efforts to preserve US legal 'flexibility' (as several of the memos called it) in handling detention and interrogation had the effect of preventing the development of a sustainable detention policy in Afghanistan from the outset. The failure of the US's NATO partners operating in Afghanistan to accept at least a share of responsibility for detention operations compounded the issues evolving from the US approach. While European policy-makers distanced themselves from Donald Rumsfeld's 2002 announcement that the USA would not apply the Geneva Conventions in Afghanistan, they did not develop a legal alternative with respect to detention and interrogation. Instead, from the beginning of operations in 2001 to 2005, European NATO partners in Afghanistan handed over all captured persons to US forces.

The Abu Ghraib scandal that broke out in April 2004 changed that. In 2005 NATO tried to develop a unified detention policy for Afghanistan, but that attempt failed. Instead, all NATO partners except the USA agreed on the standard to transfer detainees within ninety-six hours to Afghan authorities. Some NATO nations sought assurances from the Afghan government

[17] Office of the Secretary of Defense, Action Memo: 'Counter-Resistance Techniques', 27 November 2002, printed in Greenberg and Dratel (eds), *The Torture Papers*, 237.

[18] The Legal Adviser Department of State, 'Memorandum to the Counsel of the President', 2 February 2002, printed in Greenberg and Dratel (eds), *The Torture Papers*, 133.

[19] Adam Roberts, 'Review Essay: Torture and Incompetence in the "War on Terror"', *Survival*, 49/1 (2007), 199–212.

stipulating that detainees had to be treated humanely and that NATO forces were allowed to visit the Afghan detention facilities. However, human rights organizations soon criticized that such assurances were no guarantee that the transferred detainees were treated lawfully. In fact, in 2009 and 2010 the UK put into place several moratoriums on the transfer of detainees into specific Afghan prisons. Canada did the same in 2010.[20] Again the Europeans were partly to blame for these difficulties: Germany and Italy had initially been responsible for the reform of the Afghan police sector and the judiciary respectively. As the reforms made little progress, responsibility was transferred in 2007 to the EU.[21] Even on the eve of the withdrawal of NATO and US forces in 2014, Afghanistan's police and judiciary have not reached the standard that the Western coalition had envisaged. Indeed, the humane treatment of detainees, aside from concerns over the potential release of detainees whom the USA still deemed to be dangerous, was one of the main sticking points between the USA and the Afghan government in the protracted negotiations over the transfer of Bagram detention facility.[22] It also led to allegations breaking in May 2013 that the UK had illegally held eighty-five prisoners for fourteen months at Camp Bastion, which it had refused to hand over to the Afghan authorities.[23] It is likely that this refusal was due to concerns regarding their treatment at the hands of the Afghan authorities, and that the UK forces acted in good faith. However, the incident was once more a stark illustration of the difficulties involved in putting the detention regime in Afghanistan on a sustainable legal basis.

The detention regime in Iraq differed from that in Afghanistan in terms of its legal foundations. At the outset of the invasion, US forces did not anticipate capturing many irregular fighters. The conflict was deemed to be an international armed conflict between two state actors. The presumption was that the Geneva Conventions applied and that captured Iraqi armed forces would qualify for POW status. Even after the start of the insurgencies, the USA as the lead nation in the multinational coalition made no statement to the effect that Iraqi insurgents were 'unlawful combatants'. Instead, individuals not belonging to the regular Iraqi armed forces captured in Iraq where held as 'security internees' under the Geneva Convention IV.[24] The relevant articles

[20] Ashley Deeks, 'Detention in Afghanistan: The Need for an Integrated Plan', Center for Strategic and International Studies Briefing Paper, Washington DC, 14 February 2008 <http://csis.org/files/media/csis/pubs/080213_deeks_afghanistan.pdf> (accessed 26 July 2013).

[21] Cf. Sibylle Scheipers, 'Closing Guantánamo: Is Europe Ready?', *Survival*, 51/1 (2009), 5–12.

[22] Mark Mazzetti, 'US and Afghans Reach Deal on Bagram Prison Transfer', *New York Times*, 23 March 2013.

[23] Haroon Siddique, Andrew Sparrow, and Emma Graham-Harrison, 'Britain May Begin Handover of Afghan Prisoners at Camp Bastion to Kabul', *Guardian*, 30 May 2013.

[24] Geneva Convention IV, articles 27, 42–3, 78, printed in Adam Roberts and Richard Guelff (eds), *Documents on the Laws of War* (Oxford: Oxford University Press, 2000), 311, 315 ff., 327.

of the Geneva Convention IV leave large discretion over the practice of security internment, stipulating merely that interned persons have a right to appeal against their internment and that their internment ought to be subject to a periodical review every six months.[25] However, the application of the Geneva Conventions did not prevent US authorities from introducing coercive interrogation techniques in Iraq. In 2003, the use of techniques amounting to torture, including sleep deprivation, threatening detainees with dogs, and the use of stress positions, was approved for use on specific detainees deemed to 'possess critical intelligence'.[26]

After the Iraqi interim government had acquired formal sovereignty over Iraq in July 2004, the security internment option under GC IV was formally no longer available. However, the option of security internment was extended via UN Security Council Resolution 1546 (2004), which authorized the multinational forces in Iraq 'to take all necessary measures to contribute to the maintenance of security and stability in Iraq'.[27] The UN Security Council extended this authorization in resolutions 1637, 1723, and 1790 so that it applied until the end of 2008. After the formal transfer of sovereignty in 2004, security internment without trial was complemented by the possibility of prosecution within the Iraqi judicial system. However, frustration about low conviction rates in Iraqi courts reportedly irritated US forces to the point that some killed insurgents rather than capturing them if killing them was a lawful option.[28] But the low conviction rates were not the only aspect that advised against the option of prosecution within the Iraqi judicial system. With the levels of violence growing in Iraq, coalition forces were increasingly eager to regain the initiative, and the first step towards that was intelligence. Hence in the years 2004–7 they captured large numbers of random individuals primarily with a view to gathering intelligence. Most of them had to be released after

[25] Geneva Convention IV, article 78, printed in Roberts and Guelff (eds), *Documents*, 327.

[26] Lt Gen. Ricardo S. Sanchez, 'Memorandum for Commander US Central Command: CJTF-7 Interrogation and Counter-Resistance Policy', 14 September 2003 <http://www.aclu.org/files/FilesPDFs/september%20sanchez%20memo.pdf> (accessed 29 July 2013). The memo apparently tries to bridge the gap between the official Geneva-based framework in Iraq and the administration's 'new thinking' on torture, by reminding interrogators that 'CJTF-7 is operating in a theater of war in which the Geneva Conventions are applicable'. It requires that the use of the most extreme techniques on 'enemy prisoners of war' needs high-level authorization, but does not completely rule it out, which indicates that the use of torture and inhumane treatment had by that time become detached from the category of 'unlawful combatants' and was also used on 'security internees' and POWs.

[27] UN Security Council Resolution 1546 (2004), S/RES/1546 (2004), 4.

[28] Commanders reporting on this practice insisted that it never amounted to the denial of quarter, but covered a grey zone where capture may have been possible but killing was equally lawful. As Robert Chesney pointed out, the troops in question were apparently under the misconception that the transfer of detainees to the Iraqi authorities for prosecution was mandatory, when in fact the option of security internment continued to exist: Robert M. Chesney, 'Iraq and the Military Detention Debate: Firsthand Perspectives from the Other War, 2003–2010', *Virginia Journal of International Law*, 51/3 (2011), 573.

seven days, when a compulsory review of their internment was due. Still, commanders latched on to these broad sweeps as one of their few options to counter the growing insurgencies.[29] Throughout the Iraq war a total number of approximately 100,000 detainees passed through the US internment system. At its peak in 2007, US forces held 26,000 individuals as security internees.

The beginning of the 'surge' and General Petraeus's arrival in Iraq in 2007 led to changes in the detention system. Now the issue of detention and internment was explicitly viewed through the lens of counter-insurgency doctrine. 'Detainee alienation' was a factor that was seen as contributing to the radicalization of irregular fighters in Iraq.[30] The new operational approach also called into question the wisdom of holding 'moderate' internees in the detention facilities alongside the 'hardliners'. Commanders realized that the detention facilities, overcrowded and housing a mix of different categories of persons, had developed into hotspots of radicalization in their own right. As a result, they put an emphasis on releasing prisoners and on amnesties, as well as a renewed impetus on strengthening the Iraqi judicial system as an alternative to security internment. In addition, those security internees still held by US forces who were deemed 'moderates' were channelled into a rehabilitation programme that offered literacy classes and job training, but also religious re-education.[31] Many of those released entered the 'Sons of Iraq' programme upon their release.

The legal option of security internment formally ended on 1 January 2009. After that date, all detainees had to be processed through the Iraqi judicial system. However, the agreement between the US and the Iraqi government did not stipulate a date until when the coalition forces had to transfer all security internees to the Iraqi authorities. US forces retained custody of approximately 200 detainees deemed to be affiliated with al-Qaeda. They were handed over only upon the withdrawal of US forces in 2011.[32] Among them were controversial figures such as the Lebanese Ali Musa Daqduq, a suspected Hezbollah fighter who had reportedly killed US troops in Iraq. Daqduq was put on trial in an Iraqi court, but was acquitted on grounds of insufficient evidence and subsequently released.[33]

Although the detention regime in Iraq was on legally more solid grounds than the detention system in Afghanistan and in Guantánamo Bay, particular

[29] Chesney, 'Iraq and the Military Detention Debate', 579.

[30] Mason Brooks and Drew Miller, 'Inside the Detention Camps: A New Campaign in Iraq', *Joint Forces Quarterly*, 52/1 (2009), 132.

[31] Brooks and Miller, 'Inside the Detention Camps', 130.

[32] Chesney, 'Iraq and the Military Detention Debate', 624 ff.

[33] Michael Gordon, 'Against US Wishes, Iraq Releases Man Accused of Killing American Soldiers', *New York Times*, 16 November 2012.

events in Iraq brought the fundamental flaws of detention policies in the 'war on terror' into full view. The Abu Ghraib scandal that broke in April 2004 was the defining point at which many of the USA's allies, but also policymakers and citizens within the USA and commentators around the world, realized that these policies were misguided and unsustainable. The abuse of detainees in Abu Ghraib also demonstrated that manipulating the international torture regime led to the proverbial 'slippery slope': it proved that it was impossible to confine the use of certain techniques to specific detainees or to specifically authorized personnel.[34]

The Abu Ghraib scandal set off a chain of events, including the declassification and release of several US administration memos, such as the notorious Bybee memo, which narrowed the definition of torture as part of the Bush administration's attempt to prove that it was handling detention and interrogation in a 'lawful' way. This largely backfired, and the administration was forced to withdraw the Bybee memo in 2004.[35]

The year 2004 also saw the first US Supreme Court judgments regarding 'unlawful combatants' and Guantánamo inmates. In *Hamdi* v. *Rumsfeld* and *Rasul* v. *Bush* the court ruled that 'unlawful combatants' held within the USA or in Guantánamo could apply for a writ of habeas corpus before US courts. In 2006, the Supreme Court ruled in *Hamdan* v. *Rumsfeld* that common article 3 of the Geneva Conventions did apply to detainees held as 'unlawful combatants'.

Finally, in May 2004 the first reports over the secret CIA programme of 'extraordinary renditions' emerged.[36] Subsequent reports by the Council of Europe, the US Congressional Research Service, and the European Parliament confirmed the existence of such a programme in the framework of which the CIA apprehended terrorism suspects abroad and transferred them 'to locations where not too many questions will be asked about the conditions of their detention and interrogation'.[37] The reports about the CIA rendition programme showed once again that the USA was not the sole author of the

[34] 'At the very least . . . it is clear that the pictures of abuse that emerged from Abu Ghraib would have been less likely without the Haynes memo and the culture of ill-considered aggression it embraced' (Sands, *Torture Team*, 270).

[35] Bybee's successor as head of OLC, Jack Goldsmith, wrote in his account of his short tenure in the job that he had already decided to rescind the memo in late 2003 and was working on a legal opinion to replace the Bybee memo by the time he resigned from the post in the summer of 2004: Jack Goldsmith, *The Terror Presidency: Law and Judgment inside the Bush Administration* (New York: W. W. Norton, 2007), 165 ff.

[36] Stephen Grey, 'America's Gulag', *New Statesman*, 17 May 2004; see also Stephen Grey, *Ghost Plane: The True Story of the CIA Rendition and Torture Program* (New York: St Martin's Griffin, 2007); Jane Mayer, *The Dark Side: The Inside Story of How the War on Terror Turned into a War on American Ideals* (New York: Doubleday, 2008).

[37] Roberts, 'Review Essay', 200.

disastrous detention and interrogation policies in the 'war on terror': many of its European allies had acquiesced or even actively supported it.

In spite of the growing criticism starting in 2004, the legal and physical situation of detainees in the 'war on terror' improved only slowly. The military was no longer authorized to use the notorious techniques outlined in the 2002 memos after the Detainee Treatment Act in 2005 had been adopted, but Vice-President Dick Cheney still publicly stated in October 2005 that 'waterboarding is a no-brainer'.[38] The CIA was still exempt from legal restrictions on the treatment and interrogation of detainees. It was only after President Barak Obama took office in January 2009 that the exemption for the CIA was abandoned. Although the debacle of torture and detention in the 'war on terror' was the first item on Obama's political agenda, he has so far failed to close the prison at Guantánamo Bay, which in 2014 still held 166 inmates.

It is unclear how many detainees were subjected to which interrogation techniques and for how long. Interrogation logs for 'prominent' detainees such as Khalid Sheikh Mohammed (the suspected 'mastermind' behind the 9/11 attacks) and Mohammed al-Qahtani (the alleged twentieth hijacker) have been released.[39] They show that waterboarding was used repeatedly and in combination with other techniques. It is also unclear to what extent torture led to cases of death. Some cases have been reported, but it is plausible to assume that there are more.[40]

'This war has been lawyered to death': Inconsistencies and Paradoxes in the Concept of the 'Unlawful Combatant'

The 'unlawful combatants' narrative operated as a series of exclusions of irregular fighters from the law. The most prominent of these was a temporal exclusion: the Bush administration memos declared that the 'war on terror' marked the onset of a 'new paradigm' in which the law of armed conflict was no longer applicable. Next there was the spatial exclusion of detainees held at Guantánamo, where detainees were deemed not to have any constitutional rights. Also crucially important were personal and institutional exemptions, in particular with respect to CIA personnel, who were consistently declared exempt from legal and political constraints on the treatment and

[38] Mark Tran, 'Cheney Endorses Simulated Drowning', *Guardian*, 27 October 2006.

[39] Qahtani's interrogation log was published by *Time Magazine* and can be found at <http://www.time.com/time/2006/log/log.pdf> (accessed 30 July 2013).

[40] On 4 November 2002, the Iraqi Manadel al-Jamadi died under CIA interrogation in a prison in Afghanistan. On 26 November 2003, General Abed Hamed Mowhoush, head of the Iraqi Air Defence, died as a result of the use of coercive interrogation techniques at an unknown detention centre in Iraq. Cf. Grey, *Ghost Plane*, 156; Alex J. Bellamy, 'No Pain, No Gain? Torture and Ethics in the War on Terror', *International Affairs*, 82/1 (2006), 121. The case of Mowhoush is particularly interesting, as he clearly was not an 'unlawful combatant'.

interrogation of detainees.[41] The final dynamic in the 'unlawful combatants' narrative was the narrowing-down of the legal definition of torture so as to cover only those acts that intentionally inflicted the most severe physical or mental suffering. These exclusionary mechanisms were bolstered and underpinned by attempts to circumscribe the applicability of the law, in particular international law, and by increasing the President's executive power.

The dynamics of exclusions led many commentators to label Guantánamo Bay a 'legal black hole'.[42] As Philippe Sands wrote on the detention regime in the 'war on terror' more generally: 'Legal opinions became expressions of policy ... Whether this was aided by ideology or driven by other ambitions matters not: a violation of the law is a violation of the law.'[43] Critics of the detention policy also widely referred to the CIA programme of renditions as 'extralegal' renditions. The Italian philosopher Giorgio Agamben wrote on the status of 'unlawful combatants': 'Neither prisoners nor persons accused, but simply "detainees", they are the object of a pure de facto rule of a detention that is indefinite not only in the temporal sense but in its very nature as well, since it is *entirely removed from the law* and from judicial oversight.'[44]

Hence the Bush administration memos as well as their critics treated the problem of 'unlawful combatants' as one of exclusion from the law, of illegality, and of extralegality. And yet, ironically, the memos were written by lawyers, who used legal arguments and referred to legal sources, however erroneous the evolving legal opinions may have been. As Fleur Johns put it: 'The detention camps of Guantánamo Bay are above all works of legal representation and classification. They are spaces where law and liberal proceduralism have been speaking and operating in excess.'[45] In fact, while the drafting history of the so-called torture memos is often decried as a process of political expediency usurping the law, the second charge frequently levelled against the memos is that they adopted a 'narrowly legalistic approach'.[46] This perception resonates with the self-perception of some of the leading figures of the Bush administration, who were convinced that the 'torture memos' reflected their commitment to the law. At the 2004 G8 summit in Savannah, Bush said when questioned by the media about the memos released in 2004: 'The instructions went out to our people to adhere to the law.... We're a nation

[41] On the use of such exclusionary measures in times of emergencies, see Oren Gross and Fionnuala Ní Aoláin, *Law in Times of Crisis: Emergency Powers in Theory and Practice* (Cambridge: Cambridge University Press, 2006), 171 ff.

[42] e.g., Johan Steyn, 'Guantánamo Bay: The Legal Black Hole', *International and Comparative Law Quarterly*, 53/1 (2004), 1–15; see also Sands, *Torture Team*, 22.

[43] Sands, *Torture Team*, 275 ff.

[44] Giorgio Agamben, *State of Exception* (Chicago: Chicago University Press, 2005), 3 ff., emphasis added.

[45] Fleur Johns, *Non-Legality in International Law: Unruly Law* (Cambridge: Cambridge University Press, 2013), 70.

[46] Roberts, 'Detainees', 266.

of law. We adhere to laws. We have laws on the books. You might look at those laws, and that might provide comfort for you. And those were the instructions out of—from me to the government.'[47] This pledge of a commitment to the law ran so deep in various testimonies that it is difficult to dismiss it as mere cynical rhetoric. John Yoo, for instance, who was known to have co-authored the notorious Bybee memo on torture, stated in 2002: 'What the Administration is trying to do is create a new legal regime.'[48] Jack Goldsmith, Bybee's successor as head of OLC, confirmed this sentiment when he wrote in his memoir: 'These men and women did not believe they were breaking the law, and indeed they took extraordinary steps to ensure that they didn't.' He came to the conclusion that 'this war has been lawyered to death'.[49] Critics of the detention regime in the 'war on terror' picked up on this—however ill-guided—commitment to the law on the part of the Bush administration when they expressed shock over 'how legal advice can lead to abuse'.[50] What many observers and critics found particularly unsettling was that most of the drafters of the 'torture memos' had an immaculate academic pedigree, with many of them holding degrees from Ivy League law schools.

There was a further sense in which the detention regime led to an 'excess' of law, as Johns put it. As we have seen, the detention regime in Afghanistan was particularly fraught with inconsistencies and uncertainties, especially after 2004, when the USA's NATO partners were no longer willing to hand over detainees to US forces. As the issue of detention had become highly controversial at the time, NATO was unable to negotiate a unified detention policy, which led to the need for each NATO contributor to negotiate a separate agreement with the Afghan government. According to Adam Roberts, this led to a legal situation of 'almost baroque complexity': 'The Afghan government would be entitled to be bemused by the complexity of the legal arrangements for classification and treatment of detainees, consisting as they do of an extraordinary variety of different documents from different states.'[51] Herein lies an additional contradiction, as it was the self-professed aim of the Western coalition in Afghanistan to strengthen the rule of law in the country. How, one wonders, were Afghan officials supposed to know or 'learn' what the rule of law means, when the coalition forces themselves were so obviously unable to reach agreement on that question?

The paradoxes of legality/illegality/extra-legality of the detention regime were frequently attributed to a mixture of 'fear and ideology' that characterized the political climate, in particular in the USA, after 9/11.[52] There was clearly fear that the USA could be subject to new attacks and also a realization

[47] Quoted in Sands, *Torture Team*, 19.
[48] Quoted in Sands, *Torture Team*, 88.
[49] Goldsmith, *Terror Presidency*, 69.
[50] Goldsmith, *Terror Presidency*, 29.
[51] Roberts, 'Review Essay', 207.
[52] Sands, *Torture Team*, 271.

that the government did not have enough intelligence on the threat emerging from al-Qaeda and affiliated terrorist groups, which led to frantic demands for 'enhanced interrogation' as a source of new information. As to ideology, it is correct that there is a certain political–legal, intellectual tradition in the USA that has been described as 'sovereigntist'. Its main tenet is a deep-seated scepticism towards international law as an infringement on constitutional law-making and executive power.[53] Leading figures behind the 'torture memos' such as John Yoo were confessed 'sovereigntists'. Yet these explanations overlook the extent to which the detention regime in the 'war on terror' and its approach to 'unlawful combatants' in particular had its roots in a long-standing legal tradition of the exclusion of irregular fighters. This leads us to the second paradox.

The central assumption that lay at the heart of the Bush administration memos was that the 'war on terror' was a new kind of conflict. The White House Executive Order of 7 February 2002 stated that the 'war against terrorism ushers in a new paradigm': the USA was no longer facing '"regular" armed forces fighting on behalf of States', but rather 'groups with broad international reach [that] commit horrific acts against innocent civilians, sometimes with the direct support of States'.[54] This assessment echoed the various memos that had preceded the Executive Order. In a memo of 9 January 2009, for instance, John Yoo emphasized 'the novel nature of this conflict' and went on to argue that

> it seems to us overwhelmingly likely that an armed conflict between a Nation State and a transnational terrorist organization, or between a Nation State and a failed State harboring and supporting a transnational terrorist organization, could not have been within the contemplation of the drafters of Common Article 3. These would have been simply unforeseen and, therefore, not provided for.[55]

Finally, Alberto Gonzales stated in a memo of 25 January 2002 that 'this is a new type of warfare'.[56]

[53] See Stephanie Carvin, *Prisoners of America's Wars: From the Early Republic to Guantánamo* (New York: Columbia University Press, 2010), 157 ff. On the 'sovereigntist' tradition, see also Sibylle Scheipers, *Negotiating Sovereignty and Human Rights: International Society and the International Criminal Court* (Manchester: Manchester University Press, 2009). This phenomenon has also often been referred to as 'American exceptionalism'. The basic 'sovereigntist' discourse, however, is not confined to the US legal tradition; it also plays a certain role in France, for instance.

[54] 'Memorandum for the Vice President', 7 February 2002, printed in Greenberg and Dratel (eds), *The Torture Papers*, 134. Note that this assumption was shared and echoed by some European policymakers, e.g. John Reid, 'Twenty-First Century Warfare—Twentieth Century Rules', *RUSI Journal*, 151/3 (2006), 14–16.

[55] 'Memorandum for William J. Haynes II, General Counsel, Department of Defense', 9 January 2002, printed in Greenberg and Dratel (eds), *The Torture Papers*, 39, 46 ff.

[56] 'Memorandum for the President', 25 January 2002, printed in Greenberg and Dratel (eds), *The Torture Papers*, 120.

Yet, in explaining the ways in which this 'new type of warfare' differs from the established norm, many of these lawyers and policymakers cited arguments that echoed century-old stereotypes towards irregular fighters. In a 2003 paper on 'The Status of Terrorists' co-authored with James Ho, John Yoo, contended, for instance, that al-Qaeda members are by definition undisciplined as they 'are not under the control of a nation state that will force them to obey the laws of war'.[57] This charge goes back to the origins of the marginalization of irregular fighters during the French Revolutionary and Napoleonic Wars.[58] Further, Yoo and Ho wrote that al-Qaeda fighters:

> operate covertly by intentionally concealing themselves among the civilian population; they deliberately attempt to blur the lines between civilians and combatants. Most importantly, they have attacked purely civilian targets with the aim of inflicting massive civilian casualties.[59]

Here Yoo and Ho echo an argument against irregular fighters that Francis Lieber had already used one and a half centuries earlier. While it is, of course, true that al-Qaeda targets civilians, it is problematic and misleading to blame such tactics exclusively on irregular fighters.[60]

Another standard characterization of the 9/11 attacks and other terrorist incidents was to label them as acts of 'cowardice'.[61] Again, the denunciation of irregular fighters as 'cowards' is a familiar theme in the trajectory of their marginalization. It was most prominent during the American Civil War—that is, at a time when new technologies had transformed the meaning of courage on the battlefield.[62] With the so-called Revolution in Military Affairs (RMA) starting in the 1990s and its promise of increased standoff capabilities for Western militaries, it is plausible that a similar sense of disorientation over the meaning of courage and cowardice in war had spread among policymakers and commentators in the USA. Shortly after the 9/11 attacks, talk-show host Bill Maher called into question the moral coding of terrorist attacks as 'cowardly' and decried Western standoff tactics as cowardly instead; in doing so, he was perceived as breaking a political taboo and was subsequently sacked by his network ABC.[63]

[57] John C. Yoo and James C. Ho, 'The Status of Terrorists', Boalt Working Papers in Public Law, Paper 25 (Berkeley, CA: University of California, Berkeley, 2003), 10.

[58] See Chapter 2.

[59] Yoo and Ho, 'The Status of Terrorists', 10.

[60] Cf. Chapter 1.

[61] A day after the 9/11 attacks, for instance, Bush himself stated: 'Make no mistake, the United States will hunt down and punish those responsible for these cowardly acts' ('Bush: US Military on "High Alert"', *CNN.com*, 12 September 2001 <http://edition.cnn.com/2001/US/09/11/bush.second.statement/> (accessed 24 August 2013).

[62] Cf. Chapter 3.

[63] Maher said: 'We have been the cowards, lobbing cruise missiles from 2,000 miles away. That's cowardly. Staying in the airplane when it hits the building, say what you want about it, it's not

Finally, the legal and political narrative on 'unlawful combatants' after the 9/11 attacks borrowed some features from the nineteenth- and early twentieth-century discourses on colonial warfare. This is most clearly reflected in the characterization of the Taliban as a 'tribal' entity, diametrically opposed to the modern Western conception of the state.[64] The Bybee memo of 22 January 2002, for instance, argued that 'the Taliban militia in essence represented only an ethnically Pashtun movement, a "tribal militia"'.[65] According to the memo, this was the primary reason why Afghanistan was to be classified as a 'failed state' in which the Geneva Conventions did not apply and why the Taliban could not qualify as 'quasi-regular' armed forces. Tribalism was incompatible with statehood.

Against the background of this recourse to arguments stemming from discourses of colonialism, Frédéric Mégret has argued that it is plausible to draw a rather straight line from the Western imperialist image of the 'savage' to the 'war on terror's' concept of the 'unlawful combatant'.[66] According to Mégret, because the exclusion of irregular fighters is firmly entrenched in the law of armed conflict, 'the rhetoric of the Bush Administration is often merely mimicking the law. Indeed, the US authorities' case is often not a case to simply violate or do away with the law, as much as it is a characteristically strict, almost legalistic interpretation of the law.'[67] This argument is convincing, although it has to be broadened to encompass not only references to colonial discourses on irregular fighters, but also arguments that stem from the trajectory of the marginalization of the irregular more generally.

The contradiction in this context consists in the fact that the Bush administration insisted on the novelty of the 'war on terror' as the main argument for their treatment of irregular fighters, and yet it harked back to century-old legal arguments to justify the marginalization of irregular fighters as 'unlawful combatants'. This seems to be an instance of what could be called 'double historical amnesia': it combined the erroneous argument that, with the 9/11 attacks, war had changed beyond recognition with a complete unawareness of

cowardly' (quoted in Celestine Bohlen, 'In New War on Terrorism, Words Are Weapons, Too', *New York Times*, 29 September 2001).

[64] Patrick Porter summarizes the US depiction of the Taliban as a predominantly 'tribal' actor as follows: 'The "tribal mind" honours no agreements, respects no other principles outside the tribe, and has an irrational, instinctive mind of "warrior pride", not of Enlightenment reason' (Patrick Porter, *Military Orientalism: Eastern War through Western Eyes* (London: Hurst, 2009), 154). Note that Porter goes on to argue that the depiction of the Taliban was an 'ethnographic fiction' on the part of US commentators and policy-makers. Cf. Giustozzi, *Koran, Kalashnikov and Laptop*, 47.

[65] 'Application of Treaties and Laws to al Qaeada and Taliban Detainees', printed in Greenberg and Dratel (eds), *The Torture Papers*, 96 ff.

[66] Frédéric Mégret, 'From "Savages" to "Unlawful Combatants": A Post-Colonial Look at International Humanitarian Law's "Other"', in Anne Orford (ed.), *International Law and its 'Others'* (Cambridge: Cambridge University Press, 2006), 265–317.

[67] Mégret, 'From "Savages" to "Unlawful Combatants"', 304.

the historical rootedness of certain policies and arguments designed as a response to this 'new war'.[68]

The final paradox in the 'unlawful combatants' narrative is that, despite the claim that the 'war on terror' represented a new kind of war, it was strictly speaking not a spontaneous reaction to the 9/11 attacks. Rather, the claim of a new era of warfare had its roots in various arguments that had already become popular in the 1990s concerning an alleged fundamental 'transformation of war'. At the end of the cold war, a number of scholars predicted that war was undergoing profound changes. Various labels were attached to this alleged 'transformation of war'. It was described as the rise of the so-called new wars, as a dramatic increase in 'low-intensity conflict' or, alternatively, as the emergence of 'fourth-generation warfare' or, briefly, '4GW'.[69] These accounts varied with respect to the causes for the paradigm shift as well as to specific details in the analysis, but they all have three central assumptions in common. The first is that the new paradigm of war is characterized by the predominance of 'non-state' actors and by the state's loss of its monopoly on the use of armed force. The concomitant trend is the emergence of so-called failed states—that is, states that either are actively undermined by the activities of non-state actors or passively provide an operating environment for such actors (or a mix of both). Secondly, these accounts put forward the argument that, despite the obvious military weakness of non-state actors, in particular when pitted against powerful Western militaries, they pose a serious threat to Western states and societies. They achieve this disproportionate potential by using 'asymmetric' tactics and by turning the West's strengths against itself.[70] Thirdly, it was argued that the West's defence policies and institutions are woefully ill equipped to deal with this new threat emerging from non-state actors and failed states, as they are still overwhelmingly geared towards 'old wars', meaning interstate wars. Specifically, the West was said to be prone to rely on high-tech solutions when faced with an asymmetric opponent who is unlikely to confront Western militaries on a traditional battlefield.[71]

[68] See also Sibylle Scheipers, 'Irregular Fighters: Is the Law of Armed Conflict Outdated?', *Parameters*, 43/4 (2013), 1–12.

[69] See, among others, Martin van Creveld, *The Transformation of War* (New York: Free Press, 1991); Mary Kaldor, *New and Old Wars: Organized Violence in a Global Era* (Stanford, CA: Stanford University Press, 2012); William S. Lind et al., 'The Changing Face of War': Into the Fourth Generation', *Marine Corps Gazette*, 73/10 (1989), 22–6; Thomas X. Hammes, *The Sling and the Stone: On War in the 21st Century* (St Paul, MN: Zenith Press, 2006).

[70] 4GW theorists referred to this phenomenon by using the metaphor of a 'judo throw'; cf. Lind et al., 'The Changing Face of War', 26.

[71] Kaldor, *New and Old Wars*, 152 ff.; Hammes, *Sling and the Stone*, 190 ff.

The torture memos, in particular those written in early 2002 on the question of the status of al-Qaeda members and the Taliban, echoed these three assumptions. They repeatedly cite the fact that al-Qaeda is a 'non-state' actor as the main reason why the Geneva Conventions do not apply to them.[72] They insist that, because Afghanistan is a failed state, the Taliban do not count as a quasi-regular armed force.[73] The assumption that al-Qaeda's and the Taliban's status as 'non-state' actors made them a serious and deadly threat was never mentioned or explained explicitly, but the sense of urgency that emanates from the memos speaks for itself. Finally, as demonstrated above, one of the core convictions of the drafters of the torture memos was that the Western concept of war was outdated, in particular when it came to the applicability of the law of armed conflict.[74]

This is, of course, not to say that the 'new wars' and 4GW theses had in some way predetermined the path taken by the US administration after 9/11. On the contrary, many of the scholars associated with these theses had suggested a fundamentally different approach to dealing with irregular fighters.[75] Moreover, there is no empirical evidence that the authors of the torture memos had read the works on the 'new wars' and on 4GW, although there are strong indications that the latter had some impact on policymakers and military

[72] '*Al Qaeda's status as a non-State actor renders it ineligible to claim the protections of the treaties specified by the WCA* [War Crimes Act]. Al Qaeda is not a State. It is a non-governmental terrorist organization composed of members from many nations, with ongoing operations in dozens of nations.... Non-governmental organizations cannot be parties to any of the international agreements here governing the laws of war. Al Qaeda is not eligible to sign the Geneva Conventions—and even if it were eligible, it has not done so' ('Application of Treaties and Laws to al Qaeda and Taliban Detainees', printed in Greenberg and Dratel (eds), *The Torture Papers*, 48, emphasis in original).

[73] 'A "failed State" is generally characterized by the collapse or near-collapse of State authority.... The Executive can readily find that at the outset of this conflict, when the country was largely in the hands of the Taliban militia, there was no functioning central government in Afghanistan that was capable of providing the most basic services to the Afghan population, of suppressing endemic internal violence, or of maintaining normal relationships with other governments. Afghanistan, consequently, was without the status of a State for purposes of treaty law, and the Taliban militia could not have qualified as the *de facto* government of Afghanistan. Rather, the Taliban militia would have had the status only of a violent faction or movement contending with other factions for control of that country' ('Application of Treaties and Laws', printed in Greenberg and Dratel (eds), *The Torture Papers*, 53).

[74] While the 'new war' and 4GW theorists never made this claim with respect to the law of armed conflict, others who followed in the footsteps of their analysis of the transformation of war did. See Michael Gross, *Moral Dilemmas of Modern War: Torture, Assassination and Blackmail in an Age of Asymmetric Conflict* (Cambridge: Cambridge University Press, 2010).

[75] The solution Kaldor proposed is 'cosmopolitan law enforcement', which is fundamentally rooted in international humanitarian and human rights law: Kaldor, *New and Old Wars*, 133. Creveld in his *Transformation of War* came close to predicting the post-9/11 detainee disaster when he wrote: 'An armed force that violates the war convention for very long will disintegrate. This will be all the more the case if it is powerful, hence unable to convince others—and itself—of the imperative need for it to break the rules' (Creveld, *Transformation of War*, 205).

planners' perceptions and approaches in the 'war on terror'.[76] The intention here is to call into question the claim made by policymakers and lawyers involved in the 'production' of the 'unlawful combatants' narrative, that the newness of the events of 9/11 led to a sudden disruption and disorientation, after which a new legal framework had to be designed and implemented. What in fact happened was that they took recourse to the established concepts of a 'new paradigm' and used it as a lens to make sense of the events of 9/11 and their strategic context. The 'new paradigm' was to a certain extent a self-fulfilling prophecy. The decisions policymakers and lawyers took were in no way predetermined by the 'new wars' and 4GW theses; however, the sense of revolutionary change in the field of war and the fear that the West was not sufficiently well equipped to deal with the threat emerging from 'non-state' actors and 'failed states' was a direct result of these theses.

What is most disturbing is that the perception of the 'war on terror' as a new war seems to persist. The Obama administration came into office with the pledge to close Guantánamo Bay and to put the detention regime in the 'war on terror' on a more solid legal basis. At a roundtable event at the American Heritage Foundation in May 2013, then Deputy Assistant Secretary of Defense for Rule of Law and Detainee Policy William Lietzau insisted that the administration had been successful with respect to the second goal, even if it had not yet achieved the closure of the Guantánamo prison. He further argued that the Obama administration's detainee policies represented the 'gold standard' of compliance with the law of armed conflict. Yet, when he talked about the trajectory of detainee policy it became clear that, for the Obama administration, at least one of the underlying assumptions of the 'unlawful combatants' narrative was still relevant: Lietzau stated that the detainee policy during the 'war on terror' had developed into an extremely contested issue 'perhaps because of some early missteps, but I believe more profoundly because this is a *new type of war*'.[77]

[76] See, e.g., Antulio J. Echevarria II, *Fourth Generation Warfare and Other Myths* (Carlisle, PA: Strategic Studies Institute, 2005).

[77] Statement by William K., 'The Past, Present and Future of US Detainee Policy', American Heritage Foundation, Washington DC, 13 May 2013 <http://www.heritage.org/events/2013/05/detainees> (accessed 13 August 2013). Lietzau had already been concerned with detainee policy within the Bush administration. In July 2013, he announced that he would leave his post in the Pentagon. Reportedly the reason for his decision to leave was disagreements between Lietzau and the White House over the transfer of Guantánamo detainees. According to the *New York Times*, Lietzau 'has frequently defended prisoner-of-war-style indefinite detention without trial, saying it is a moral, lawful and humane part of warfare. His leading role within the administration in defending its continued use in the open-ended war against Al Qaeda has drawn fire from many of the same critics angry at President Obama for failing to close the prison at Guantánamo' (Charlie Savage, 'US to Send 2 at Guantánamo Back to Algeria, Saying Security Concerns Are Met', *New York Times*, 26 July 2013).

Irregular Fighters as the Subject of Targeted Killings

Targeted killings are defined as:

> The intentional killing of a specific civilian or unlawful combatant who cannot reasonably be apprehended, who is taking a direct part in hostilities, the targeting done at the direction of the state, in the context of an international or non-international armed conflict.[78]

Targeted killings are thus a practice that is explicitly geared towards irregular fighters, as regular opponents are usually fought by more conventional means. Targeting regular opponents would also not cause legal issues, as it would represent 'straightforward examples of acts of war'.[79]

Drone strikes and targeted killings more generally became a military instrument of choice for the US administration in the framework of the 'war on terror'. While attempts by states to kill certain individuals have always taken place, targeted killings as an explicit tactical tool with a specific legal justification are a new phenomenon. The first state that openly discussed the practice and its legal basis was Israel. Israel had started to use targeted killings against Palestinian irregular fighters in the 1970s. It ramped up its campaign after the beginning of the Second Intifada in 2000. In 2006, the Israeli Supreme Court issued a ruling according to which the practice per se was neither lawful nor unlawful. Rather, its lawfulness depended on the conditions of specific cases.[80]

In the US cold war context, after dubious CIA practices had come to light in the course of the Watergate scandal, subsequent administrations had adopted a prohibition against assassinations. Executive Orders 11905 (1976) and 12333 (1981) prohibited political assassinations carried out by any state official. The strength of these orders had already been undermined in the 1990s, when the Clinton administration made various (obviously failed) attempts to kill Osama bin Laden.[81] However, the real turning point in US policy on targeted killings came with the attacks of 9/11 and the subsequent difficulties in prevailing against irregular fighters in Afghanistan and Iraq. In 2003, during the invasion of Iraq, US forces tried but failed to kill Saddam Hussein in a so-called decapitation strike. After the outbreak of the insurgencies in Iraq, US forces increased their reliance on this practice. On 7 June 2006, a US airstrike killed Abu Musab al-Zarqawi, the head of AQM.

[78] Gary Solis, quoted in David Whetham, 'Drones and Targeted Killing: Angels or Assassins?', in Bradley Jay Strawser (ed.), *Killing by Remote Control: The Ethics of an Unmanned Military* (Oxford: Oxford University Press, 2013), 75 ff.

[79] Whetham, 'Drones and Targeted Killing', 76.

[80] See Nils Melzer, *Targeted Killing in International Law* (Oxford: Oxford University Press, 2008), 32 ff.

[81] Daniel Byman, 'Do Targeted Killings Work?', *Foreign Affairs*, 85/2 (2006), 95–111.

Since 2006, coalition forces have also increasingly relied on targeted killings in Afghanistan. Prominent targets have included members of the Taliban's *Shura Council*, al-Qaeda, the Haqqani network, and various other extremist groups.[82] The Obama administration's decision to start withdrawing forces as of 2010 and complete the withdrawal from Afghanistan by mid-2014 reportedly led to a spike in drone strikes in the country.[83]

But targeted killings were not confined to the theatres of war in Afghanistan and Iraq. In 2004, the USA initiated a programme of drone surveillance and strikes in Pakistan in order to gain control over Taliban insurgent forces and suspected al-Qaeda members who had increasingly gained a foothold in the FATA. The most high-level, targeted killing incident was undoubtedly the killing of Osama bin Laden by a US Special Forces commando on 2 May 2011. Other high-level cases include a successful CIA drone strike against Haithan al-Yemeni, a suspected member of the al-Qaeda leadership on 10 May 2005, and the attempted killing of Osama bin Laden's suspected second in command, Ayman al-Zawahiri, on 13 January 2006. While the latter attempt failed to kill al-Zawahiri, it caused eighteen unintended deaths and triggered heavy international criticism. The US government never publicly acknowledged responsibility for the attacks on al-Yemeni and on al-Zawahiri. The political and legal context of the USA's targeted killings in the Afghan–Pakistani border region is compounded by the fact that the Pakistani government has never officially acknowledged its authorization of the programme, although it appears that it tacitly acquiesced to them. Approximately 360 drone strikes have been carried out in Pakistan between 2004 and 2013. The overall number of deaths caused by the strikes was approximately 3,000. Between 260 and 880 of these deaths were civilians who had not taken part directly in hostilities.[84] In 2010, there was a sharp drop in the number of civilian deaths as a result of increased precautionary measures owing to international criticism and as a result of the use of a new type of drone and weaponry.[85] However, of the thousands of intended deaths as a result of drone attacks in Pakistan, only a small fraction were high-level targets such as members of the al-Qaeda leadership. The overwhelming majority were low-level militants.[86]

[82] Alex S. Wilner, 'Targeted Killings in Afghanistan: Measuring Coercion and Deterrence in Counterterrorism and Counterinsurgency', *Studies in Conflict and Terrorism*, 33/3 (2010), 316.

[83] Helene Cooper and Mark Landler, 'Targeted Killing Is New US Focus in Afghanistan', *New York Times*, 31 July 2010.

[84] International Crisis Group, *Drones: Myths and Reality in Pakistan* (Brussels: International Crisis Group, 2013), 8.

[85] Avery Plaw, 'Counting the Dead: The Proportionality of Predation in Pakistan', in Strawser (ed.), *Killing by Remote Control*, 150.

[86] Plaw, 'Counting the Dead', 143.

The USA also conducted targeted killing operations in other countries. On 3 November 2002, a CIA drone strike killed six suspected al-Qaeda members in Yemen. The Yemeni government had reportedly authorized the strikes. Since 2011 Yemen has become the predominant target of drone strikes. In 2012 the USA carried out more drone strikes in Yemen than anywhere else. The summer of 2013 saw a sharp rise in drone attacks in Yemen, which were apparently related to intelligence alerts over new attacks planned by al-Qaeda.[87] Since 2011 the USA has reportedly also conducted between three and nine drone strikes in Somalia.

The effectiveness of the practice of drone strikes and of targeted killings is contested. Some commentators arrive at a wholly negative assessment. In an op-ed in the *New York Times*, David Kilcullen and Andrew Exum called for a moratorium on drone strikes in Pakistan. They argued that, instead of achieving US strategic aims, the strikes actually alienate and radicalize the local populations and exacerbate the problems in the Afghan–Pakistani border region.[88] Others pointed out that drone strikes disrupt social structures and institutions of the local communities. Suspected local informants are often pursued and killed by local communities, which can spark a cycle of violence and counter-violence. Moreover, local inhabitants in Pakistan have become reluctant to participate in large gatherings, as drone strikes were reportedly often targeted at weddings, funerals, and other get-togethers. A particularly problematic result is that the traditional councils of elders, the *jirgas*, no longer convene for fear of attracting an attack.[89] A further argument against drone strikes focuses on the elasticity of legal justifications for them: if the USA sanctions a definition of legitimate targets that is excessively inclusive, this could serve as a justification for other states or even non-state actors to justify attacks on civilians.[90]

However, some commentators are more optimistic about targeted killings. Daniel Byman, for instance, presented a cautious but positive assessment of the capacity of targeted killings to disrupt terrorist activities and unravel terrorist networks. However, he admitted that drone strikes pose difficult problems related to their legality and legitimacy, and called for more

[87] Yalda Hakim, 'Why Drone Attacks in Yemen Are Like "Trying to Hit a Ghost"', BBC News, 10 August 2013 <http://www.bbc.co.uk/news/world-middle-east-23653488> (accessed 17 August 2013).

[88] David Kilcullen and Andrew McDonald Exum, 'Death from Above, Outrage Down Below', *New York Times*, 16 May 2009. The ICG disputes the claim that drone strikes led to a direct radicalization among the local population; see International Crisis Group, *Drones*, 25.

[89] International Crisis Group, *Drones*, 11; James Cavallaro, Stephan Sonnenberg, and Sarah Knuckey, *Living under Drones: Death, Injury and Trauma to Civilians from US Drone Practices in Pakistan* (Stanford, CA: Stanford Law School; New York: NYU School of Law, 2012), 95 ff.

[90] Anthony Dworkin, *Drones and Targeted Killing: Defining a European Position* (London: European Council on Foreign Relations, 2013), 3.

transparency.[91] Others were even more optimistic and regarded targeted killings as the tactic of choice against elusive terrorists and insurgents.[92]

Targeted killings belong to the same strategic, political, and moral complex of issues as 'unlawful combatants'. However, in contrast to the debate over the detention regime in the 'war on terror', legal opinions and justifications underpinning US decisions regarding targeted killings have not been widely published yet. The debate has only just begun to become public.[93] President Obama's speech at the National Defense University on 23 May 2013 broke new ground insofar as he devoted it almost entirely to the topics of detention and targeted killings.[94] No doubt, repeated calls for more transparency, both from within the USA and internationally, played a role in prompting this turn. However, rather than presenting a policy plan for achieving more transparency regarding targeted killings, Obama pledged only to work toward this aim. One option that Obama reportedly considered was to abandon the collaboration between DOD and CIA regarding drone strikes and to assign responsibility for the drone programme solely to the DOD. Yet, as of July 2014, the CIA is still conducting the majority of US drone strikes.[95] The large involvement of the CIA in targeted killings is an 'irregular' response to irregular challenges, inasmuch as the CIA is not a part of the US regular armed forces and often claims to operate in a different legal context from the latter, as we have already seen with regard to the detention and interrogation of 'unlawful combatants'. This resort to 'irregular' means and actors should not be surprising, as it has been a staple of anti-irregular campaigns throughout the history of modern war. However, calls for the 're-regularization' of the drone programme reflect a certain degree of uneasiness about this irregularization on the part of the US administration.

While most of the legal justifications for the US targeted killing policy have not been published yet, the contours of the debate that has emerged allow us to compare the US approach to targeted killing with its approach to detainee policy. Strikingly, the publicly available legal justifications for targeted killings differ considerably from the legal arguments regarding detention. First, there is no claim that the 'war on terror' is a new war in which the law of armed conflict has become obsolete. On the contrary, the main strand of legal arguments on targeted killing

[91] Byman, 'Do Targeted Killings Work?'.

[92] Wilner, 'Targeted Killings in Afghanistan'.

[93] Melzer, *Targeted Killing*, 42 ff.

[94] A transcript of the speech can be found at 'Obama's Speech on Drone Policy', *New York Times*, 23 May 2013.

[95] Greg Miller, 'CIA Remains behind Most Drone Strikes, Despite Effort to Shift Campaign to Defense', *Washington Post*, 25 November 2013.

> relies on the claim that the US is engaged in an armed conflict with al-Qaeda, the Taliban, and associated forces... While administration officials admit that the international laws governing such a conflict against an external non-state groups are unclear, they contend that the rules should be understood *by analogy with more traditional forms of conflict* to allow the targeting of all members of enemy forces wherever they are found.[96]

This explanation is based on an 'expansive' model of armed conflict.[97] In the words of an International Crisis Group (ICG) report:

> the broad focus on the 'transnational' nature of modern-day terrorism and the inclusion of a wide range of al-Qaeda 'associates' among those who pose a threat to US national security dilute legal limits on the US's ability to strike suspected terrorists abroad. It also undermines the idea that war is an exceptional state of affairs and entrenches the idea of a war with no temporal or geographical constraints.[98]

But the legal thinking on which targeted killing is based not only depicts the 'war on terror' as a war without geographical boundaries and without a clear end. Rather, it also characterizes all 'terrorists' as legitimate military targets.[99] In cases of so-called signature strikes, meaning strikes aimed at a particular pattern of activities rather than at particular persons, everyone carrying out this activity becomes a legitimate target. Hence the legal construct on which targeted killing is founded is based on a series of analogies and inclusions rather than exclusions, as is the case in the debate over detainee policy. Ultimately, therefore, the legal arguments surrounding targeted killings are incompatible with those on which the detention regime is based. The clash between the different legal arguments raises difficult questions. Is the 'war on terror' a new war or a rather a form of a conventional, regular, armed conflict fought by different actors? Are irregular fighters lawful targets or unlawful combatants or both? It would be hard to argue that the two different approaches represent complementary alternatives, as is sometimes said of the 'kill or capture' alternative. Because the underlying assumptions for these legal arguments are contradictory, they are ultimately incompatible.

In spite of the differences regarding the legal foundations of the detention regime and the practice of targeted killing, they have a common political

[96] Dworkin, *Drones and Targeted Killing*, 5 (emphasis added). According to Dworkin, the alternative justification relies on arguments of self-defence, but this justification is even fuzzier than the first one and seems to represent a minority opinion within the US administration. The leaked DOJ paper 'Lawfulness of a Lethal Operation Directed against a US Citizen: Who Is a Senior Operational Leader of Al-Qa'ida or an Associated Force' (8 November 2011) mentions both arguments <http://msnbcmedia.msn.com/i/msnbc/sections/news/020413_DOJ_White_Paper.pdf> (accessed 17 August 2013).

[97] Dworkin, *Drones and Targeted Killing*, 6.

[98] International Crisis Group, *Drones*, 15.

[99] Melzer, *Targeted Killing*, 429.

effect in that they both work as a display of power towards the opponents of the USA and the West more generally. As we have seen in the previous chapter, torture was often used in the framework of colonial warfare, not only in order to 'make people talk', but to make them listen and understand, in particular to understand which side ultimately wielded power in a conflict. Throughout the 'war on terror', US policy on detention and interrogation was characterized by a certain ambivalence over information policies. In 2002, Donald Rumsfeld publicly announced that the USA would not apply the Geneva Conventions in Afghanistan. In 2004, the Bush administration released the so-called torture memos in spite of the international outcry that they were bound to provoke. Part of the reason for this 'mismanagement' of information was probably the erroneous assessment of international sympathy and international understanding for these legal measures. But it is also plausible that the USA was trying to communicate with at least two different audiences at once: its international allies *and* its international opponents. While the former were to be reassured, the latter were to be intimidated.[100]

Viewed from this perspective, there is not such a great difference between the premeditated and bureaucratically implemented acts of coercive interrogation at places such as Guantánamo and the seemingly gratuitous and 'amateurish' instances of torture in Abu Ghraib. Many commentators regarded Abu Ghraib as an instance of 'bad apples' mimicking certain techniques that had been authorized only for use in certain prisons by certain interrogators. Abu Ghraib was seen as a perfect example for the 'slippery slope' of torture.[101] Yet what was striking about Abu Ghraib was the number of photographs and videos that those in charge of the facility took and the fact that these pictures often depicted the torturer as well the victim. Alex Danchev noted that the pictures were 'almost triumphal'.[102] They were displays of power just in the same way as public half-knowledge of 'enhanced interrogation' techniques was a display of power intended to intimidate the opponent and to break his or her will to continue fighting.

Targeted killings can be understood in the same context, especially when conducted by drone strikes. Constant drone surveillance coupled with occasional strikes reportedly has a deeply intimidating effect on the affected communities. One interviewee of the *Living under Drones* project described this effect as follows:

[100] Cf. Bob Woodward, *Obama's Wars* (New York: Simon and Schuster, 2010), 55.

[101] e.g. Yuval Ginbar, *Why Not Torture Terrorists?: Moral, Practical, and Legal Aspects of the 'Ticking Bomb' Justification for Torture* (Oxford: Oxford University Press, 2008), ch. 9.

[102] Alex Danchev, 'Review Article: Bad Apples, Dead Souls: Understanding Abu Ghraib', *International Affairs*, 84/6 (2008), 1272.

> If I am walking in the market, I have this fear that maybe the person walking next to me is going to be a target of the drone. If I'm shopping, I'm really careful and scared. If I'm standing on the road and there is a car parked next to me, I never know if that is going to be the target. Maybe they will target the car in front of me or behind me. Even in mosques, if we're praying, we're worried that maybe one person who is standing with us praying is wanted. So, wherever we are, we have this fear of drones.[103]

Evidence from the experience of targeted killing practices in the Palestinian territories confirms this effect. According to the former head of the Israeli human rights NGO Shin Bet, Avi Dichter, 'When a Palestinian child draws a picture of the sky, he doesn't draw it without a helicopter in it.'[104] Hence Uwe Steinhoff is convincing when he draws a parallel between the policies of intimidation used in the 'war on terror' and in colonial warfare.[105]

Rebels: Useful Auxiliaries, but not 'Unlawful Combatants'?

In the 'war on terror' the USA and its allies relied from the very start on the support of local auxiliary forces. Most prominent among these was the Northern Alliance in Afghanistan. It consisted principally of three groups: Jamiat, led by Ahmad Shah Massoud, Jubesh-i Milli under the leadership of Abdul Rashid Dostum, and Hezbi Whadat under Karim Khalili. These groups had already fought the Taliban during the Afghan Civil War and were hence seen as natural allies of the USA after the 9/11 attacks and the decision to invade Afghanistan.[106] The Northern Alliance and other anti-Taliban militias (sometimes referred to as the Southern Alliance) played a decisive role in the early fighting on the ground in Afghanistan. Abdul Rashid Dostum and his forces, for instance, were instrumental in the fall of Mazar-i-Sharif in November 2001. Indigenous auxiliaries were also heavily involved in the fighting along Highway 4 in December 2001 and in Operation 'Anaconda' in March 2002. According to one observer: 'In Afghanistan, the Northern and Southern Alliances, eventually combined with American and Canadian troops that fought during Anaconda, together provided significant ground forces that ultimately shouldered an essential load of old-fashioned close combat against surviving, actively resisting opponents.'[107]

[103] Quoted in Cavallaro, Sonnenberg, and Knuckey, *Living under Drones*, 98.

[104] Quoted in Byman, 'Do Targeted Killings Work?', 100.

[105] Uwe Steinhoff, 'Killing them Safely: Extreme Asymmetry and its Discontents', in Strawser (ed.), *Killing by Remote Control*, 207.

[106] Amin Saikal, 'Afghanistan's Ethnic Conflict', *Survival*, 40/2 (1998), 114–26.

[107] Stephen Biddle, 'Afghanistan and the Future of Warfare', *Foreign Affairs*, 82/2 (2003), 44.

However, apart from the fact that the Taliban had harboured Osama bin Laden and other members of al-Qaeda, and that the Northern Alliance had fought against the Taliban, there had been few differences in the behaviour of the various sides in the Afghan Civil War. In fact, the US administration had not only supported and armed the mujahidin during the Afghan–Soviet war in the 1980s; during the Civil War in the 1990s the Clinton administration had also had a positive outlook on the Taliban.[108] All sides in the Civil War had committed war crimes, and the forces gathered under the umbrella of the Northern Alliance were no exception. Hezbi Wahdat, for instance, was reportedly responsible for the massacre of 2,000 Taliban prisoners during the Civil War.[109] This pattern of violence persisted even after the US invasion of Afghanistan and the elevation of the Northern Alliance to allies of the West. Abdul Rashid Dostum's forces were accused of having committed a massacre among Taliban prisoners in Dasht-i-Leili in December 2001. His forces had reportedly dumped containers full of Taliban prisoners in the desert near Shar-i-Kot and left them to suffocate and starve. In spite of allegations against Abdul Rashid Dostum, the Bush administration refused to investigate the incident at the time.[110]

At the time when the collaboration between US forces and the Northern Alliance started in the autumn of 2001, human rights organizations warned 'that [Northern Alliance] troops routinely indulge in looting, raping, and killing in their attempts to capture the whole of Afghanistan'.[111] Nevertheless, the irregular fighters of the Northern Alliance and other anti-Taliban forces were nowhere labelled as 'unlawful combatants'. Instead, they were called 'rebels' and 'militias', and their leaders were persistently referred to by their military ranks such as 'General Dostum' and 'General Massoud', although the provenance of such titles was obscure.

During the invasion of Iraq, the coalition forces' most important allies were irregulars from the Kurdish provinces of northern Iraq called *peshmergas*. As with the Northern Alliance, there had been allegations of war crimes committed by the *peshmergas*.[112] The second important irregular force who came to be a Western ally were the so-called Awakening Councils and the Sons of Iraq, who turned away from the Sunni insurgency as of 2007. While these fighters obviously had fought as 'enemy combatants' before the Awakening, winning them over was seen as the most important turning point in the trajectory of

[108] Saikal, 'Afghanistan's Ethnic Conflict', 21.

[109] Saikal, 'Afghanistan's Ethnic Conflict', 120.

[110] 'The Truth about Dasht-i-Leili', 13 July 2009.

[111] BBC News, 'Should the West Support the Northern Alliance?' <http://news.bbc.co.uk/2/hi/talking_point/1615191.stm> (accessed 18 August 2013).

[112] During the 1991 Kurdish uprising in Iraq, the *peshmergas* allegedly massacred Iraqi regular forces who were trying to surrender. See Kurt Schork, '*Hors de Combat*', Crimes of War <http://www.crimesofwar.org/a-z-guide/hors-de-combat/> (accessed 18 August 2013).

the Iraq war. There were reports that the withdrawal of US forces could leave the Sons of Iraq programme without financial funding or a political base in the newly formed Iraqi state and that Sunni irregulars could turn back to AQM.[113] The rise of the Islamic State of Iraq and the Levant (ISIS), an AQM splinter group, which took control of large parts of Iraq in June 2014, has confirmed this assessment.

The question of Western support for irregular fighters became even more relevant beyond the 'war on terror'. In February 2011, a violent uprising started in Libya with the aim to oust the country's dictatorial leader Muammar Gaddafi. In March 2011, the UN Security Council authorized the imposition of a no-fly zone and air strikes to protect civilians in the country. NATO forces received the mandate to implement these measures. In the same month, Obama authorized covert support for the anti-Gaddafi armed groups. In August 2011, the US and other governments recognized the Transitional National Council (TNC), a political coalition formed by the anti-Gaddafi movement, as the governing authority of Libya. In October armed groups captured and killed Gaddafi. Subsequently, the TNC declared that Libya was officially 'liberated'.

Once again, the official terminology referred to the anti-Gaddafi armed groups as 'rebels'. The 'rebels' enjoyed the West's full military and political support, and few questions were asked about their political credentials or about their military conduct as irregular fighters. On the contrary, they were de facto regularized by the West's decision to recognize the TNC. However, a UN investigation into war crimes committed during the Libyan revolution came to the conclusion that both sides, the Gaddafi forces and the armed opposition groups, were responsible for war crimes, even if it also stated: 'The commission received fewer reports of facts that would amount to the commission of international crimes by forces connected with the opposition.'[114] In September 2012, an Amnesty International Report 'warned Libya's leaders that unless they rein in armed militias and bring them to account Libya risks reproducing the same human rights violations that led to the "17 February Revolution" against Mu'ammar al-Gaddafi's repressive rule last year'.[115] Earlier that month the West had already received a wake-up call regarding the new political and military situation in the country when a heavily armed group of Jihadi extremists attacked the US diplomatic mission in Benghazi and killed

[113] Amnesty International Deutschland, 'Ein Funke genügt', June 2009 <http://www.amnesty.de/journal/2009/juni/ein-funke-genuegt> (accessed 18 August 2013).

[114] *Report of the International Commission of Inquiry to Investigate All Alleged Violations of International Human Rights Law in the Libyan Arab Jamahiriya* (Geneva: Office of the Human Rights Council, 2011), 7.

[115] Amnesty International, 'Libya's Leadership Urged to Prioritize Human Rights', 24 September 2012 <http://www.amnestyusa.org/news/news-item/libya-s-leadership-urged-to-prioritize-human-rights> (accessed 19 August 2013).

four of its staff, including the ambassador, J. Christopher Stevens. Subsequently evidence emerged that the attack on the embassy had been carried out with weapons that Qatar had transferred to Libyan rebels with US authorization in 2011.[116]

The question of support for irregular armed groups also developed into the main sticking point in debates over the West's possible reactions to the civil war in Syria, which began in 2012. Whereas the West was quick to deliver non-lethal support to the Syrian 'rebels', the question of arms transfers was more controversial, owing to fears that the weapons could fall into the hands of groups affiliated with al-Qaeda. As of the summer of 2012, the CIA was reportedly involved in covert transfers of heavy weapons to armed fighters in Syria.[117] In May 2013, the EU lifted its arms embargo to Syria, thereby opening up the possibility for its member states to transfer arms to the 'rebels'. However, human rights groups such as Amnesty International cautioned against this possibility:

> Amnesty International has gathered evidence of abuses by some of the many armed opposition groups operating in Syria, some of which amount to war crimes, including summary killings of captured members of the security forces, paramilitary militias, suspected informers and others . . . because of the existence of such abuses by some armed opposition groups, Amnesty International believes that arms transfers to any armed opposition group in Syria should be withheld until the removal of any substantial risk of serious violations of human rights and IHL [International Humanitarian Law].[118]

After evidence emerged in June 2013 that Bashar al-Assad's forces had used chemical weapons against the opposition forces and civilians, the US administration decided to deliver arms openly to the opposition forces in Syria. However, the measure remained contested within the USA as well as internationally.

The West's treatment of auxiliaries in the 'war on terror' is on the surface entirely disconnected from the problem of 'unlawful combatants'. The armed groups that the West supports as auxiliaries or as revolutionary movements in their own right are referred to as 'rebels' or 'militias'. The political rhetoric thus paints a black and white picture with 'bad' 'unlawful combatants', on the one

[116] James Risen, Mark Mazzetti, and Michael S. Schmitt, 'US Approved Arms for Libya Rebels Fell into Jihadis' Hands', *New York Times*, 5 December 2012 <http://www.nytimes.com/2012/12/06/world/africa/weapons-sent-to-libyan-rebels-with-us-approval-fell-into-islamist-hands.html?pagewanted=all&_r=0> (accessed 19 August 2013).

[117] Eric Schmitt, 'CIA Said to Aid in Steering Arms to Syrian Opposition', *New York Times*, 21 June 2012 <http://www.nytimes.com/2012/06/21/world/middleeast/cia-said-to-aid-in-steering-arms-to-syrian-rebels.html?pagewanted=all> (accessed 19 August 2013).

[118] Amnesty International, 'EU Declaration on Arms Transfer to Syria', 31 May 2013 <http://www.amnesty.org/fr/library/asset/MDE24/027/2013/fr/dcfff54d-76bc-4230-a6ca-3b6a2c1ec79f/mde240272013en.html> (accessed 19 August 2013).

hand, and 'good' 'rebels', on the other. From a legal point of view, it is unclear what distinguishes the 'bad' irregular fighters from the 'good' ones other than political expediency. However, one of the basic principles of the modern law of armed conflict is that it applies universally and that it does not discriminate against certain groups on the grounds of the political aims that they are fighting for. But the political rhetoric also obscures that the reality is often not quite as black and white as some want to believe. Cases in which US auxiliaries all of a sudden turned into 'unlawful combatants' are instructive in this respect. The Afghan militia leader Ziauddin, for instance, whose forces participated in the battle of Shah-i-Kot, and his friend Abdullah Mujahid, local police chief in Gardez in Paktia, who also had cooperated with the Americans in the years 2001 and 2002, were suddenly declared 'unlawful combatants'. After their relations with the Americans soured, both were arrested in 2003. Ziauddin ended up in the US prison at Bagram air base and Mujahid in Guantánamo.[119] Within months, both had supposedly crossed the large dividing line that allegedly distinguished good 'rebels' from 'bad' 'unlawful combatants'. In reality, both Ziauddin and Mujahid had been neither 'white' nor 'black'—they had been 'grey', meaning they had been irregular fighters, who can be a strategic asset, but who are notoriously difficult to control and whose loyalty is often difficult to retain.

The debate on weapons transfers to Syrian armed groups is only minimally more enlightened than the naïve take on auxiliaries in Afghanistan or on the Libyan 'rebels'. While it does reflect the insight that arms transfers are difficult to control and that a variety of different groups are likely to operate in a civil war scenario, it is so far entirely focused on the fear that arms could fall into the hands of groups related to al-Qaeda. This focus obscures the problem that *all* irregular groups are difficult to control and will pursue their own political objectives and renegotiate their alliances if push comes to shove. But not only irregular fighters tend to reassess their alliances. In reality, the same holds also for their patrons if they perceive a shift in the strategic context. The experience of US support for the mujahidin in Afghanistan in the 1980s should serve as a warning that support for irregular groups can have hugely adverse long-term ramifications, even if it fits short-term strategic objectives.

Conclusion

The preceding chapter discussed the development of the concept of the irregular during in the twenty-first century. The largest part of the chapter

[119] Farah Stockman, 'US behind Afghan Warlords Rise and Fall: At Guantánamo Unruly Chieftains Join Combatants', *Boston Globe*, 12 August 2007.

was devoted to the notion of the 'unlawful combatant', which emerged predominantly from questions regarding the detention of irregular fighters in the 'war on terror'. It showed that the concept of the 'unlawful combatant' is characterized by internal inconsistencies and paradoxes and sits uneasily with parallel legal narratives describing irregular fighters as objects of targeted killing operations and as auxiliaries of the West.

Internally, the 'unlawful combatants' concept is characterized by three paradoxes in particular. First, although the detention regime in the 'war on terror' has frequently been denounced as featuring 'legal black holes' and as sacrificing adherence to the law for political expediency, the concept of the 'unlawful combatant' is the result of intensive legal argument and writing. Rather than being void of the law, it seems to have been 'lawyered to death', in the words of Jack Goldsmith. Second, the US administration's main argument for introducing the concept of the 'unlawful combatant', which was based on the idea that the 'war on terror' was a 'new war' and that the law of armed conflict was hence outdated, was actually based upon the century-old arguments already underpinning the legal marginalization of irregular fighters. Finally, although those responsible for the detainee regime in the 'war on terror' claimed that they had been taken by surprise by this 'new' kind of war, their arguments on its nature echoed the analyses published in the 1990s on the 'new wars' and on 4GW. To a certain extent, the 'newness' of the 'war on terror' hence turned out to be a self-fulfilling prophecy.

But the 'unlawful combatants' concept is not only internally inconsistent; it also sits uneasily with the treatment and characterization of irregular fighters in other legal and strategic contexts. On the one hand, the legal arguments underpinning targeted killing seem to be inconsistent with the 'unlawful combatants' concept, in that they consist of a series of analogies and inclusions rather than relying on the exclusion of irregular fighters from the remit of the law. Their aim is to show that 'unlawful combatants' are in fact lawful targets. However, the most obvious inconsistency exists between the depiction of irregular fighters as 'unlawful combatants', on the one hand, and as useful auxiliaries, on the other. While political rhetoric suggests that these two manifestations have nothing to do with each other, the reality has repeatedly disproven this assessment. If it is the professed aim of the USA and its allies to put the detention and the targeting of irregular fighters on a sounder legal basis, these inconsistencies will have to be addressed.

Conclusion

The status of the irregular fighter in the twenty-first century is both contested and marginalized. In order to understand how this status evolved, this study has analysed the emergence and the development of the concept of the irregular throughout the history of modern war. The concept of *'la guerre irregulière'* originated in the mid-eighteenth century. It differentiated between 'regular warfare' and 'irregular warfare' at the tactical and operational levels. However, irregular warfare in this context was not yet the conceptual opposite of regular warfare. Rather, it was regarded as a tactical complement to regular warfare. Regular and irregular warfare were expected to occur alongside one another. Irregular warfare hence originated from the 'private–public partnership' arrangements that characterized the state in early modern Europe. Its formalization, mainly in the context of mid-eighteenth-century efforts to spell out a doctrine for irregular warfare, signified an attempt by the state to appropriate and regulate it.

The acknowledged coexistence of regular and irregular warfare as tactically complementary components of war ended with the French Revolution. Irregular warfare ceased to be the tactical counterpart of regular warfare. Rather, it became its conceptual opposite. The notions of the regular soldier and the irregular fighter became morally charged, and a moral hierarchy developed, equating the irregular fighter with the illegitimate combatant. The historical constellation in which this transformation came about was characterized by the nationalization of war and the emerging cult of the *levée en masse*. The double meaning of the term *levee*, which could signify both an official levy and a spontaneous popular uprising, made it necessary for the revolutionary government to distinguish between the 'proper' *levée en masse* and the counter-revolutionary insurgencies, which were portrayed as illegitimate. At the same time, from the point of view of the military, the inclusion of a wider group of the population as soldiers through the nationalization of war had the paradoxical effect of leading to a greater professionalization of the armed forces. The result of this professionalization was that the differences between the regular army and irregular guerrilla bands became more pronounced. The negative attributes of the irregular, such as their indiscipline and their propensity to loot, became more visible. Yet, the attempts to blame such negative traits exclusively on irregular fighters were inaccurate. Napoleon's armies may have increased the standard of discipline and drill on the battlefield, but off the battlefield they were just as prone to loot and plunder as their irregular counterparts. Thus the evolving moral hierarchy, which depicted the irregular as morally inferior, was a reflection of the regular armies' ambitions rather than their empirical reality. By presenting the irregular as the conceptual opposite of the regular, regular armies were, at least at the discursive level, purged of all negative traits.

This conceptual trend was to continue despite the fact that contemporaries of the period of restoration did not anticipate the lasting impact of the French Revolutionary and Napoleonic Wars on modern war. Their aim was to restore the status quo ante. Yet the second half of the nineteenth century brought the realization that irregular fighters never really stopped being involved in wars in Europe and North America. At the same time, negotiations concerning their status made visible the ideological conflicts over the question of where to draw the line between the regular and the irregular, as well as between the irregular fighter and non-fighting citizens. The republican tradition, which envisaged war would be conducted with the involvement of the citizens of a state, clashed with other approaches that either regarded legal restraints in warfare as superfluous or insisted that war was the sole preserve of states and their regular armed forces, but not of the citizens of a state.[1] The ideological clash brought a desire for legal clarification, leading to the drafting of the Lieber Code in 1863 but also to the failed attempt to codify the law of armed conflict at the 1874 Brussels conference. Both draft texts as well as the 1880 Oxford Manual were not only largely exclusionary towards irregular fighters but also reflected the punitive approach towards 'civilians' that had become common practice in the field by that time. It is important to remember that the term 'civilians' in today's sense was not yet defined at the time of the early attempts to codify the law of armed conflict. Protections for those ostensibly not involved in fighting were dependent on their behaviour and even their political *Weltanschauung*. Occupation forces demanded absolute loyalty, which often put 'civilians' in an impossible situation, as they were simultaneously under pressure from irregular fighters operating in their vicinity. Regular forces were effectively entitled to pass on security risks to 'civilians'.

The 1907 Hague Convention reinforced this trend. At the same time, it introduced some inclusive measures towards irregular fighters. It stipulated

[1] Cf. Karma Nabulsi, *Traditions of War: Occupation, Resistance, and the Law* (Oxford: Oxford University Press, 1999).

that militias and volunteer corps and even spontaneous risings en masse for the defence against an invading army were legitimate, provided they fulfilled strict conditions, which essentially required them to resemble regular armies as closely as possible. In addition, the Martens clause introduced an appeal to humanitarian conscience not to let all restraint slip towards those groups of persons that were not subject to explicit legal protections.

The Second World War brought a new turning point in the trajectory of the concept of the irregular. Nazi German occupation policy epitomized a climax in both the exclusionary approach towards irregular fighters and the punitive approach to 'civilians'. Nazi propaganda fused the demarcation of the irregular as an illegitimate combatant as well as the denunciation of the 'civilian' as a potentially treacherous enemy with racist and, on the eastern front, also anti-Bolshevist stereotypes. This was the enabling ideational context of the war of annihilation in the east and in the south-east and of numerous massacres and atrocities in occupied western Europe. However, the Second World War also led to two important changes in the assessment of irregular fighters. First, from the perspective of the Allies it led to a partial reappraisal of the irregular as the morally vindicated resistance fighter. And, secondly, it once again reincarnated the concept of the irregular as the strategically useful auxiliary. British strategic planning in the early phases of the war produced a particularly explicit conceptual approach to the irregular as auxiliary, but the use of irregulars as auxiliaries was also common on the Soviet and on the German sides.

The horrific levels of violence against irregular fighters and civilians that characterized the Second World War both in Europe and in Asia showed the need for the codification of new legal protections. The main innovation in the 1949 Geneva Conventions was the unequivocal protection of civilians in occupied territory. It was obvious to the drafters of the Conventions that the potentially punitive approach towards civilians was no longer viable. However, the exclusionary approach towards irregular fighters persisted. The criteria for lawful belligerency remained almost unchanged. At the same time, the Geneva Conventions introduced inclusive aspects, even though they remained weak. Common article 3 to the Conventions spelled out minimum protections applicable in 'armed conflicts not of an international character'—that is, common article 3 effectively restricted protections for irregular fighters to civil war contexts. However, Europe's two major colonial powers, Britain and France, saw common article 3 with apprehension, as they had a strong interest in shielding their colonial policies from legal interference.

Until the second half of the twentieth century, war in the colonies did not have a large impact on the conceptualization of the irregular fighter as a figure of the codified law of armed conflict. Anti-irregular warfare in the colonies mostly drew upon the same toolkit of established practices that were used in the centre of the international system. Specific practices that were characteristic for particular actors and particular periods seemed to travel from the centre to the peripheries and back. However, anti-irregular warfare in the colonies was distinct in that it led more easily to exterminatory policies. Racism and the so-called standard of civilization may have played a role in justifying unrestrained violence; however, the extreme anti-irregular measures, and a very wide definition of irregular fighters to include many groups of civilians, reflected above all the desire on the part of the colonizers to 'produce' suitable colonial subjects, even if that involved the total destruction of the indigenous inhabitants' physical and social existence. Extermination was the 'civilizing mission' of the colonial territories carried to extremes. However, apart from the levels of violence used to suppress resistance in the colonies, there was no qualitative difference between anti-irregular warfare in the centre and in the peripheries. If anything, actors in the peripheries were more easily stigmatized as 'irregular' even if they did not necessarily or exclusively use irregular tactics. If irregular fighters in Europe and North America were denounced as uncivilized because they were irregular, this causality was stood on its head in the colonies: here colonial authorities stigmatized colonial 'subjects' as irregular fighters, because they regarded them as uncivilized.

Britain and France solved the problem posed by the potential applicability of common article 3 to wars in their colonies by declaring that these wars were 'domestic' affairs to be tackled within the framework of emergency legislation. With common article 3, the 'rebel' had become a figure in civil wars who was potentially able to claim minimum protections. If wars in the colonies after the Second World War were not to be categorized as civil wars, the terminology had to change. The irregular fighter now became the (communist) 'terrorist'. Undoubtedly the wars of decolonization, as they were referred to in hindsight, were fought in a context of heightened legal awareness. This awareness resulted in a characteristic ambivalence on the part of the colonial authorities: on the one hand, the aim was to subdue colonial uprisings effectively and to produce 'good' colonial subjects in the process. On the other, there was a need to act with restraint, owing to the changed international context, making it necessary to create more flexibility for their repressive acts by keeping the legal situation as vague as possible. This led to double-bind communications in which colonial authorities preached moderation, but failed to prosecute and punish transgressions of their orders towards the civilian population and the opposing forces. As the most ruthless means of suppression were ostensibly no longer available, the internment and detention in camps with the aim of rehabilitating former fighters through a process of recantation and re-education became more important. Although the notion of 'rehabilitation' gave these measures a civilized veneer, they often included torture as their prime instrument.

The process of the liberation of the colonies and the inclusion of the newly decolonized states into the law-making process opened up the possibility of renegotiating the status of irregular fighters. This happened mainly with the codification of the 1977 Additional Protocol I to the Geneva Conventions. Additional Protocol I is more inclusive than the Geneva Conventions regarding combatant privilege. It states that 'the armed forces of a Party to a conflict consist of all organized armed forces, groups and units which are under a command responsible to that Party for the conduct of its subordinates, even if that Party is represented by a government or an authority not recognized by an adverse Party'. The Protocol effectively abandons the distinction between the regular and the irregular and restricts its criteria for privileged combatancy to the requirement that a combatant 'carries his arms openly: (a) during each military engagement, and (b) during such time as he is visible to the adversary while he is engaged in a military deployment preceding the launching of an attack in which he is to participate'.[2] Additional Protocol I reflects the realization that irregular fighters are often severely constrained in their choice of tactics and methods by the political situation in which they find themselves and against which they fight.

Despite this tendency towards more inclusiveness towards irregular fighters, the figure of the irregular remained contested even after Additional Protocol I came into force. A number of states, including the USA and Israel, refused to ratify the Protocol because they worried that it was too lenient towards 'terrorists'. The latest obvious manifestation of the contested character of the irregular fighter was, of course, the debate over the concept of the 'unlawful combatant'. This concept arose because of external events, but also because of the internal dynamics of the legal concept of the irregular fighter. The 'unlawful combatants' narrative would not have emerged without the 9/11 attacks or an incident of a similar nature and magnitude. Yet the way in which policy-makers and lawyers in the USA reacted to this event was heavily influenced by the previous assumptions underlying the legal concept of the irregular and its strong tendencies towards exclusion. The 'unlawful combatants' narrative reproduced the pre-Second World War view of the irregular fighter, which consists in the presentation of the irregular as the conceptual opposite of the regular combatant.

[2] Additional Protocol I, articles 43 (1) and 44 (3), printed in Adam Roberts and Richard Guelff (eds), *Documents on the Laws of War* (Oxford: Oxford University Press, 2000), 444 ff.

Is the Law of Armed Conflict Outdated?

As we have seen, the 'unlawful combatants' narrative was based on the argument that the law of armed conflict is outdated. But this argument was not confined to the 'unlawful combatants' narrative. Rather, many critical voices in Europe and in the USA argued that al-Qaeda terrorists should be prosecuted under criminal law rather than be placed outside the law; they deplored that the boundaries between criminal law and the law of armed conflict had become too rigid to allow for a flexible approach to the problem of terrorism.

The underlying assumption of both arguments is that the law of armed conflict was made for a particular kind of war—namely, regular, interstate war. In as much as the wars in the twenty-first century no longer fit that template, the law in its current form is difficult to apply, if not completely obsolete. However, this study has shown that this assumption could not be further from the truth. The law of armed conflict was drafted precisely as a response to the challenges posed by irregular fighters. At all time in the history of the codification of the law of armed conflict, the issue of irregular fighters gave the greatest impetus for new conventions and protocols. At the same time, irregular fighters were the most contested subjects of the law of armed conflict. This constellation resulted in the emergence of a double-negative approach to irregular fighters. First, irregulars were defined *ex negativo*, as all those combatants who do not belong to the regular armed forces or to one of the other narrowly confined groups of lawful combatants. Secondly, the law failed to spell out specific standards for their treatment other than the minimum protections entailed in common article 3 of the Geneva Conventions and article 75 of Additional Protocol I, which has by now developed into customary law. This double-negative approach to irregular fighters is the reason why policymakers and lawyers in the twenty-first century find it difficult to categorize irregular fighters and to agree on standards for their treatment. In sum, the law's approach to irregular fighters is not particularly clear, and it is fraught with strong exclusionary measures towards that category of combatants. This is not because the law was only to address regular interstate war, but because it propagates that regular form of war as an ideal type.

What implications does the preceding analysis have for policymakers and military practitioners who face the difficult task of dealing with irregular fighters in current and future conflicts? The first step is to realize that what has changed in the twenty-first century is not the nature of armed conflict or the types of actors involved in it. Irregular fighters have played a role at all times in the history of modern war. Their occurrence is not new, and it certainly does not signify the erosion of the state system or of the foundations of the organization of political and military power. However, what is new are the West's typical strategic aims in these conflicts. The West no longer fights irregulars in order to prevail in a conflict over the domestic organization of the state and the structure of society, as the Union did in the American Civil War. Neither do Western coalitions these days aim to dominate the peripheries of the international system by force and to 'police' post-colonial resistance movements with the occasional unrestrained application of military force *pour encourager les autres*, as the colonial powers did. The aim of most Western wars and military operations since the end of the cold war was 'stabilization'—in other words, the achievement of a sustainable political situation of relative peace and stability, which stops short of making too many prescriptions regarding the design of political institutions. In other words, even if the 'war on terror' was begun with the confessed aim of democratization, this aim was soon perceived as unrealistic, and stabilization became paramount.

In most cases, those branded as 'irregulars' will have to be reintegrated into society and into the political system. Of course, this does not apply to those who have committed or planned mass atrocities. They will have to be prosecuted under either domestic criminal law or international law. However, the vast majority of irregular fighters are low- to mid-level militants or warlords who do not pose an acute threat to the West. They will have to be integrated in the post-war political process in order to enable a sustainable political situation to emerge. Large-scale, exclusionary policies towards irregular fighters such as indefinite detention, torture, and inhumane treatment are counterproductive with respect to the aim of political stabilization. They alienate and destroy societies and corrupt their political institutions. Moreover, this exclusionary treatment sends ambivalent messages to the governments in the target states and encourages them to copy dubious detention policies. Against this background, it is no surprise that, in both Afghanistan and Iraq, prisons have been the preferred targets of extremist attacks in recent years, as they remain symbols of the normative double-bind situation in those countries. In sum, there can be no doubt that large-scale, exclusionary measures towards irregular fighters are obstacles on the way to political stabilization.

How should the West treat irregular fighters in order to further the aim of stabilization and societal reconciliation? The Islamic legal tradition of *akham al-bughat* is instructive in this respect. *Akham al-bughat* is a body of moral and legal thought that developed between the eighth and the seventeenth centuries (second and eleventh centuries in the Islamic calendar). It stipulates that rebellions can be legitimate, if the rebels have a legitimate reason to rebel, a *ta'wil*. The notion of *ta'wil* points to the fact that both the ruler and the rebels share a common religious and normative background; it underlines that the difference between the ruler and the rebels arose because of their divergent interpretations of some of the rules and orders emerging from that common background:

> the term *ta'wil* does not necessarily mean a religious interpretation on a particular point of law. Rather, a *ta'wil* could be either a disagreement on a legal point or simply a grievance (*mazlama*). Therefore, a refusal to obey a command which the rebels believe to be illegal or unjust qualifies as a proper *ta'wil*.[3]

The assumption that rebels can have a legitimate reason to resist the authority of the ruler has an important impact on the question of how rebellions ought to be fought within the framework of *akham al-bughat*. It is legitimate for the ruler to fight against rebellions, but the guiding principle in the conduct of hostilities ought to be reconciliation: 'The notion that rebels are to be reconciled implies that war not only is a means of discipline, but is to be conducted in such a way that any legitimate grievances of the rebels may be redressed.'[4] This has two implications: first, captured rebels must be treated with respect. Executions, torture, and inhumane treatment are prohibited. Secondly, the rebels ought to be granted some latitude with respect to the tactics they use, although this latitude is not unrestricted: 'those who fight "from below" must have room to maneuver, since the forces of the establishment typically enjoy overwhelming advantages. Yet this latitude cannot be, and is not, absolute. Rebels who claim the mantle of the Shari'a are not free to utilize means that the Shari'a cannot approve.'[5]

Of course, this is not to say that we should abandon the established law of armed conflict and substitute it with *akham al-bughat*. Yet *akham al-bughat* offers interesting principles and insights that seem to be particularly well suited to the current strategic situation and its emphasis on political stabilization and reconciliation. Rather than relying on the exclusion of irregular fighters, *akham al-bughat* stresses inclusion and leniency. To a certain extent, Additional Protocol I's approach to irregular fighters echoes these principles. Yet, as we have seen, the more inclusive aspects of the law of armed conflict have not been particularly prominent in the West's policies in recent years. Although the West's professed strategic aim in recent conflicts in Iraq and Afghanistan was stabilization, its policies towards irregular fighters—with the exception of limited programmes such as the Sons of Iraq scheme—belied this objective.

How to explain this misfit between ends and means? The law of armed conflict is definitely not outdated in the sense that it was not made with a view to irregular fighters. As argued above, if it is outdated, that is because it reflects approaches to the problem of irregular fighters that no longer fit the West's

[3] Khaled Abou El Fadl, *Rebellion and Violence in Islamic Law* (Cambridge: Cambridge University Press, 2001), 243.

[4] John Kelsay, *Arguing the Just War in Islam* (Cambridge, MA: Harvard University Press, 2007), 103.

[5] Kelsay, *Arguing the Just War in Islam*, 193.

strategic aims in the twenty-first century. The law's approach to irregular fighters bears the mark of the strategic interests of major states at the time of its codification. At the beginning of the twentieth century, Germany and Russia had an interest in excluding irregular fighters from legal protections because of their strategic situations as great powers with large land armies. After the Second World War, the colonial powers had an interest in restricting protections for combatants in non-international conflicts. It is these strategic interests behind the law and its more exclusionary tendencies that are obsolete, not the law itself.

The charge frequently levelled against the USA is that its strategic interest dictated its legal approach to the 'war on terror'. However, viewed from the perspective outlined above, this charge has to be turned on its head, in the sense that it was the law, or, more accurately, a particular exclusionary strand within the law of armed conflict, that framed the US strategic and tactical approach to irregular fighters and its detention policies in particular. An approach geared to the strategic interest of stabilization properly conceived would have brought the treatment of irregular fighters in line with the aim of political stabilization. This would have meant relying on the more inclusive aspects of the law such as those outlined in Additional Protocol I. Moreover, it would have implied that additional protections would have had to be implemented as a matter of policy in cases where legal protections are too weak or too vague. Of course, the law as such is not to blame for the disasters of the detention policy in the 'war on terror'. It offers minimum protections for irregular fighters, but it leaves open the possibility of additional restraint and moderation. However, in the 'war on terror' policymakers and lawyers tried to hide behind the law in order to justify their disastrous strategic decisions.

Should the law of armed conflict be revised? Ideally, a revision of the law ought to achieve two things: first, to provide a clarification of what protections apply to irregular fighters and, second, to make these protections more comprehensive.[6] However, it is unlikely that substantial revisions of the law of armed conflict will emerge any time soon. Local initiatives such as the Copenhagen process, which emerged because of the confusion over detention policies in Afghanistan, are a starting point. The Copenhagen Principles and Guidelines address questions such as the length of detention, habeas corpus rights, and treatment standards. However, they fail to qualify which legal text they use as a foundation—the Geneva Conventions or Additional Protocol I. This is not surprising, as the Copenhagen process includes both signatories and non-signatories of Additional Protocol I. In addition, among the

[6] For a more comprehensive discussion of the questions that ought to be addressed, see Adam Roberts, 'Detainees: Misfits in Peace and War', in Sibylle Scheipers (ed.), *Prisoners in War* (Oxford: Oxford University Press, 2010), 276 ff.

signatories, different reservations to Additional Protocol I apply. Nonetheless, the Principles and Guidelines are intended to provide a common ground to secure future military interoperability and to prevent detention policies in joint operations from ending up in chaos, as happened in Afghanistan. The price to be paid is that the Principles often remain vague and refer to the 'applicable law' without specifying which law is meant.[7] However, as already noted, in the absence of a revision of the law of armed conflict and of its approach to irregular fighters, nothing in the law of armed conflict prevents policymakers and lawyers from granting additional protections to irregular fighters if this is necessary for the achievement of their strategic aims.

Cooperating with Irregular Fighters: The State and its Auxiliaries

The irregular fighter in its incarnation as the auxiliary of regular forces has played as important a role in the history of modern war as its 'illegitimate' alter ego. Colonialism would have been unthinkable without the colonial powers' reliance on various forms of colonial armed forces, among them often irregular auxiliaries. But equally important, irregular auxiliaries never ceased to play a role in conflicts in Europe and North America. The irregular fighter's incarnation as an auxiliary to regular forces has its roots in mid-eighteenth-century strategic theory on *la guerre irregulière*. With the French Revolution it was complemented and to a certain extent eclipsed by the irregular as the illegitimate combatant. While the latter was subject to intense legal and philosophical arguments, the former remained under-theorized throughout the history of modern war. In a way, it remained stuck in the context of its eighteenth-century origin: *la guerre irregulière* defined irregular fighters in an auxiliary role as a tactical complement to regular warfare. It thus ignored the fact that the use of auxiliaries in fact often had strategic as well as political motives and implications. Examples range from Confederate irregulars during the American Civil War to the use of native auxiliaries in colonial wars. The conceptual blind spot regarding the strategic and political implications of using auxiliaries operates as the main mechanism that upholds the artificial distinction between irregular fighters and auxiliaries.

However, in most cases of their empirical occurrence, irregular fighters were both illegitimate combatants and useful auxiliaries, depending on the perspective of the respective parties to the conflict. For Napoleon the Spanish

[7] *The Copenhagen Process on the Handling of Detainees in International Military Operations: Principles and Guidelines* (Copenhagen: Department of International Law) <http://um.dk/en/~/media/UM/English-site/Documents/Politics-and-diplomacy/Copenhagen%20Process%20Principles%20and%20Guidelines.pdf> (accessed 5 September 2013).

guerrillas were 'brigands', and for Wellesley they were auxiliaries. Similarly, for the Germans, the resistance fighters during the Second World War were 'partisans', and for the Allies, they were auxiliaries. In recent years auxiliaries gained renewed prominence in the framework of Western strategies in Afghanistan, Iraq, Libya, and Syria. How should we assess the cooperation of Western regular armed forces with irregular auxiliaries?

Without doubt, irregular auxiliaries are often useful, as they alleviate the West's endemic manpower problems and balance out the West's reluctance to commit sufficient troop numbers to conflicts in the peripheries of the international system. Reliance on auxiliaries reflects the same underlying trends and issues as the use of private security contractors. However, auxiliaries have the added advantage of being much cheaper than the latter. They usually need only weapons supplies and logistical support and do not charge for their combat role. Moreover, they are even less visible in terms of the political costs of war: their deaths will hardly register among Western publics. Yet they have important downsides. Their abilities and skills will vary. At the lower end of the spectrum, auxiliaries may therefore turn out to be a liability on the battlefield rather than an asset. They are hard to control and potentially not very disciplined. They may fight in ways that their Western supporters find hard to tolerate. British commanders in the French and Indian War as well as US commanders in Afghanistan who had to cooperate with local auxiliaries had to face situations in which the latter committed atrocities. At times, regular officers in charge of auxiliaries tolerated and exploited undisciplined behaviour, as it had an intimidating effect on local opponents and the local population, while at the same time allowing regular officers to deflect responsibility from the regular side. Finally, there is the question of the auxiliaries' long-term loyalty. The example of occupied Yugoslavia during the Second World War, where both Mihailovic and Tito attempted to collaborate with both the Allies and with the Germans, is a stark reminder of the pitfalls of relying on irregular auxiliaries.

However, if cooperation with auxiliaries carries risks from the regular side's point of view, it is even more perilous from the point of view of the irregular fighters. Auxiliaries often are more dependent on their 'state sponsor' than vice versa. The regular side can suddenly drop and sideline individual irregular leaders and warlords if the perception of the local situation changes, as happened to General Ziauddin and Abdullah Mujahid in Afghanistan. More importantly, the regular sponsor can also suddenly withdraw all support from the auxiliaries if the strategic situation changes so as to render the latter's services redundant. This usually leaves the auxiliaries in an exposed situation, often with disastrous consequences. The 1944 Warsaw Uprising, the situation of the Italian resistance in 1944, and the situation of the Shiite and Kurdish resistance in Iraq in 1991 are cases in point.

The strategic debates over support for 'rebels' in places such as Libya and Syria show little awareness of the deeper issues related to cooperation between states and irregular auxiliaries. Many would acknowledge that cooperation with auxiliaries is potentially problematic. However, concerns often boil down to avoiding support for groups that have links with al-Qaeda. But the problems related to cooperating with irregular auxiliaries are much more fundamental than those that could be solved by 'vetting' their different political perspectives and affiliations. Cooperating with auxiliaries is a difficult and potentially dangerous undertaking for both sides. The risks involved are manifold and inherent in the nature of the cooperation. They cannot simply be solved by good intelligence. However, the artificial and paradigmatic, yet almost unspoken, distinction between the irregular as the illegitimate fighter and the irregular as the useful auxiliary smothers any meaningful debate over the use of auxiliaries in twenty-first-century conflict. It is necessary to overcome this artificial distinction in order to enable policymakers and military practitioners to make more informed strategic decisions on the cooperation with auxiliaries.

Irregular Fighters and the Identity Crises of Regular Armed Forces

Throughout the history of modern war, periods of heightened attention to the concept of the irregular were at the same time periods during which the identity of regular armed forces underwent a crisis. As noted above, the French Revolutionary and Napoleonic Wars and the concomitant birth of the modern conscripted mass army necessitated the construction of new identity narratives. This was all the more important since the social and political situation in which the new citizen–soldiers found themselves was characterized by internal paradoxes: the political narrative of sacrifice for the nation clashed with the population's perception of conscription as the 'blood tax', and the aspiration to drill and discipline were countered by the need to forage in occupied territory. Against the background of these innate paradoxes and uncertainties in the social and political situation of the new citizens' army, it was difficult to articulate a coherent identity narrative. Taking recourse to an external 'other' thus became all the more important. That 'other' was the irregular fighter. It was convenient to deflect from the regular army's internal problems by projecting them entirely on the irregular.

In the second half of the nineteenth century the identity of regular armed forces, especially infantrymen, faced a new challenge, this time because of technological innovations. The introduction of long-range rifles called into question established notions of courage on the battlefield. Prior to this innovation courageous behaviour was perceived as making someone almost invulnerable; this mindset proved suicidal under the new battlefield conditions. This not only led to the adaptation of battlefield tactics, but also had repercussions in different social contexts. It affected the cultural perception and role of leadership in that officers now found it increasingly difficult to lead by example for fear of being singled out by snipers. More fundamentally, it led to a clash between societal ideas about valorous conduct and actual battlefield conditions. Under these circumstances, it was again the irregular fighter who served as a smokescreen for the externalization of the internal issues of regular armed forces. Especially in the American Civil War the contemporary image of the irregular oscillated between a denunciating portrayal of them as cowards and an admirable one of them as archaic warriors who refused to succumb to the maelstrom of mechanized mass warfare. If the regular soldier increasingly saw himself as the object of impersonalized forces and constraints, the irregular had retained agency, even in the face of new battlefield conditions.

The Second World War and the wars of decolonization were followed by a trend towards a greater inclusion of irregular fighters owing to shifting discourses over the political legitimacy of armed resistance against political oppression and a new emphasis on national self-determination. Also the revival of the irregular as the auxiliary of regular armed forces further facilitated the move towards more inclusive legal parameters for irregulars. At the same time, the identity of Western regular armed forces remained relatively stable. Mass armies and conscription, the latter in particular in continental Europe, continued to be relevant in the strategic context of the cold war. Even if the introduction of thermonuclear weapons was a major innovation in weapons technology after the Second World War, owing to the sheer unthinkability of their use mass land forces remained the military instrument of choice in cold war strategic planning.

The beginning of the twenty-first century brought a renewed emphasis on an exclusionary approach towards irregular fighters. This approach was defended by pointing out that Western states had become more vulnerable to irregular attacks and that stern measures needed to be taken to re-level the battlefield. However, it is worthwhile to ask how this development had been affected by the relationship between the regular and irregular in this context. It is plausible to argue that the renewed emphasis on marginalizing irregulars is an indicator of a crisis of the identity of regular armed forces rather than a reaction to a perceived preponderance of 'new wars' or 'asymmetric' warfare. Most Western states' armed forces underwent deep transformations after the end of the cold war. Small professional armies supported by high-tech weapons systems, including robotics and drones, replaced the conscripted mass armies of the twentieth century. The concomitant challenges to the established social and political identities of Western armed forces are twofold. First, the increasing technologization of the battlefield precipitates questions about soldiers' identities: is the UAV pilot who operates drones thousands of miles away from the battlefield, in the safety of his bunker, actually a soldier or rather a technician? Has technology decreased the risks for twenty-first-century armed forces to such an extent that core qualities within the traditional understanding of soldiering such as courage have come under pressure?[8] To a certain extent the post-cold-war situation is a reverse image of the trend prevailing during the American Civil War: if valorous behaviour in the traditional sense had become suicidal on the battlefield of the American Civil War, large parts of the Western armed forces may no longer need courage in order to perform in twenty-first-century wars. Blaming terrorists for their 'cowardice' hence is to a certain extent an externalization of these internal uncertainties.

Secondly, Western armed forces have become more professionalized than ever before in the modern history of regular armies. They are capable of performing at much higher professional standards than the citizen–soldiers of the twentieth century.[9] Against this background, some scholars have predicted that the image of the soldier will gradually give way to that of the warrior. In contrast to the former, the latter is deliberately set apart from society, and his behaviour is regulated by a separate set of norms, which Christopher Coker calls the 'warrior ethos'.[10] Others are concerned about this trend, in as much as the growing professionalization opens up a—potentially politically dangerous—gap between the soldier, who is no longer a citizen–soldier, and the rest of society.[11] It also leaves open fundamental questions about his or her political identity. The image of the 'warrior' sits uneasily with his or her rootedness in Western democratic societies.[12] Even more threatening to the identity of Western professional armies is the fact that, despite their unsurpassed professionalism, they often fail to prevail against lesser, 'irregular' opponents. In most cases, the main reasons for this failure are political rather than military. They often stem from the fact that few Western governments are willing to commit adequate resources to wars in the periphery of the international system. Yet, branding the irregular as an 'unlawful combatant' again offers a convenient way to deflect from this problem: if the latter prevails, it must be due to his or her unfair ways of

[8] Christopher Coker, *The Warrior Ethos: Military Culture and the War on Terror* (London: Routledge, 2007), 130.

[9] Anthony King, *The Combat Soldier: Infantry Tactics and Cohesion in the Twentieth and Twenty-First Centuries* (Oxford: Oxford University Press, 2013).

[10] Coker, *The Warrior Ethos*.

[11] Hew Strachan, 'Heroic Warfare and the Problem of Mass Armies, France 1871–1914', in Sibylle Scheipers (ed.), *Heroism and the Changing Character of War: Toward Post-Heroic Warfare?* (Basingstoke: Palgrave, 2014), 47–63.

[12] Andreas Herberg-Rothe, 'The Democratic Warrior and the Emergence of World Order Conflicts', in Scheipers (ed.), *Heroism*, 286–304.

fighting and not because of the Western mismatch between strategic aspirations and actual military commitment. Paradoxically, in blaming the irregular for the problems Western armed forces encounter in twenty-first-century wars, their political masters have made it even less likely that the latter will prevail in twenty-first-century wars.

Bibliography

Adekson, J. 'Bayo, 'The Algerian and the Mau Mau Revolts: A Comparative Study in Revolutionary Warfare'. *Comparative Strategy*, 3/1 (1981), 69–92.

Agamben, Giorgio, *State of Exception* (Chicago: Chicago University Press, 2005).

Ahmad, Abu Talib, 'The Impact of the Japanese Occupation on Colonial and Anti-Colonial Armies in Southeast Asia', in Tobias Rettig and Karl Hack (eds), *Colonial Armies in Southeast Asia* (London: Routledge, 2009), 213–38.

Alexander, Amanda, 'The Genesis of the Civilian', *Leiden Journal of International Law*, 20/2 (2007), 359–76.

Amnesty International, 'EU Declaration on Arms Transfer to Syria', 31 May 2013 <http://www.amnesty.org/fr/library/asset/MDE24/027/2013/fr/dcfff54d-76bc-4230-a6ca-3b6a2c1ec79f/mde240272013en.html> (accessed 19 August 2013).

Amnesty International, 'Libya's Leadership Urged to Prioritize Human Rights', 24 September 2012 <http://www.amnestyusa.org/news/news-item/libya-s-leadership-urged-to-prioritize-human-rights> (accessed 19 August 2013).

Amnesty International Deutschland, 'Ein Funke genügt', June 2009 <http://www.amnesty.de/journal/2009/juni/ein-funke-genuegt> (accessed 18 August 2013).

Anderson, David, 'British Abuse and Torture in Kenya's Counter-Insurgency, 1952–1960', *Small Wars & Insurgencies*, 23/4–5 (2012), 700–19.

Anderson, David, *Histories of the Hanged: Britain's Dirty War in Kenya and the End of Empire* (London: Weidenfeld & Nicolson, 2005).

Angevine, Robert G., *The Railroad and the State: War, Politics and Technology in Nineteenth-Century America* (Stanford, CA: Stanford University Press, 2004).

Armstrong, John, 'Introduction', in John Armstong (ed.), *The Soviet Partisans in World War II* (Madison: University of Wisconsin Press, 1964), 1–43.

Arquilla, John, *Insurgents, Raiders, and Bandits: How Masters of Irregular Warfare have Shaped our World* (Lanham, MD: Ivan R. Dee, 2011).

Arreguín-Toft, Ivan, *How the Weak Win Wars: A Theory of Asymmetric Conflict* (Cambridge: Cambridge University Press, 2005).

Ashley, Richard K., 'Untying the Sovereign State: A Double Reading of the Anarchy Problematique', *Millennium: Journal of International Studies*, 17/2 (1995), 227–62.

Asprey, Robert B., *War in the Shadows: The Guerrilla in History* (London: Macdonald and Jane's, 1976).

Axtell, James, *The European and the Indian: Essays in the Ethnohistory of Colonial North America* (Oxford: Oxford University Press, 1981).

Bartelson, Jens, *A Genealogy of Sovereignty* (Cambridge: Cambridge University Press, 1995).

Bartov, Omer, *The Eastern Front, 1941–45: German Troops and the Barbarisation of Warfare* (Houndmills: Palgrave, 2001).

Bartov, Omer, *Hitler's Army: Soldiers, Nazis, and War in the Third Reich* (Oxford: Oxford University Press, 1992).

Bayly, Christopher, and Harper, Tim, *Forgotten Wars: The End of Britain's Asian Empire* (London: Penguin, 2008).

BBC News, 'Should the West Support the Northern Alliance?' <http://news.bbc.co.uk/2/hi/talking_point/1615191.stm> (accessed 18 August 2013).

Beckett, Ian F., *Modern Insurgencies and Counterinsurgencies: Guerrillas and their Opponents since 1750* (London: Routledge, 2001).

Bell, David A., *The First Total War: Napoleon's Europe and the Birth of Modern Warfare* (London: Bloomsbury, 2007).

Bellamy, Alex J., 'No Pain, No Gain? Torture and Ethics in the War on Terror', *International Affairs*, 82/1 (2006), 121–48.

Bellamy, Christopher, '"Tools of Ill-Omen": The Shifted Conflict Paradigm and Reduced Role of Conventional Military Power', *Cambridge Review of International Affairs*, 15/1 (2002), 149–57.

Bennett, Huw, *Fighting the Mau Mau: The British Army and Counter-Insurgency in the Kenya Emergency* (Cambridge: Cambridge University Press, 2012).

Bennett, Huw, '"A Very Salutary Effect": The Counter-Terror Strategy in the Early Malayan Emergency, June 1948 to December 1949', *Journal of Strategic Studies*, 32/3 (2009), 415–44.

Best, Geoffrey, *Humanity in Warfare: The Modern History of the International Law of Armed Conflicts* (London: Weidenfeld & Nicolson, 1980).

Best, Geoffrey, *War and Law since 1945* (Oxford: Oxford University Press, 1994).

Best, Geoffrey, *War and Society in Revolutionary Europe, 1770–1870* (Leicester: Leicester University Press, 1982).

Biddle, Stephen, 'Afghanistan and the Future of Warfare', *Foreign Affairs*, 82/2 (2003), 31–46.

Biddle, Stephen, Friedman, Jeffrey A., and Shapiro, Jacob N., 'Testing the Surge: Why did Violence Decline in Iraq in 2007', *International Security*, 37/1 (2012), 7–40.

Black, Jeremy, 'Book Review of "The First Total War: Napoleon's Europe and the Birth of Warfare as We Know It"', *English Historical Review*, 123/502 (2008), 765–6.

Black, Jeremy, 'Introduction', in Jeremy Black (ed.), *European Warfare, 1453–1815* (Houndmills: Palgrave Macmillan, 1999), 1–22.

Blackburn, Robin, 'Haiti, Slavery, and the Age of Democratic Revolution', *William and Mary Quarterly*, 63/4 (2006), 643–74.

Blanning, T. C. W., *The Origins of the French Revolutionary Wars* (London: Longman, 1986).

Blocher, Joseph, 'The Guantanamo Three-Step', *Yale Law Journal*, 117/1 (2007), 1–7.

Blood, Philip W., *Hitler's Bandit Hunters: The SS and the Nazi Occupation of Europe* (Dulles, VA: Potomac Books, 2006).

Bohlen, Celestine, 'In New War on Terrorism, Words Are Weapons, Too', *New York Times*, 29 September 2001.

Bois, Paul, *Paysans de L'Ouest: Des structures économiques et sociales aux options politiques depuis l'époque révolutionnaire dans la Sarthe* (Le Mans: M. Vilaire, 1960).

Boot, Max, *Invisible Armies: An Epic History of Guerrilla Warfare* (New York: Norton & Company, 2013).

Bowen, Don R., 'Guerrilla War in Western Missouri, 1861–1865: Historical Extensions of the Relative Deprivation Hypothesis', *Comparative Studies in Society and History*, 19/1 (1977), 30–51.

Boycott-Brown, Martin, 'Guerrilla Warfare *avant la lettre*: Northern Italy, 1792–97', in Charles J. Esdaile (ed.), *Popular Resistance in the French Wars: Patriots, Partisans and Land Pirates* (Basingstoke: Palgrave Macmillan, 2005), 45–66.

Brahimi, Alia, 'Religion in the War on Terror', in Hew Strachan and Sibylle Scheipers (ed.), *The Changing Character of War* (Oxford: Oxford University Press, 2011), 184–201.

Branch, Daniel, *Defeating Mau Mau, Creating Kenya: Counterinsurgency, Civil War, and Decolonization* (Cambridge: Cambridge University Press, 2009).

Branche, Raphaëlle,'The French in Algeria: Can there be Prisoners of War in a "Domestic" Operation?', in Sibylle Scheipers (ed.), *Prisoners in War* (Oxford: Oxford University Press, 2010), 173–86.

Broers, Michael, *Europe under Napoleon, 1799–1815* (London: Arnold, 1996).

Broers, Michael, *Napoleon's Other War: Bandits, Rebels and their Pursuers in the Age of Revolution* (Oxford: Peter Lang, 2010).

Brooks, Mason, and Miller, Drew, 'Inside the Detention Camps: A New Campaign in Iraq', *Joint Forces Quarterly*, 52/1 (2009), 129–33.

Brown, Howard G., *Ending the French Revolution: Violence, Justice, and Repression from the Terror to Napoleon* (Charlottesville, VA: University of Virginia Press, 2006).

Browning, Christopher R., *Ordinary Men: Reserve Battalion 101 and the Final Solution in Poland* (London: Penguin, 2001).

Brownlee, Richard S., *Gray Ghosts of the Confederacy: Guerrilla Warfare in the West, 1861–1865* (Baton Rouge, LA: Louisiana State University Press, 1984).

Brumwell, Stephen, *Redcoats: The British Soldier and the War in the Americas, 1755–1763* (Cambridge: Cambridge University Press, 2006).

B'tselem, 'Human Shields' <http://www.btselem.org/english/Human_Shields/Index.asp> (accessed 15 June 2013).

Bührer, Tanja, *Die Kaiserliche Schutztruppe für Deutsch-Ostafrika: Koloniale Sicherheitspolitik und transkulturelle Kriegführung, 1885–1918* (Munich: Oldenbourg, 2011).

'Bush: US Military on "High Alert" ', *CNN.com*, 12 September 2001 <http://edition.cnn.com/2001/US/09/11/bush.second.statement/> (accessed 24 August 2013).

Byman, Daniel, 'Do Targeted Killings Work?', *Foreign Affairs*, 85/2 (2006), 95–111.

Callwell, C. E., *Small Wars: Their Principle and their Practice* (Lincoln, NB: University of Nebraska Press, 1996 [1896]).

Campbell, Courtney S., 'Moral Responsibility and Irregular War', in James Turner Johnson and John Kelsay (eds), *Cross, Crescent, and Sword: The Justification and*

Limitation of War in Western and Islamic Tradition (New York: Greenwood Press, 1990), 103–28.

Carvin, Stephanie, *Prisoners of America's Wars: From the Early Republic to Guantánamo* (New York: Columbia University Press, 2010).

Cavallaro, James, Sonnenberg, Stephan, and Knuckey, Sarah, *Living under Drones: Death, Injury and Trauma to Civilians from US Drone Practices in Pakistan* (Stanford, CA: Stanford Law School; New York: NYU School of Law, 2012).

Chambers, John Whiteclay, II, 'American Views of Conscription and the German Nation in Arms in the Franco-Prussian War', in Daniel Moran and Arthur Waldron (eds), *The People in Arms: Military Myth and the National Mobilization since the French Revolution* (Cambridge: Cambridge University Press, 2003), 75–99.

Chandler, David G., *The Campaigns of Napoleon* (London: Folio Society, 2002), ii.

Checkel, Jeffrey T., 'The Constructivist Turn in International Relations Theory', *World Politics*, 50/2 (1998), 324–48.

Chesney, Robert M., 'Iraq and the Military Detention Debate: Firsthand Perspectives from the Other War, 2003–2010', *Virginia Journal of International Law*, 51/3 (2011), 549–636.

Chiari, Bernhard, *Alltag hinter der Front: Besatzung, Kollaboration und Widerstand in Weißrussland, 1941–1944* (Düsseldorf: Droste Verlag, 1998).

Clausewitz, Carl von, *On War*, ed. and trans. Michael E. Howard and Peter Paret (Princeton: Princeton University Press, 1976).

Cobb, Richard, *The People's Armies* (New Haven: Yale University Press, 1987).

Coil, George L., 'War Crimes of the American Revolution', *Military Law Review*, 82 (1978), 171–98.

Coker, Christopher, *The Warrior Ethos: Military Culture and the War on Terror* (London: Routledge, 2007).

Colley, Linda, *Captives: Britain, Empire and the World, 1600–1850* (London: Jonathan Cape, 2002).

Collier, Paul, and Höffler, Anke, 'Greed and Grievance in Civil War', *Oxford Economic Papers*, 56/4 (2004), 563–95.

Conway, Stephen, 'From Fellow-Nationals to Foreigners: British Perceptions of the Americans, circa 1739–1783', *William and Mary Quarterly*, 59/1 (2002), 65–100.

Cooper, Helene, and Landler, Mark,'Targeted Killing Is New US Focus in Afghanistan', *New York Times*, 31 July 2010.

Corni, Gustavo, 'Italy', in Bob Moore (ed.), *Resistance in Western Europe* (Oxford: Berg, 2000), 157–88.

Creveld, Martin van, *Supplying War: Logistics from Wallenstein to Patton* (Cambridge: Cambridge University Press, 1977).

Creveld, Martin van, *The Transformation of War* (New York: Free Press, 1991).

Dams, Carsten, and Stolle, Michael, *Die Gestapo: Herrschaft und Terror im Dritten Reich* (Munich: Beck, 2009).

Danchev, Alex, 'Review Article: Bad Apples, Dead Souls: Understanding Abu Ghraib', *International Affairs*, 84/6 (2008), 1271–80.

Davies, Norman, *Rising '44: The Battle for Warsaw* (London: Penguin, 2005).

Davis, Merian Helen, 'Restaging *mise en valeur*: "Postwar Imperialism" and the Plan de Constantine', *Review of Middle East Studies*, 44/2 (2010), 176–86.

Deeks, Ashley, 'Detention in Afghanistan: The Need for an Integrated Plan', Center for Strategic and International Studies Briefing Paper, Washington, 14 February 2008 <http://csis.org/files/media/csis/pubs/080213_deeks_afghanistan.pdf> (accessed 26 July 2013).

Denbeaux, Mark, and Denbeaux, Joshua, *Report on Guantánamo Detainees* (South Orange, NJ: Seton Hall University School of Law, 2008).

Downes, Alexander B., *Targeting Civilians in War* (Ithaca, NY: Cornell University Press, 2008).

Dubois, Laurent, *Avengers of the New World: The Story of the Haitian Revolution* (Cambridge, MA: Harvard University Press, 2004).

Dunlap, Charles C., 'Lawfare amid Warfare', *Washington Times*, 3 August 2007.

Dworkin, Anthony, *Drones and Targeted Killing: Defining a European Position* (London: European Council on Foreign Relations, 2013).

Earl, Hilary, *The Nuremberg* SS-Einsatzgruppen *Trial, 1945–1958: Atrocity, Law and History* (Cambridge: Cambridge University Press, 2009).

Echevarria, Antulio J., II, *Fourth Generation Warfare and Other Myths* (Carlisle, PA: Strategic Studies Institute, 2005).

Edelstein, Dan, 'War and Terror: The Law of Nations from Grotius to the French Revolution', *French Historical Studies*, 31/2 (2008), 229–62.

El Fadl, Khaled Abou, *Rebellion and Violence in Islamic Law* (Cambridge: Cambridge University Press, 2001).

Elkins, Caroline, *Britain's Gulags: The Brutal End of Empire in Kenya* (London: Pimlico, 2005).

Ellis, John, *From the Barrel of a Gun: A History of Guerrilla, Revolutionary and Counter-Insurgency Warfare, from the Romans to the Present* (London: Greenhill Books, 1995).

Ellis, John, *A Short History of Guerrilla Warfare* (New York: St Martin's Press, 1976).

Esdaile, Charles J., *Fighting Napoleon: Guerrillas, Bandits and Adventurers in Spain, 1808–1814* (New Haven: Yale University Press, 2004).

Esdaile, Charles J., 'Patriots, Partisans and Land Pirates in Retrospect', in Charles J. Esdaile (ed.), *Popular Resistance in the French Wars: Patriots, Partisans and Land Pirates* (Basingstoke: Palgrave Macmillan, 2005), 1–24.

Esdaile, Charles J., and Enviz, Leonor Hernández, 'The Anatomy of a Research Project: The Sociology of the Guerrilla War in Spain, 1808–14', in Charles J. Esdaile (ed.), *Popular Resistance in the French Wars: Patriots, Partisans and Land Pirates* (Basingstoke: Palgrave Macmillan, 2005), 115–36.

Evans, Martin, *Algeria: France's Undeclared War* (Oxford: Oxford University Press, 2012).

Evans, Martin, 'French Resistance and the Algerian War', *History Today*, 41/7 (1991), 38–45.

Fanon, Frantz, *The Wretched of the Earth* (London: Penguin, 2001 [1961]).

Faucheux, Marcel, *L'Insurrection vendéenne de 1793: Aspects économique et socieaux* (Paris: Imprimerie nationale, 1964).

Fellman, Michael, *Inside War: The Guerrilla Conflict in Missouri during the American Civil War* (Oxford: Oxford University Press, 1989).

Fenn, Elizabeth A., 'Biological Warfare in Eighteenth-Century North America: Beyond Jeffery Amherst', *Journal of American History*, 86/4 (2000), 1552–80.

Field Manual FM 34–52: Intelligence Interrogation, Headquarters, Department of the Army, Washington DC, 28 September 1992, sections 1–11, D-1 <https://www.fas.org/irp/doddir/army/fm34-52.pdf> (accessed 25 July 2013).

Fick, Carolyn E., *The Making of Haiti: The Saint Domingue Revolution from Below* (Knoxville, TN: University of Tennessee Press, 1990).

Finkelstein, Norman G., and Birn, Ruth Bettina, *A Nation on Trial: The Goldhagen Thesis and Historical Truth* (New York: Henry Holt and Company, 1998).

Fisher, Noel C., *War at Every Door: Partisan Politics and Guerrilla Violence in East Tennessee, 1860–69* (Chapel Hill, NC: University of North Carolina Press, 1997).

Foot, M. R. D., *SOE in France* (London: Frank Cass, 2006).

Foot, M. R. D., *SOE: The Special Operations Executive, 1940–1946* (London: Pimlico, 1999).

Forrest, Alan, '*La Patrie en danger*: The French Revolution and the First *Levée en masse*', in Daniel Moran and Arthur Waldron (eds), *The People in Arms: Military Myth and National Mobilization since the French Revolution* (Cambridge: Cambridge University Press, 2003), 8–32.

Forrest, Alan, 'The Ubiquitous Brigand: The Politics and Language of Repression', in Charles J. Esdaile (ed.), *Popular Resistance in the French Wars: Patriots, Partisans and Land Pirates* (Basingstoke: Palgrave Macmillan, 2005), 25–43.

Förster, Jürgen, 'The Wehrmacht and the War of Extermination against the Soviet Union', *Yad Vashem Studies*, 14 (1981), 7–34.

Galen Last, Dick van, 'The Netherlands', in Bob Moore (ed.), *Resistance in Western Europe* (Oxford: Berg, 2000), 189–222.

Gann, Lewis H., *Guerrillas in History* (Stanford, CA: Hoover Institution Press, 1971).

Geggus, David Patrick, *Haitian Revolutionary Studies* (Bloomington, IN: Indiana University Press, 2002).

Ginbar, Yuval, *Why Not Torture Terrorists?: Moral, Practical, and Legal Aspects of the 'Ticking Bomb' Justification for Torture* (Oxford: Oxford University Press, 2008).

Giustozzi, Antonio, *Koran, Kalashnikov and Laptop: The Neo-Taliban Insurgency in Afghanistan* (New York: Columbia University Press, 2008).

Goeschel, Christian, and Wachsmann, Nikolaus,'Before Auschwitz: The Formation of the Nazi Concentration Camps, 1933–9', *Journal of Contemporary History*, 45/3 (2010), 515–34.

Goldhagen, Daniel J., *Hitler's Willing Executioners—Ordinary Germans and the Holocaust* (London: Abacus, 1997).

Goldsmith, Jack, *The Terror Presidency: Law and Judgment inside the Bush Administration* (New York: W. W. Norton, 2007), 165 ff.

Gong, Gerrit W., *The Standard of Civilization in International Society* (Oxford: Oxford University Press, 1984).

Gordon, Michael, 'Against US Wishes, Iraq Releases Man Accused of Killing American Soldiers', *New York Times*, 16 November 2012.

Grab, Alexander, 'State Power, Brigandage and Rural Resistance in Napoleonic Italy', *European History Quarterly*, 25/1 (1995), 39–70.

Grey, Stephen, 'America's Gulag', *New Statesman*, 17 May 2004.

Grey, Stephen, *Ghost Plane: The True Story of the CIA Rendition and Torture Program* (New York: St Martin's Griffin, 2007).

Greenberg, Karen J. (ed.), *The Torture Debate in America* (Cambridge: Cambridge University Press, 2008).

Greenberg, Karen J., and Dratel, Joshua L. (eds), *The Torture Papers: The Road to Abu Ghraib* (Cambridge: Cambridge University Press, 2005).

Grenier, John, *The First Way of War: American War Making on the Frontier, 1607–1814* (Cambridge: Cambridge University Press, 2005).

Grimsley, Mark, *The Hard Hand of War: Union Military Policy towards Southern Civilians 1861–1865* (Cambridge: Cambridge University Press, 1995).

Gross, Michael, *Moral Dilemmas of Modern War: Torture, Assassination and Blackmail in an Age of Asymmetric Conflict* (Cambridge: Cambridge University Press, 2010).

Gross, Oren, and Aoláin, Fionnuala Ní, *Law in Times of Crisis: Emergency Powers in Theory and Practice* (Cambridge: Cambridge University Press, 2006).

Gumz, Jonathan, 'German Counterinsurgency Policy in Independent Croatia, 1941–1944', *Historian*, 61/1 (1998), 33–50.

Gumz, Jonathan E., 'Reframing the Historical Problematic of Insurgency: How the Professional Military Literature Created a New History and Missed the Past', *Journal of Strategic Studies*, 32/4 (2009), 553–88.

Gumz, Jonathan, 'Wehrmacht Perceptions of Mass Violence in Croatia, 1941–1942', *Historical Journal*, 44/4 (2001), 1015–38.

Guzzini, Stefano, 'Structural Power: The Limits of Neo-Realist Power Analysis', *International Organization*, 47/3 (1993), 443–78.

Hack, Karl, 'Everyone Lived in Fear: Malaya and the British Way of Counter-Insurgency', *Small Wars & Insurgencies*, 23/4–5 (2012), 671–99.

Hack, Karl, '"Iron Claws of Malaya": The Historiography of the Malayan Emergency', *Journal of Southeast Asian Studies*, 30/1 (1999), 99–125.

Hack, Karl, with Rettig, Tobias, 'Imperial Systems of Power, Colonial Forces and the Making of Modern Southeast Asia', in Tobias Rettig and Karl Hack (eds), *Colonial Armies in Southeast Asia* (London: Routledge, 2009), 3–36.

Hakim, Yalda, 'Why Drone Attacks in Yemen Are Like "Trying to Hit a Ghost"', BBC News, 10 August 2013 <http://www.bbc.co.uk/news/world-middle-east-23653488> (accessed 17 August 2013).

Hammes, Thomas X., *The Sling and the Stone: On War in the 21st Century* (St Paul, MN: Zenith Press, 2006).

Hartigan, Richard Shelly (ed.), *Lieber's Code and the Laws of War* (Chicago: Precedent, 1983).

Hartley, Janet, 'The Patriotism of the Russian Army in the "Patriotic" or "Fatherland" War of 1812', in Charles J. Esdaile (ed.), *Popular Resistance in the French Wars: Patriots, Partisans and Land Pirates* (Basingstoke: Palgrave Macmillan, 2005), 181–200.

Hashim, Ahmed, *Iraq's Sunni Insurgency* (Oxford: Oxford University Press for the International Institute for Strategic Studies, 2008) [Adelphi Paper 48/402].

Hattaway, Herman, and Jones, Archer, *How the North Won: A Military History of the American Civil War* (Champaign, IL: University of Illinois Press, 1983).

Hazareesingh, Sudhir, 'Republicanism, War and Democracy: The Ligue du Midi in France's War against Prussia, 1870–1871', *French History*, 17/1 (2003), 48–78.

Heer, Hannes, and Naumann, Klaus (eds), *Vernichtungskrieg—Verbrechen der Wehrmacht 1941–1944* (Frankfurt: Hamburger Edition, 1995).

Herberg-Rothe, Andreas, 'The Democratic Warrior and the Emergence of World Order Conflicts', in Sibylle Scheipers (ed.), *Heroism and the Changing Character of War: Toward Post-Heroic Warfare?* (Basingstoke: Palgrave, 2014), 286–304.

Heuser, Beatrice, 'Small Wars in the Age of Clausewitz: The Watershed between Partisan War and People's War', *Journal of Strategic Studies*, 33/1 (2010), 139–62.

Hill, Alexander, *The War behind the Eastern Front: The Soviet Partisan Movement in North-West Russia 1941–44* (London: Frank Cass, 2005).

Hippler, Thomas, *Citizens, Soldiers and National Armies: Military Service in France and Germany, 1789–1830* (London: Routledge, 2007).

Hobsbawm, Eric J., *Bandits* (Harmondsworth: Pelican Books, 1972).

Hodes, Cyrus, and Sedra, Mark, *The Search for Security in Post-Taliban Afghanistan* (Oxford: Oxford University Press for the International Institute for Strategic Studies, 2007) [Adelphi Paper 47/391].

Horne, Alistair, *A Savage War of Peace: Algeria 1954–1962* (New York: New York Review of Books, 2006).

Horne, John, 'Defining the Enemy: War, Law, and the Levée en masse from 1870 to 1945', in Daniel Moran and Arthur Waldron (eds), *The People in Arms: Military Myth and National Mobilization since the French Revolution* (Cambridge: Cambridge University Press, 2003), 100–23.

Horne, John, and Kramer, Alan, *German Atrocities, 1914: A History of Denial* (New Haven: Yale University Press, 2002).

Howard, Michael, *The Franco-Prussian War: The German Invasion of France, 1870–1871* (London: Routledge, 1981).

Hull, Isabel V., *Absolute Destruction: Military Culture and the Practices of War in Imperial Germany* (Ithaca, NY: Cornell University Press, 2005).

International Committee of the Red Cross, 'Project of an International Declaration concerning the Laws and Customs of War', Brussels, 27 August 1874 <http://www.icrc.org/ihl.nsf/135-FULL?Open Document> (accessed 17 August 2013).

International Committee of the Red Cross, 'The Laws of War on Land', Oxford, 9 September 1880 <http://icrc.org/ihl.nsf/140-FULL?OpenDocument> (accessed 17 August 2013).

International Crisis Group, *Drones: Myths and Reality in Pakistan* (Brussels: International Crisis Group, 2013).

Jackson, Robert, *Contemporary State Terrorism: Theory and Practice* (London: Routledge, 2010).

James, C. L. R., *The Black Jacobins* (London: Penguin, 1980).

Jeffery, Keith, *MI6: The History of the Secret Intelligence Service, 1909–1949* (London: Bloomsbury, 2010).

Jennings, Francis, *Empire of Fortune: Crown, Colonies, and Tribes in the Seven Years War in America* (New York: Norton & Company, 1988).

Johns, Fleur, *Non-Legality in International Law: Unruly Law* (Cambridge: Cambridge University Press, 2013).

Johnson, James Turner, *Can Modern War Be Just?* (New Haven: Yale University Press, 1986).

Judd, Dennis, and Surridge, Keith, *The Boer War: A History* (London: I. B. Tauris, 2002).

Judt, Tony, 'Introduction', in Tony Judt (ed.), *Resistance and Revolution in Mediterranean Europe, 1939–1948* (London: Routledge, 1989), 1–28.

Kaldor, Mary, *New and Old Wars: Organized Violence in a Global Era* (Stanford, CA: Stanford University Press, 2012).

Kalyvas, Stathis N., *The Logic of Violence in Civil War* (Cambridge: Cambridge University Press, 2006).

Kanter, Sanford, 'Exposing the Myth of the Franco-Prussia War', *War and Society*, 4/1 (1986), 13–30.

Kedward, H. R., *In Search of the Maquis: Rural Resistance in Southern France, 1942–1944* (Oxford: Oxford University Press, 1993).

Keegan, John, *A History of Warfare* (New York: Vintage, 1993).

Keene, Edward, *Beyond the Anarchical Society: Grotius, Colonialism and Order in World Politics* (Cambridge: Cambridge University Press, 2002).

Kelsay, John, *Arguing the Just War in Islam* (Cambridge, MA: Harvard University Press, 2007).

Kennedy, Paul, and Andreopoulos, George J., 'The Laws of War: Some Concluding Reflections', in Michael Howard, George J. Andreopoulos, and Mark R. Shulman (eds), *The Laws of War: Constraints on Warfare in the Western World* (New Haven: Yale University Press, 1994), 214–25.

Kilcullen, David, and McDonald Exum,Andrew, 'Death from Above, Outrage Down Below', *New York Times*, 16 May 2009.

King, Anthony, *The Combat Soldier: Infantry Tactics and Cohesion in the Twentieth and Twenty-First Centuries* (Oxford: Oxford University Press, 2013).

Kirchhoff, Hans, 'Denmark', in Bob Moore (ed.), *Resistance in Western Europe* (Oxford: Berg, 2000), 93–124.

Kirkpatrick, David, 'Logistics of the American Civil War', *RUSI Journal*, 152/5 (2007), 76–81.

Klinkhammer, Lutz, 'Der Partisanenkrieg der Wehrmacht 1941–1944', in Rolf-Dieter Müller and Hans-Erich Volkmann (eds), *Die Wehrmacht: Mythos und Realität*, (Munich: Oldenbourg, 1999), 76–81.

Knight, Betsy, 'Prisoner Exchange and Parole in the American Revolution', *William and Mary Quarterly*, 48/2 (1991), 201–22.

Kramer, Alan, 'Editorial', *Mittelweg 36*, 20/4 (2011), 36.

Kuß, Susanne, *Deutsches Militär auf kolonialen Kriegsschauplätzen: Eskalation der Gewalt zu Beginn des 20. Jahrhunderts* (Berlin: Ch. Links, 2010).

Lagrou, Pierre, 'Belgium', in Bob Moore (ed.), *Resistance in Western Europe* (Oxford: Berg, 2000), 27–64.

Laqueur, Walter, *Guerrilla Warfare: A Historical and Critical Study* (New York: Transaction, 1998).

Laqueur, Walter, 'The Origins of Guerrilla Doctrine', *Journal of Contemporary History*, 10/3 (1975), 341–82.

'Lawfulness of a Lethal Operation Directed against a US Citizen: Who Is a Senior Operational Leader of Al-Qa'ida or an Associated Force', 8 November 2011 <http://msnbcmedia.msn.com/i/msnbc/sections/news/020413_DOJ_White_Paper.pdf> (accessed 17 August 2013).

Lieb, Peter, 'Repercussions of Eastern Front Experiences on Anti-Partisan Warfare in France 1943–1944', *Journal of Strategic Studies*, 31/5 (2008), 797–823.

Lieb, Peter, 'Suppressing Insurgencies in Comparison: The Germans in the Ukraine, 1918, and the British in Mesopotamia, 1920', *Small Wars & Insurgencies*, 23/4–5 (2012), 627–47.

Lietzau, William K., 'The Past, Present and Future of US Detainee Policy', American Heritage Foundation, Washington DC, 13 May 2013, podcast <http://www.heritage.org/events/2013/05/detainees> (accessed 13 August 2013).

Lieven, Michael, '"Butchering the Brutes All Over the Place": Total War and Massacre in Zululand, 1879'. *History* 84/276 (1999), 614–32.

Lind, William S., et al., 'The Changing Face of War: Into the Fourth Generation', *Marine Corps Gazette*, 73/10 (1989), 22–6.

Linderman, Gerald F., *Embattled Courage: The Experience of Combat in the American Civil War* (New York: Free Press, 1987).

Long, Austin, *On 'Other War': Lessons from Five Decades of RAND Counterinsurgency Research* (Santa Monica, CA: RAND, 2006).

Lucas, Colin, 'The Problem of the Midi in the French Revolution', *Transactions of the Royal Historical Society*, 5th ser., 28 (1978), 1–25.

Lumans, Valdis O., *Himmler's Auxiliaries: The Volksdeutsche Mittelstelle and the German National Minorities of Europe, 1933–1945* (Chapel Hill, NC: University of North Carolina Press, 1993).

Luther, Martin, 'Wider die räuberischen und mürderischen Rotten der Bawren', *Weimarer Ausgabe*, 18 (Weimar: Böhlau, 1908 [1525]), 357–61.

Lynn, John A., *Battle: A History of Combat and Culture* (New York: Basic Books, 2003).

Mackey, Robert R., *The Uncivil War: Irregular Warfare in the Upper South, 1861–1865* (Norman, OK: University of Oklahoma Press, 2004).

McPherson, James M., *Battle Cry of Freedom: The American Civil War* (Oxford: Oxford University Press, 1988).

Madley, Benjamin, 'From Africa to Auschwitz: How German South West Africa Incubated Ideas and Methods Adopted and Developed by the Nazis in Eastern Europe', *European History Quarterly*, 35/3 (2005), 429–64.

Mahon, John K., 'Anglo-American Methods of Indian Warfare, 1676–1794', *Mississippi Valley Historical Review*, 45/2 (1958), 254–75.

Maslov, A. A., 'Concerning the Role of Partisan Warfare in Soviet Military Doctrine of the 1920s and 1930s', *Journal of Slavic Military Studies*, 9/4 (1996), 885–94.

Maslowski, Peter, 'To the Edge of Greatness: The United States, 1783–1865', in Williamson Murray, MacGregor Knox, and Alvin Bernstein (eds), *The Making of Strategy: Rulers, States and War* (Cambridge: Cambridge University Press, 1994), 205–41.

Masur, Louis P., *The Civil War: A Concise History* (Oxford: Oxford University Press, 2011).

Mawdsley, Evan, 'Anti-German Insurgency and Allied Grand Strategy', *Journal of Strategic Studies*, 31/5 (2008), 695–719.

Mayer, Jane, *The Dark Side: The Inside Story of How the War on Terror Turned into a War on American Ideals* (New York: Doubleday, 2008).

Mazower, Mark, *Hitler's Empire: Nazi Rule in Occupied Europe* (London: Allen Lane 2008).

Mazower, Mark, *Inside Hitler's Greece: The Experience of Occupation* (New Haven: Yale University Press, 1993).

Mazzetti, Mark, 'US and Afghans Reach Deal on Bagram Prison Transfer', *New York Times*, 23 March 2013.

Mégret, Frédéric, 'From "Savages" to "Unlawful Combatants": A Post-Colonial Look at International Humanitarian Law's "Other"', in Anne Orford (ed.), *International Law and its 'Others'* (Cambridge: Cambridge University Press, 2006), 265–317.

Melzer, Nils, *Targeted Killing in International Law* (Oxford: Oxford University Press, 2008).

Meron, Theodor, 'Lieber's Code and the Principles of Humanity', *Columbia Journal of Transnational Law*, 36/2 (1997), 269–81.

Michel, Henri, *The Shadow War: Resistance in Europe 1939–1945* (London: Corgi, 1972).

Milazzo, Matteo J., *The Chetnik Movement and the Yugoslav Resistance* (Baltimore, MD: Johns Hopkins University Press, 1975).

Miller, Dawn M., '"Raising the Tribes": British Policy in Italian East Africa, 1938–41', *Journal of Strategic Studies*, 22/1 (1999), 96–123.

Miller, Greg, 'CIA Remains behind Most Drone Strikes, Despite Effort to Shift Campaign to Defense', *Washington Post*, 25 November 2013.

Mitze, Katja, '"Seit der babylonischen Gefangenschaft hat die Welt nichts derart erlebt": Französische Kriegsgefangene und Franctireurs im Deutsch-Französischen Krieg 1870/71', in Rüdiger Overmans (ed.), *In der Hand des Feindes: Kriegsgefangenschaft von der Antike zum Zweiten Weltkrieg* (Cologne: Böhlau Verlag, 1999), 235–54.

Moland, Arnfin, 'Norway', in Bob Moore (ed.), *Resistance in Western Europe* (Oxford: Berg, 2000), 223–48.

Moran, Daniel, 'Introduction: The Legend of the *Levée en masse*', in Daniel Moran and Arthur Waldron (eds), *The People in Arms: Military Myth and National Mobilization since the French Revolution* (Cambridge: Cambridge University Press, 2003), 1–7.

Morris, Donald R., *The Washing of the Spears: The Rise and Fall of the Zulu Nation* (London: Pimlico, 1994).

Mountcastle, Clay, *Punitive War: Confederate Guerrillas and Union Reprisals* (Lawrence, KS: University Press of Kansas, 2009).

Müller, Eberhard, 'Autonome Bewegungen des Volkskriegs in Sowjetrußland nach der Revolution von 1917', in Gerhard Schulze (ed.), *Partisanen und Volkskrieg: Zur Revolutionierung des Krieges im 20. Jahrhundert* (Göttingen: Vandehoeck & Ruprecht, 1985), 36–56.

Münkler, Herfried, *The New Wars* (Cambridge: Polity, 2005).

Murphy, David E., *What Stalin Knew: The Enigma of Barbarossa* (New Haven: Yale University Press, 2005).

Nabulsi, Karma, 'Evolving Concepts of Civilians and Belligerents: One Hundred Years after the Hague Peace Conferences', in Simon Chesterman (ed.), *Civilians in War* (Boulder, CO: Lynne Rienner, 2001), 9–24.

Nabulsi, Karma, *Traditions of War: Occupation, Resistance, and the Law* (Oxford: Oxford University Press, 1999).

Neff, Stephen C., *War and the Law of Nations: A General History* (Cambridge: Cambridge University Press, 2005).

Newman, Edward, 'The "New Wars" Debate: A Historical Perspective Is Needed', *Security Dialogue*, 35/2 (2004), 173–89.

Nolte, Hans-Heinrich, 'Partisan War in Belorussia, 1941–1944', in Roger Chickering, Stig Förster, and Bernd Greiner (eds), *A World at Total War: Global Conflict and the Politics of Destruction* (Cambridge: Cambridge University Press, 2005), 261–76.

'Obama's Speech on Drone Policy', *New York Times*, 23 May 2013.

Oman, Charles, *A History of the Peninsular War* (London: Greenhill, 1995–9), vi.

Osiel, Mark, *The End of Reciprocity: Terror, Torture, and the Law of War* (Cambridge: Cambridge University Press, 2009).

Overmans, Rüdiger, 'Die Kriegsgefangenenpolitik des Deutschen Reiches 1939–1945', in Jörg Echternkamp (ed.), *Das Deutsche Reich und der Zweite Weltkrieg, Band 9/2: Die Deutsche Kriegsgesellschaft: Ausbeutung, Deutungen, Ausgrenzung* (Munich: Deutsche Verlags-Anstalt, 2005), 729–875.

Pakenham, Thomas, *The Boer War* (New York: Random House, 1979).

Pakenham, Thomas, *The Scramble for Africa* (London: Weidenfeld & Nicolson, 1991).

Paret, Peter, 'Conscription and the End of the *Ancien Régime* in France and Prussia': in Peter Paret (ed.), *Understanding War: Essays on Clausewitz and the History of Military Power* (Princeton: Princeton University Press, 1992), 53–374.

Paret, Peter, 'The Relationship between the American Revolutionary War and European Military Thought and Practice', in Peter Paret (ed.), *Understanding War: Essays on Clausewitz and the History of Military Power* (Princeton: Princeton University Press, 1992), 26–38.

Paret, Peter, *The Vendée, 1789–1796* (Princeton: Center for International Studies, 1961).

Parker, David, 'Had a Distinct Template for "Western Way of War" Been Established before 1800?', in Hew Strachan and Sibylle Scheipers (eds), *The Changing Character of War* (Oxford: Oxford University Press, 2011), 48–63.

Parker, Geoffrey, 'The "Military Revolution," 1560–1660—a Myth?', *Journal of Military History*, 48/2 (1976), 195–214.

Parker, Geoffrey, *The Military Revolution: Military Innovation and the Rise of the West 1500–1800* (Cambridge: Cambridge University Press, 1988).

Pelzer, Erich, '"Il ne sera fait aucun prisonnier anglais ou hanovrien": Zur Problematik der Kriegsgefangenen während der Revolutions- und Empirekriege (1792–1815)', in Rüdiger Overmans (ed.), *In der Hand des Feindes: Kriegsgefangenschaft von der Antike bis zum Zweiten Weltkrieg* (Cologne: Böhlau Verlag, 1999), 189–210.

Percy, Sarah, *Mercenaries: The History of a Norm in International Relations* (Oxford: Oxford University Press, 2007).

Petitfrère, Claude, 'The Origins of the Civil War in the Vendée', *French History*, 2/2 (1988), 187–207.

Plaw, Avery, 'Counting the Dead: The Proportionality of Predation in Pakistan', in Bradley Jay Strawser (ed.), *Killing by Remote Control: The Ethics of an Unmanned Military* (Oxford: Oxford University Press, 2013), 126–53.

Popple, Simon, 'From "Brother Boer" to "Dirty Boer": Colonizing the Colonizers through the Popular Representations of the Boer in the British Illustrated Journal 1899–1902', *Journal of War & Culture Studies*, 5/2 (2012), 137–56.

Porch, Douglas, 'The Dangerous Myths and Dubious Promise of COIN', *Small Wars & Insurgencies*, 22/2 (2011), 239–57.

Porch, Douglas, *Counterinsurgency: Exposing the Myths of the New Way of War* (Cambridge: Cambridge University Press, 2013).

Porter, Patrick, *Military Orientalism: Eastern War through Western Eyes* (London: Hurst, 2009).

Prada, Antonio Moliner, 'Popular Resistance in Catalonia: *Somatens* and *Miquelets*, 1808–14', in Charles J. Esdaile (ed.), *Popular Resistance in the French Wars: Patriots, Partisans and Land Pirates* (Basingstoke: Palgrave Macmillan, 2005), 91–114.

Price, Richard, 'A Genealogy of the Chemical Weapons Taboo', *International Organization*, 49/1 (1995), 73–103.

Reid, John, 'Twenty-First Century Warfare—Twentieth Century Rules', *RUSI Journal*, 151/3 (2006), 14–16.

Reid, Richard J., *Warfare in African History* (Cambridge: Cambridge University Press, 2012).

Report of the International Commission of Inquiry to Investigate All Alleged Violations of International Human Rights Law in the Libyan Arab Jamahiriya (Geneva: Office of the Human Rights Council, 2011).

Rieber, Alfred J., 'Civil Wars in the Soviet Union', *Kritika: Explorations in Russian and Eurasian History*, 4/1 (2003), 129–62.

Rink, Martin, 'The Partisan's Metamorphosis: From Freelance Military Entrepreneur to German Freedom Fighter, 1740 to 1815', *War in History*, 17/1 (2010), 6–36.

Risen, James, Mazzetti, Mark, and Schmitt, Michael S., 'US Approved Arms for Libya Rebels Fell into Jihadis' Hands', *New York Times*, 5 December 2012.

Roberts, Adam, 'The Civilian in Modern War', in Hew Strachan and Sibylle Scheipers, *The Changing Character of War* (Oxford: Oxford University Press, 2011), 357–80.

Roberts, Adam, 'Detainees: Misfits in Peace and War', in Sibylle Scheipers (ed.), *Prisoners in War* (Oxford: Oxford University Press, 2010), 263–80.

Roberts, Adam, 'Review Essay: Torture and Incompetence in the "War on Terror"', *Survival*, 49/1 (2007), 199–212.

Roberts, Adam, and Guelff, Richard (eds), *Documents on the Laws of War* (Oxford: Oxford University Press, 2000).

Roberts, Michael, 'The Military Revolution, 1560–1660', in Michael Roberts (ed.), *Essays in Swedish History* (London: Weidenfeld & Nicolson, 1967), 195–225.

Rothenberg, Gunther, 'The Age of Napoleon', in Michael Howard, George J. Andropoulos, and Mark R. Shulman (eds), *The Laws of War: Constraints on Warfare in the Western World* (New Haven: Yale University Press, 1994), 86–97.

Rothenberg, Gunther, *The Art of Warfare in the Age of Napoleon* (London: Batsford, 1977).

Rothenberg, Gunther, 'The Origins, Causes, and Extension of the Wars of the French Revolution and Napoleon', *Journal of Interdisciplinary History*, 18/4 (1988), 771–93.

Russell, Peter E., 'Redcoats in the Wilderness: British Officers and Irregular Warfare in Europe and America, 1740 to 1760', *William and Mary Quarterly*, 35/4 (1978), 629–52.

Saikal, Amin, 'Afghanistan's Ethnic Conflict', *Survival*, 40/2 (1998), 114–26.

Sands, Philippe, *Torture Team: Deception, Cruelty and the Compromise of Law* (London: Allen Lane, 2008).

Savage, Charlie, 'US to Send 2 at Guantánamo back to Algeria, Saying Security Concerns Are Met', *New York Times*, 26 July 2013.

Savory, Reginald, 'The Convention of Écluse, 1759–1762: The Treatment of the Sick and Wounded, Prisoners of War, and Deserters of the British and French Armies during the Seven Years War', *Journal of the Society for Army Historical Research*, 42/170 (1964), 68–77.

Scheipers, Sibylle, 'Closing Guantánamo: Is Europe Ready?', *Survival*, 51/1 (2009), 5–12.

Scheipers, Sibylle, 'Irregular Fighters: Is the Law of Armed Conflict Outdated?', *Parameters*, 43/4 (2013), 1–12.

Scheipers, Sibylle, *Negotiating Sovereignty and Human Rights: International Society and the International Criminal Court* (Manchester: Manchester University Press, 2009).

Scheipers, Sibylle, 'The Status and Protections of POWs and Detainees', in Hew Strachan and Sibylle Scheipers (eds), *The Changing Character of War* (Oxford: Oxford University Press, 2011), 394–409.

Schmider, Klaus, 'Auf Umwegen zum Vernichtungskrieg? Der Partisanenkrieg in Jugoslawien, 1941–1944', in Rolf-Dieter Müller and Hans-Erich Volkmann (eds), *Die Wehrmacht: Mythos und Realität* (Munich: Oldenbourg, 1999), 901–22.

Schmider, Klaus, *Partisanenkrieg in Yugoslawien 1941–1944* (Hamburg: E. S. Mittler & Sohn, 2002).

Schmitt, Carl, *The* Nomos *of the Earth in the International Law of the* Jus Publicum Europaeum (New York: Telos Press, 2003 [1950]).

Schmitt, Carl, *Theory of the Partisan: Intermediate Commentary on the Concept of the Political* (New York: Telos Press, 2007 [1963]).

Schmitt, Eric, 'CIA Said to Aid in Steering Arms to Syrian Opposition', *New York Times*, 21 June 2012.

Schofield, Philip, 'British Politicians and French Arms: The Ideological War of 1793–1795', *History*, 77/250 (1992), 183–201.

Schork, Kurt, '*Hors de Combat*', Crimes of War <http://www.crimesofwar.org/a-z-guide/hors-de-combat/> (accessed 18 August 2013).

Scotti-Douglas, Vittorio, 'Regulating the Irregulars: Spanish Legislation on *la guerrilla* during the Peninsular War', in Charles J. Esdaile (ed.), *Popular Resistance in the French Wars: Patriots, Partisans and Land Pirates* (Basingstoke: Palgrave Macmillan, 2005), 137–60.

Selesky, Harold E., 'Colonial America', in Michael Howard, George J. Andropoulos, and Mark R. Shulman (eds), *The Laws of War: Constraints on Warfare in the Western World* (New Haven: Yale University Press, 1994), 59–85.

Shandley, Robert R. (ed.), *Unwilling Germans? The Goldhagen Debate* (Minneapolis, MN: University of Minnesota Press, 1998).
Shepherd, Ben, *Terror in the Balkans: German Armies and Partisan Warfare* (Cambridge, MA: Harvard University Press, 2012).
Shepherd, Ben, *War in the Wild East: The German Army and Soviet Partisans* (Cambridge, MA: Harvard University Press, 2004).
Showalter, Dennis, 'Infantry Weapons, Infantry Tactics, and the Armies of Germany, 1849–64', *European Studies Review*, 4/2 (1974), 119–40.
Shy, John, 'The American Military Experience: History and Learning', *Journal of Interdisciplinary Research*, 1/2 (1971), 205–28.
Shy, John, 'The American Revolution: The Military Conflict Considered as a Revolutionary War', in Stephen J. Kurtz and James H. Hutson (eds), *Essays on the American Revolution* (Chapel Hill, NC: University of North Carolina Press, 1973), 121–56.
Siddique, Haroon, Sparrow, Andrew, and Graham-Harrison, Emma, 'Britain May Begin Handover of Afghan Prisoners at Camp Bastion to Kabul', *Guardian*, 30 May 2013.
Slepyan, Kenneth, *Stalin's Guerrillas: Soviet Partisans in World War II* (Lawrence, KS: University of Kansas Press, 2006).
Slim, Hugo, *Killing Civilians: Methods, Madness and Morality in War* (London: Hurst, 2007).
Slomp, Gabriella, 'Theory of the Partisan: Carl Schmitt's Neglected Legacy', *History of Political Thought*, 26/3 (2005), 502–19.
Smith, Adam I. P., '"On the Altar of the Nation": Narratives of Heroic Sacrifice in the American Civil War', in Sibylle Scheipers (ed.), *Heroism and the Changing Character of War: Toward Post-Heroic Warfare?* (Basingstoke: Palgrave, 2014), 33–46.
Smith, Iain R., and Stucki, Andreas, 'The Colonial Development of Concentration Camps (1868–1902)', *Journal of Imperial and Commonwealth History*, 39/3 (2011), 417–37.
Smith, M. L. R., 'Guerrillas in the Mist: Reassessing Strategy and Low Intensity Warfare', *Review of International Studies*, 29/1 (2009), 19–37.
Stafford, David, *Britain and European Resistance, 1940–1945: A Survey of the Special Operations Executive, with Documents* (London: Macmillan, 1980).
Stahel, David, *Operation Barbarossa and Germany's Defeat in the East* (Cambridge: Cambridge University Press, 2009).
Starkey, Armstrong, *European and Native American Warfare, 1675–1815* (London: UCL Press, 1998).
Statiev, Alexander, 'Was *Smuglianka* a Lunatic or a Siguranţa's Agent-Provocateur? Peculiarities of the Soviet Partisan Struggle in the Western Borderlands', *Journal of Strategic Studies*, 31/5 (2008), 743–70.
Steele, Ian K., *Betrayals: Fort William Henry & the 'Massacre'* (Oxford: Oxford University Press, 1990).
Steele, Ian K., *Warpaths: Invasions of North America* (Oxford: Oxford University Press, 1994).
Steyn, Johan, 'Guantánamo Bay: The Legal Black Hole', *International and Comparative Law Quarterly*, 53/1 (2004), 1–15.

Steinhoff, Uwe, 'Killing them Safely: Extreme Asymmetry and its Discontents', in Bradley Jay Strawser (ed.), *Killing by Remote Control: The Ethics of an Unmanned Military* (Oxford: Oxford University Press, 2013), 179–207.

Stockman, Farah, 'US behind Afghan Warlords Rise and Fall: At Guantánamo Unruly Chieftains Join Combatants', *Boston Globe*, 12 August 2007.

Stoneman, Mark R., 'The Bavarian Army and French Civilians in the War of 1870–1871: A Cultural Interpretation', *War in History*, 8/3 (2001), 271–93.

Strachan, Hew, 'Heroic Warfare and the Problem of Mass Armies, France 1871–1914', in Sibylle Scheipers (ed.), *Heroism and the Changing Character of War: Toward Post-Heroic Warfare?* (Basingstoke: Palgrave, 2014), 47–63.

Streit, Christian, *Kein Kameraden: Die Wehrmacht und die sowjetischen Kriegsgefangenen, 1941–1945* (Bonn: Verlag J. H. W. Dietz, 1991).

Stucki, Andreas, 'Aufbruch ins Zeitalter der Lager? Zwangsumsiedlung und Deportation in der spanischen Antiguerilla auf Kuba, 1868–98', *Mittelweg 36*, 20/4 (2011), 20–34.

Surridge, Keith, 'An Example to Be Followed or a Warning to Be Avoided? The British, Boers, and Guerrilla Warfare, 1900–1902', *Small Wars & Insurgencies*, 23/4–5 (2012), 608–26.

Sutherland, Daniel E., *A Savage Conflict: The Decisive Role of Guerrillas in the American Civil War* (Chapel Hill, NC: University of North Carolina Press, 2009).

Sutherland, Donald M. G., *The French Revolution and Empire: The Quest for a Civic Order* (Oxford: Blackwell, 2003).

Taylor, Lynne, 'The Parti Communiste Français and the French Resistance in the Second World War', in Tony Judt (ed.), *Resistance and Revolution in Mediterranean Europe, 1939–1948* (London: Routledge, 1989), 53–79.

The Copenhagen Process on the Handling of Detainees in International Military Operations: Principles and Guidelines (Copenhagen: Department of International Law) <http://um.dk/en/~/media/UM/English-site/Documents/Politics-and-diplomacy/Copenhagen%20Process%20Principles%20and%20Guidelines.pdf> (accessed 5 September 2013).

Thompson, Janice E., *Mercenaries, Pirates, and Sovereigns: State-Building and Extraterritorial Violence in Early Modern Europe* (Princeton: Princeton University Press, 1996).

Thornton, John K., '"I am the Subject of the King of Congo": African Political Ideology and the Haitian Revolution', *Journal of World History*, 4/2 (1993), 181–214.

Thornton, John K., *Warfare in Atlantic Africa, 1500–1800* (London: Routledge, 1999).

Thornton, Rod, *Asymmetric Warfare: Threat and Response in the 21st Century* (Cambridge: Polity, 2007).

Thurston, Robert W., 'Cauldrons of Loyalty and Betrayal: Soviet Soldiers' Behavior, 1941 and 1945', in Robert W. Thurston and Bernd Bonwetsch (eds), *The People's War: Responses to World War II in the Soviet Union* (Urbana, IL: University of Illinois Press, 2000), 235–57.

Tilly, Charles, *Coercion, Capital and European States, AD 990–1990* (Oxford: Blackwell, 1990).

Tilly, Charles, *The Vendée: A Sociological Analysis of the Counter-Revolution of 1793* (Cambridge, MA: Harvard University Press, 1976).

Tilly, Charles, 'War Making and State Making as Organized Crime', in Dietrich Rueschemeyer, Peter B. Evans, and Theda Skocpol (eds), *Bringing the State Back In* (Cambridge: Cambridge University Press, 1985), 203–21.

Tone, John Lawrence, *The Fatal Knot: The Guerrilla War in Navarre and the Defeat of Napoleon in Spain* (Chapel Hill, NC: University of North Carolina Press, 1994).

Tran, Mark, 'Cheney Endorses Simulated Drowning', *Guardian*, 27 October 2006.

Travis, David, 'Communism and Resistance in Italy, 1943–8', in Tony Judt (ed.), *Resistance and Revolution in Mediterranean Europe, 1939–1948* (London: Routledge, 1989), 80–109.

'The Truth about Dasht-i-Leili', *New York Times*, 13 July 2009.

Ullman, Harlan, and Wade, James, Jr, *Shock and Awe: Achieving Rapid Dominance* (Washington: National Defense University, 1996).

Vattel, Emmerich de, *The Law of Nations; or, Principles of the Law of Nature Applied to the Conduct and Affairs of Nations and Sovereigns* (New York: AMS Press, 1982 [1758]).

Wæver, Ole, 'Discursive Approaches', in Thomas Diez and Antje Wiener (eds), *European Integration Theory* (Oxford: Oxford University Press, 2004), 197–215.

Ward, Harry M., *Between the Lines: Banditti of the American Revolution* (Westport, CT: Praeger, 2002).

Ward, Matthew C., '"The European Method of Warring is not Practiced Here": The Failure of British Military Policy in the Ohio Valley, 1755–1759', *War in History*, 4/3 (1997), 247–63.

Watkin, Kenneth, 'Warriors without Rights? Combatants, Unprivileged Belligerents, and the Struggle over Legitimacy', *Humanitarian Policy and Conflict Research Occasional Paper* (Cambridge, MA: Harvard University, 2005).

Wawro, Geoffrey, *The Franco-Prussian War: The German Conquest of France, 1870–1871* (Cambridge: Cambridge University Press, 2003).

Westermann, Edward B., *Hitler's Police Battalions: Enforcing Racial War in the East* (Lawrence, KS: University Press of Kansas, 2005).

Whetham, David, 'Drones and Targeted Killing: Angels or Assassins?', in Bradley Jay Strawser (ed.), *Killing by Remote Control: The Ethics of an Unmanned Military* (Oxford: Oxford University Press, 2013), 69–83.

White, Richard, 'Outlaw Gangs of the Middle Border: American Social Bandits', *Western Historical Quarterly*, 12/4 (1981), 387–408.

Wieviorka, Olivier, 'France', in Bob Moore (ed.), *Resistance in Western Europe* (Oxford: Berg, 2000), 125–55.

Wilner, Alex S., 'Targeted Killings in Afghanistan: Measuring Coercion and Deterrence in Counterterrorism and Counterinsurgency', *Studies in Conflict and Terrorism*, 33/3 (2010), 307–29.

Winter, Yves, 'The Asymmetric War Discourse and its Moral Economies: A Critique', *International Theory*, 3/3 (2011), 488–514.

Woodward, Bob, *Obama's Wars* (New York: Simon and Schuster, 2010).

Yoo, John C., 'Terrorists have No Geneva Rights', *Wall Street Journal*, 26 May 2004.

Yoo, John C., and Ho, James C., 'The Status of Terrorists', *Boalt Working Papers in Public Law*, Paper 25 (Berkeley, CA: University of California, Berkeley, 2003).

Zariczniak, Larysa, 'Major Stepan Stebelski of the Ukrainian Insurgent Army: Examples of Insurgency Leadership and Tactics', *Small Wars & Insurgencies*, 22/3 (2011), 435–47.

Zimmerer, Jürgen, *Von Windhuk nach Auschwitz? Beiträge zum Verhältnis von Kolonialismus und Holocaust* (Münster: LIT Verlag, 2011).

Index